川習時期
美中霸權競逐新關係

李大中 主編

U0082032

淡江大學出版中心

主編序

自 2017 年年底以來，隨著川普政府的印太倡議逐漸成形，美國對中政策的遏制色彩日益明顯，美中關係進入霸權競逐的新階段，最後是否演變為新冷戰，成為全球關注焦點，面對此瞬息萬變的外部形勢，如何趨吉避凶，維護自身最大利益，普遍成為區域各國的首要之務。

有鑑於此，本所於 2018 年 5 月舉辦「淡江戰略學派年會暨第 14 屆紀念鈕先鍾教授國際學術研討會」，並以「蔡總統執政兩週年的台灣國家安全戰略總體檢：挑戰與展望」為研討會主軸，邀集國內外一流的學者專家進行為期兩天的學術饗宴，而本學術專書即為大家共襄盛舉的心血結晶。

本書共收錄其中 20 篇優秀論文，並將其歸類為五大主題，包括：「印太戰略與區域研究」、「戰略研究的理論與實踐」、「台灣國家安全戰略」、「中國大陸政軍發展」以及「全國全民國防教育」。集合所有撰稿人的智慧結晶，相信本書不僅有助於各界理解當前亞太安全情勢的變遷脈絡，更能提供我國國家安全政策之制定參考。

本書得以順利付梓，除感謝每位作者的貢獻之外，特別感謝董事長張家宜博士與校長葛煥昭博士對於本所的支持與肯定；其次，感謝淡江大學出版中心歐陽崇榮主任、吳秋霞總編輯與張瑜倫編輯，無論是編輯或設計，均提供專業的建議與協助；至於在本所的工作團隊方面，陳文政教授肩負執行編輯之重任，陳秀真助理提供無微不至的行政支援，博士候選人鄭智懷進行橫向聯繫，徐立中碩士生投入繁瑣的校對與排版工作，正因為眾志成城，才有最後豐美的成果。

本所創立迄今已逾 37 年，回首過去，在前人辛勤的耕耘下，不僅奠定堅實的基礎，更成為鞭策我們不斷向前的動力。展望未來，我們將持續豐富「淡江戰略學派」的內涵，並以打造「學術之重鎮、國家之干城」的目標自我期許。

淡江大學國際事務與戰略研究所 所長

2019 年 6 月 27 日

目次

印太戰略與區域研究

The Idea of Indo-Pacific Strategy and the Role of India

Srikanth Kondapalli *

Asian region today is poised to be dynamic not only in its recent economic vibrancy but also in strategic arena as well. As economic, technological and military developments in the eastern hemisphere are inexorably acquiring prominence, global attention is today focused on the region that is renamed as Indo-Pacific. Known since the 1980s as Asia-Pacific — mainly to accommodate China in the regional and international system — the Indo-Pacific idea is gathering increasing attention as other states and actors are emerging recently. While China, which emerged as the 2nd largest economy in the world by 2010, displacing Japan from this position, is still the focus of the region, such focus is increasingly becoming one of a concern, if not alarming. For, with the rise of China is also the changing "core interests" discourse with assertive positions on Taiwan, Tibet of the yore now extending stridently on South China Sea islands, Senkaku Islands and others. Also, with the Belt and Road Initiative and China intending to occupy "centre-stage" in global and regional orders, the region is set to respond to this phenomenon. These developments have led to concern among the states of the region but also including the United States. Indeed, the idea for the Indo-Pacific came in the 2000s with the United States leaders suggesting to alternatives to the Asia-Pacific as a concept as well as a security mechanism to manage regional tensions and provide security for the region in the light of the assertiveness of China.

I. Indo-Pacific

Defining geographical regions is fraught with difficulties and most times as well ahistorical. However, they do reflect to the dominant themes of the times

---

* Professor and Chairperson, Centre for East Asian Studies, Jawaharlal Nehru University, India

and a reflection of a shift in power in the region and the world at large. In 1954 the United States National Defense Education Act defined five major regions of Asia for better management of the resources and direction of policies. Thus came into being the sub-regions of East Asia (from Far East or Cathay or Oriental of the yore), Southeast Asia, South Asia and the then Soviet controlled Central Asia and the Middle East (West Asia). Containment of Soviet and Chinese communist expansion in East and Southeast Asian regions became the main policy thrust during this time resulting in the alliances of Southeast Asia Treaty Organisation in the 50s and the formation of the Association of Southeast Asian Nations grouping in the 1960s and alliances with Pakistan in South Asia. India at that time was an active member of the Non Aligned Movement advocating positive neutrality between the then Super powers or even aligning with the then Soviet Union in 1971. While the US built up relations with China to counter the then Soviet Union, it had also expanded relations with Beijing in economic and technological spheres. It was during this period that the concept of Asia-Pacific became popular mainly in the economic field with the Asia-Pacific Economic Cooperation (APEC) becoming a major driver in the regional economic integration.

Four decades of reform and opening up in China had built substantial depth to the US-China relations, propelling as well China to the 2nd largest economy in the world by 2010 and largest high-tech exporting country by 2014 displacing Japan from these two positions. Bilateral trade increased to over $500 billion, investments increased with China investing $1.2 trillion in the US Treasury Securities. However, while "the US welcomed the rise of China" as successive Presidents of the US stated, China began encroaching on the US alliances or influence regions such as South China Sea, East China Sea, Pacific and Indian Oceans. China also began "setting up a different kitchen" in the Belt and Road Initiative since 2013 in the continental and maritime regions — chipping away the US / it allies' influence.

II. China as a Factor

The last decade witnessed not only the spectacular rise of China but unprecedented forays by Beijing in global and regional dynamics. Firstly after China became the 2nd largest economy in the world displacing Japan in 2010 utilising the rules of the World Trade Organisation which it joined in 2001, it's appetite increased substantially. It began questioning the rules by terming these as "discriminatory". After building huge trade surpluses with the US, Europe, Japan, South Korea and Southeast Asian Nations, China did not meet their legitimate demands on market economy status, mounting trade deficits and currency valuation. Instead today China had proposed a grand strategy of creating a niche for itself in the form of Belt and Road Initiative since 2013 both in the continental and maritime dimensions. However, a recent Washington-based CSIS report suggested that more than 80 percent of business contracts of China-financed construction projects in these areas went to Chinese companies much to the chagrin of others. Besides, countries like India were critical of China violating the sovereignty norms in the Belt and Road Initiative projects by not only building infrastructure projects in disputed territories of Kashmir but also positioning troops.

Moreover, by centralising power and annihilating any opposition to his rule, President Xi Jinping has indicated to his passion to stay longer in power. Recent constitutional amendments to give possibly unlimited terms for the state leaders, China's parliamentary sessions recently have instituted unbridled powers to one person thus raising uncertainties on the security front in the region. In the concentration of power drive, President Xi had earlier cited to political opposition from within the communist party and also to fulfil promises of a "well-off society" and celebrating "two centennials", viz., the hundredth anniversary of the communist party in 2021 and that of the People's Republic in 2049.

Secondly, China's assertiveness on its recently revised "core interests" in South China and East China Seas and Arunachal Pradesh (termed by China as "Southern Tibet" since 2005) rattled its neighbours. China's juggernaut rolled without much resistance. Chinese navy pushed the USS Impeccable and 12 other vessels from South China Sea besides daredevil mid-air engagements. Military buildup by China of occupied reefs (about 3000 acres in South China Sea alone) and border areas increased briskly. China's armed forces marched on to Japan-administered Senkaku Islands. More than 600 transgressions were made by China's coast guards and air force while such transgressions on Indian borders went up from nearly 300 a year to more than 400 in the last four years. Chinese military leaders like Luo Yuan also began calling for preparing for "two front war". To cap these, China rejected The Hague Tribunal ruling in July 2016 on the welfare of fishermen, environmental degradation, reef build-up and imaginary -9 dashed line. Rule of law and freedom of navigation were thus threatened and the Southeast Asian Nations unity divided without a joint declaration being issued at their annual meetings.

Thirdly, side-by–side, China's strategic posture under radical shift from "keeping a low profile" to "accomplishing something". At the May 2014 CICA meeting at Shanghai, China called for Asian countries to fend for themselves suggesting extra-regional powers to withdraw but without adequate protect of any beleaguered country. The October 2017 nineteenth Communist Party congress began preparation for occupying the "Centre -stage" in global and regional orders. China also increased its defence budget to $174 billion — which represents more than 8 percent increase over the previous year and also over the GDP figures of about 6 percent. This was enough for many in Asia and beyond whose security interests were threatened. This is the context for the birth of the Indo-Pacific and represents a shift in power from the Atlantic towards the Pacific and to the Indian Oceans with the rise of India.

III. The US Role

The idea of Indo-Pacific is gaining traction in the recent times although the concept is almost a decade old coined during the Obama Administration as well as by Japanese leader Shinzo Abe.[1] This broadly refers to the maritime area between east of Africa and west of Oceania. After the initial work by the then US Secretary of State Condoleezza Rice who proposed interaction between these four countries' navies to provide timely relief measures for Southeast Asian countries affected by Tsunami disaster, in September 2007 they conducted a naval exercise in the Bay of Bengal. China saw this exercise as a part of an effort at potentially blocking its energy supplies-80 percent of more than 300 million tonnes of its imports from West Asia and Africa passing through the region. This was also the time when China's rhetoric and coercive diplomatic efforts against Taiwan President Chen Shuibian were becoming acute. China's subsequent demarché to Australia and Japan, change in government in the US and President Obama's visit to China in 2009 changed the situation in favour of Beijing.

The idea of Indo-Pacific was referred to by the then US Secretary of State Hillary Clinton in 2010 obliquely but more extensively in 2011. Clinton stated:

> *Stretching from the Indian subcontinent to the western shores of the Americas, the region spans two oceans — the Pacific and the Indian — that are increasingly linked by shipping and strategy. It boasts almost half the world's population. It includes many of the key engines of the global economy, as well as the largest emitters of greenhouse gases. It is home to several of our key allies and important emerging powers like China, India, and Indonesia.[2]*

---

1 On the subject see Rory Medclaf, "The Indo-Pacific: What's in a Name?," The American Interest, Volume 9, Number 2, October 10, 2013, https://www.the-american-interest.com/2013/10/10/the-indo-pacific-whats-in-a-name/; Medclaf, "Reimagining Asia: From Asia-Pacific to Indo-Pacific," The Asan Forum, June 26, 2015, http://www.theasanforum.org/reimagining-asia-from-asia-pacific-to-indo-pacific/; Premvir Das, "India's 'Indo-Pacific' challenges," Business Standard, April 1, 2017, http://www.business-standard.com/article/opinion/india-s-indo-pacific-challenges-117040100822_1.html.

2 Hillary Clinton, "America's Pacific Century," *Foreign Policy*, October 11, 2011, http://foreignpolicy.com/2011/10/11/americas-pacific-century/. See also Hillary Clinton, "Hillary Clinton lauds India's role in Indo-Pacific region, urges for increased participation," *India Today*, November 14, 2012, https://www.indiatoday.in/world/rest-of-the-world/story/hillary-clinton-lauds-indias-role-indo-pacific-region-talks-china-breifly-in-australia-121455-2012-11-14.

President Obama in his speech to the Australian Parliament in November 2011, suggested to a "rebalance" of the US from the Atlantic to the Asia-Pacific.[3] The rebalance is aimed at rationalizing US force structures in the trans-Atlantic and the Asia-Pacific to about 40:60 in favour of the Asia-Pacific, in addition to building ballistic missile defence (BMD) shield; "strengthening alliances" with Australia, Japan and South Korea, "deepening partnerships with emerging powers" like Singapore, Indonesia, Thailand, Vietnam and India; "building stable, productive and constructive relationship with China"; "empowering regional institutions" like the East Asian Summit; "helping to build a regional economic architecture" in the form of Trans-Pacific Partnership (TPP) and others.

Kurt Campbell, who served the Obama Administration on the East Asian affairs of the State Department, argued that the "pivot" policy is not to contain China militarily but reinforce the US influence through a combination of policies including in the economic domain. The TPP of high grade liberalization and environmental standards were proposed as a way to reinforce the US economy as well as its allies.[4]

After President Trump took over in January 2017, a number of changes took place. Increasingly, Trump had used Indo-Pacific as a concept in his speeches and documents. Trump during his speech at Da Nang in Vietnam in November 2017, proposed the Indo-Pacific strategy. Trump mentioned that his strategy was "to strengthen America's alliances and economic partnerships in a free and open Indo-Pacific, made up of thriving, independent nations, respectful of other countries and their own citizens, and safe from foreign domination and economic servitude".[5] According to Alex N. Wong, US Deputy Assistant Secretary, Bureau

---

3 The White House Office of the Press Secretary, "Remarks by President Obama to the Australian Parliament," November 17, 2011, https://obamawhitehouse.archives.gov/the-press-office/2011/11/17/remarks-president-obama-australian-parliament.

4 Kurt M. Campbell, *The Pivot: The Future of American Statecraft in Asia* (New York: Hachette Book Group, 2016).

5 The White House, "Remarks by President Trump on His Trip to Asia," November 15, 2017, https://www.whitehouse.gov/briefings-statements/remarks-president-trump-trip-asia/. See also Nikhil Sonnad, "All about 'Indo-Pacific,' the new term Trump is using to refer to Asia," *Quartz*, November 7, 2017, https://qz.com/1121336/trump-in-asia-all-about-indo-pacific-the-new-term-trump-is-using-to-refer-to-asia/.

of East Asian and Pacific Affairs, the Indo-Pacific strategy has the following components:

We want the nations of the Indo-Pacific to be free from coercion, that they can pursue in a sovereign manner the paths they choose in the region. ...... we want the societies of the various Indo-Pacific countries to become progressively more free — free in terms of good governance, in terms of fundamental rights, in terms of transparency and anti-corruption. [We want] open sea lines of communication and open airways. [We want] more open logistics — infrastructure [-] to encourage greater regional integration, encourage greater economic growth. [We want] more open investment [and] open trade. it acknowledges the historical reality and the current-day reality that South Asia, and in particular India, plays a key role in the Pacific and in East Asia and in Southeast Asia. That's been true for thousands of years and it's true today. Secondly, it is in our interest, the U.S. interest, as well as the interests of the region, that India play an increasingly weighty role in the region. India is a nation that is invested in a free and open order. It is a democracy. It is a nation that can bookend and anchor the free and open order in the Indo-Pacific region, and it's our policy to ensure that India does play that role, does become over time a more influential player in the region.[6]

The above conceptual proposals for the Indo-Pacific strategy were reiterated by the then Secretary of State Rex Tillerson in a speech to the CSIS in October 2017. Tillerson stated to the expanding relations with India in this speech and underlined that :

India and the United States must foster greater prosperity and security with the aim of a free and open Indo-Pacific. The Indo-Pacific, including the entire Indian Ocean, the Western Pacific and the nations that surround them, will be the most consequential part of the globe in the 21st century. Home to more than 3 billion people, this region is the focal point of the world's energy and trade routes.

---

6 Statement by Alex N. Wong, US Deputy Assistant Secretary, Bureau of East Asian and Pacific Affairs, April 2, 2018, https://www.state.gov/r/pa/prs/ps/2018/04/280134.htm.

Forty percent of the world's oil supply crisscrosses the Indian Ocean every day, through critical points of transit like the Straits of Malacca and Hormuz…India can also serve as a clear example of a diverse, dynamic, and pluralistic country to others. [7]

On November 11, 2017, officials of four democratic countries the United States, Japan, India and Australia got together at Manila to form a low key grouping called as Quad- the Quadrilateral Security Dialogue. They announced individually their support in varying degrees, to rule of law, freedom of navigation, maritime connectivity, maritime security, countering North Korean nuclear/ballistic missile proliferation and other related aspects. There has also a change in terminology from Asia-Pacific to Indo-Pacific as an area of their collective cooperation. It is likely that many countries in the region may find this initiative interesting and hence join while others like China had expressed concerns and may make measures to counter the Quad. For instance, China's foreign minister Wang Yi was critical of the idea of India-Pacific suggesting that the current phase of international relations do not reflect to a renewed cold war phase and that the Indo-Pacific strategy is basically an "attention-grabbing idea" that will "dissipate like ocean foam".[8] A Chinese military commentary argued that "As the Indo-Pacific Strategy is in full swing, we cannot neglect it, but we also do not need to be inexplicably nervous." It further stated:

> *The theoretical basis and behavioral models of the Indo-Pacific Strategy is a reflection of the Cold War mentality based on ideology and the demarcation of friend or enemy. This definitely goes against the economic globalization, cultural inclusion, political pluralism, shared interests and other requirements of our times.*[9]

---

7 Superior Transcriptions LLC, "Defining Our Relationship with India for the Next Century: An Address by U.S. Secretary of State Rex Tillerson," October 18, 2017, pp. 1-14, https://csis-prod. s3.amazonaws.com/s3fs-public/event/171018_An_Address_by_U.S._Secretary_of_State_Rex_Tillerson. pdf?O0nMCCRjXZiUa5V2cF8_NDiZ14LYRX3m.

8 Yi Wang cited at Bill Birtles, "China mocks Australia over 'Indo-Pacific' concept it says will 'dissipate'," March 8, 2018, http://www.abc.net.au/news/2018-03-08/china-mocks-australia-over-indo-pacific-con-cept/9529548.

9 Minwen Wu, "Where will the Indo-Pacific Strategy go?" *China Military Online*, February 23, 2018, http://

However, despite enunciating the strategy of the Indo-Pacific, a number of sinews of this strategy are not explained nor implemented so far.[10] The US, Japan and India did discuss "economically sensible projects" though.[11]

IV. Japan's Role

The role of Japan in the Indo-Pacific has become crucial. Japan was the earliest to have proposed the idea of Indo-Pacific region in a speech by Prime Minister Shinzo Abe in New Delhi in 2007 referring to the confluence of the two oceans in an "arc of freedom and prosperity". Japan has been making initiatives since the 1990s — constrained as it were by its Constitutional provisions as well as to counter an assertive China which began making inroads in the Japanese administered Senkaku Islands. The 1997 treaty revisions with the US included provisions for acting in the "surrounding areas", although this has not been defined. With China increasing first its naval "research activities" and then transgressions in the Senkaku Islands through coastguard and air force intrusions, Tokyo began amending its constitutional provisions to include "collective self-defence", going beyond the 1,000 nautical mile limits (as in its non-combat role in Iraq and Afghan campaigns), increasing the defence budget, and the like. Japan's dependence on the Indian Ocean through which an estimated 75 percent of its energy transits, has become a main concern partly due to a spike in piracy incidents in the Indian Ocean and Southeast Asia but also due to the militarization of the South China Sea by China in the recent past. China's "territorial sea" concept for South China Sea and claiming of 80 percent of this sea territory could result in Japan losing its edge in the region at a minimum and depending on China for trade transit in the longer run. While much of economic and technological aspects of Japan are still in China, the maritime dimensions are

---

english.chinamil.com.cn/view/2018-02/23/content_7949833.htm.

10 Eric Sayers, "15 Big Ideas to Operationalize America's Indo-Pacific Strategy," *War On The Rocks*, April 6, 2018, https://warontherocks.com/2018/04/15-big-ideas-to-operationalize-americas-indo-pacific-strategy/.

11 PTI, "US wants to work on economically-sensible projects in Indo-Pacific, says official," *Financial Express*, April 20, 2018, https://www.financialexpress.com/india-news/us-wants-to-work-on-economically-sensible-projects-in-indo-pacific-says-official/1139157/.

becoming a major constraint in the region. Also, as the earliest state to modernize and democratize in Asia, Japan's "arc of freedom and prosperity" are a natural choice for the Indo-Pacific region. Support to this idea is also in the Japanese interests in diversifying economic assets from an increasingly nationalist China towards the Southeast Asian states and India. This expands the role and leadership of Japan in the two oceans in the longer run. Thus Japan is in the forefront of the Bay of Bengal quadrilateral exercise in September 2007 with the US, India and Australia but also in the recent revival of the concept of Indo-Pacific and joining the trilateral Malabar exercises with the US and India in the last few years and proposing the concept of Asia-Africa Growth Corridor in 2017. According to Japan's Ministry of Foreign Affairs, "Japan will enhance "connectivity" between Asia and Africa through a free and open Indo-Pacific to promote the stability and prosperity of the regions as a whole".[12] Providing a robust role for Japan (as with the others) is the trilateral with the US and India which is taking a more concrete shape recently, beginning with lower official interactions. Nine Trilateral meetings between the US, Japan and Indian officials took place till April 2018.[13] Also, a 2+2 dialogue begun between the foreign and defence establishments with the interoperable Malabar naval exercises accommodating Japan recently.

V. India's Role

The idea of Indo-Pacific is attractive to India as it propels New Delhi's role beyond the Cold War confinement to South Asia. India had a set of policies even before the Indo-Pacific idea came into being.[14] It began a Look East policy in 1991 which has been changed to Act East policy. The Act East policy coincides with the Indo-Pacific idea to a large extent.

---

12 Ministry Of Foreign Affairs, "Priority Policy for Development Cooperation FY2017," April, 2017, pp. 1-13, http://www.mofa.go.jp/files/000259285.pdf.

13 Dipanjan Roy Choudhury, "India-US-Japan discuss South China Sea tensions; Indo-Pacific region," *The Economic Times*, April 4, 2018, https://economictimes.indiatimes.com/news/defence/india-us-japan-discuss-south-china-sea-tensions-indo-pacific-region/articleshow/63614072.cms.

14 Samir Saran and Abhijit Singh, "India's struggle for the soul of Indo-Pacific," *Lowy Institute*, May 3, 2018, https://www.lowyinstitute.org/the-interpreter/india-struggle-soul-indo-pacific.

In India, the concept of the Indo-Pacific was first utilized by a naval officer Gurpreet S. Khurana in an article on cooperation between Indian and Japanese navies.[15] India's rise in the recent times, its robust democratic processes and rule of law, geographical position in the Indian Ocean, rising middle class, sustainable economic growth rates with high domestic consumption, its integration process in the world and IT software and others have attracted global attention, specifically of the US and its allies in the recent times. India's Look East policy since the 90s and its recent rechristening towards Act East since 2015 had expanded its interactions from South Asia to an extended format. The January 2015 Obama-Modi joint vision statement on the Indian Ocean and the South China Sea caught global attention although the Indian naval doctrine is based on the Indian Ocean as a primary area of responsibility and South China Sea and the Persian Gulf as "secondary". Nevertheless, China's forays in the South China Sea alerted the Indian side which learnt through the INS Airavat incident in 2009 that freedom of navigation through the region is crucial for the protection of its more than half of global trade passing through these waters. India's alignment with the renewed Quad members is thus visible since November 2017.[16]

The new government in New Delhi since 2014 has been attempting to introduce certain new policies. One of these include the up gradation of the 1991 Look East policy into Act East policy, the latter formally announced by Prime Minister Narendra Modi in his speeches at Naypyidaw in November 2014.[17] Although Act East policy was mentioned by various others before, after the implementation of the Look East policy for over two decades, a mid-course

---

15 Gurpreet S. Khurana, "Security of Sea Lines: Prospects for India–Japan Cooperation," *Strategic Analysis*, Vol. 31, No. 1, Jan–Feb, 2007, pp. 139-153, https://www.academia.edu/7710744/Security_of_Sea_Lines_Prospects_for_India-Japan_Cooperation; TCA Raghavan, "What Indo-Pacific means for India," *The Hindustan Times, March* 31, 2018, https://www.hindustantimes.com/opinion/what-indo-pacific-means-for-india/story-VmLixgjeLnLKWV58i8e3yM.html.

16 See Rajiv Bhatia and Vijay Sakhuja, Eds. *Indo Pacific Region: Political and Strategic Prospects* (New Delhi: Vij Books, 2014) and Brahma Chellaney, "A New Order for the Indo-Pacific," *Project Syndicate*, March 9, 2018, https://www.project-syndicate.org/commentary/china-indo-pacific-security-framework-by-brahma-chellaney-2018-03.

17 PTI, "'Look East' policy now turned into 'Act East' policy: Modi," *The Hindu*, November 13, 2014, http://www.thehindu.com/news/national/look-east-policy-now-turned-into-act-east-policy-modi/article6595186.ece.

correction towards an action-oriented outlook, regional demands and aspirations and change in the context have all necessitated the new initiatives by India.[18] With over 7 percent economic growth rates, and prospects for sustaining such growth rates, the new leadership had announced a change in policy from being reactive to one of aspiring to be the "leading power" in global affairs as India's foreign secretary S.Jaishankar suggested in a speech at Singapore in 2015. This necessitates a shift in focus and policy and to provide public goods and services to the neighbourhood and beyond. To cushion such a policy is to invoke the diplomatic, commercial and military resources of the country in a coordinated manner. Many of the visits undertaken by PM Modi in the first two years of assuming office since mid-2014 are towards the east and include visits to Japan in September 2014 and November 2016, Myanmar, Singapore, Fiji, South Korea, Mongolia, Vietnam, Australia and others.

The key issues in the Act East policy, according to Minister of State Gen. V.K. Singh, in a reply to a query in the lower house of the Indian Parliament, is "to promote economic cooperation, cultural ties and develop strategic relationship with countries in the Asia-Pacific region through continuous engagement at bilateral, regional and multilateral levels thereby providing enhanced connectivity to the States of North Eastern Region with other countries in our neighbourhood".[19] According to the Ministry of Defence, in its annual report of 2016, under the 'Act East' policy which places renewed emphasis on engagement with the Asia Pacific, India has been an active participant in various bilateral as well as multilateral fora with a focus on security matters such as the East Asia Summit, ADMM – Plus and ASEAN Regional Forum (ARF). There is also a need to further improve regional responses to challenges such as transnational crime, terrorism, natural disasters, pandemics, cyber security as well as food and energy security.[20]

---

18 See Amitandu Palit, "India's Act East Policy and Implications for Southeast Asia," *Southeast Asian Studies*, 2016, pp. 81-91; Danielle Rajendram, "India's new Asia-Pacific strategy: Modi acts East," *Lowy Institute*, December, 2014, pp. 1-24, https://www.lowyinstitute.org/sites/default/files/indias-new-asia-pacific-strategy-modi-acts-east_0.pdf.

19 See Lok Sabha, "Look East and Act East Policy" *Unstarred Question,* No. 3121, March 16, 2016, http://mea.gov.in/lok-sabha.htm?dtl/26554/question+no+3121+look+east+and+act+east+policy.

20 Ministry of Defence Government of India, Annual Report 2016 (New Delhi: 2016) p. 4, http://www.mod.

Speaking to the gathering of over 30 countries at Singapore meeting on the Indian Ocean on September 1, 2016, Minister of State for External Affairs M.J. Akbar suggested that India occupies a unique position between the Phoenix Horizon towards its east with high growth rates and the Toxin Horizon of conflict, war and terrorism towards its west. The strategic choice for India then is clear of veering towards the east.[21]

VI. Political and Diplomatic Dimension

India became a sectoral dialogue partner with ASEAN in 1992, elevated this position to a dialogue partner in 1996 and a Summit level partner in 2002 and to the current "strategic partnership" with the ASEAN in 2012. Further, India elevated this position by appointing an ambassador level interaction with the ASEAN region. Also, diplomatic and political relations were further elevated with Japan, South Korea, Mongolia and Australia.[22] Apart from its summit level meetings with Russia, India has similar annual contacts with Japan since the 1990s. India and Japan further elevated these ties towards a 21st Century partnership. Apart from the 2+2 dialogue mechanism (between foreign and defence ministries), which strengthened mutual understanding on a number of issues, both today have been intensifying relations in the field of defence, bilateral economic and technological fields reinforced by comprehensive economic partnership, and other fields. PM Modi made his first visit abroad to Tokyo in September 2014 outside the South Asian region. Japan is also one of the largest investors in India with over 1300 companies basing their operations in India. Further, Japan is upgrading Indian infrastructure projects by connecting Mumbai with Delhi through a dedicated freight corridor.

---

nic.in/writereaddata/Annual2016.pdf.

21 See M.J. Akbar, "Welcome Remarks by Minister of State for External Affairs M.J. Akbar at the Inaugural Session of Indian Ocean Conference," Ministry of External Affairs Government of India, September 1, 2016, http://mea.gov.in/Speeches-Statements.htm?dtl/27355/welcome+remarks+by+minister+of+state+for+external+affairs+mj+akbar+at+the+inaugural+session+of+indian+ocean+conference+september+01+2016.

22 See Ministry of External Affairs Government of India, Annual Report 2015-16 (New Delhi: 2016), pp. 1-387, https://www.mea.gov.in/Uploads/PublicationDocs/26525_26525_External_Affairs_English_AR_2015-16_Final_compressed.pdf.

South Korea also plays a significant role in the Act East policy of India. Since 2009 relations were upgraded to include maritime cooperation, defence technology transfers and commercial cooperation in the comprehensive economic partnership arrangement. Soon after China's denial of a visa to an Indian Army General hailing from Jammu and Kashmir region in mid-2009, India considered it fit to elevate ties with Seoul by September that year. PM Modi visited Seoul in June 2015 and South Korean Presidents visited India in quick succession.

At a time when India's energy needs are increasingly to support over 7 percent economic growth rates, it is imperative for India to expand its energy basket to include renewable sources such as solar, wind and nuclear power. On the last, it had signed 13 agreements with various countries in the nuclear domain and many in the east, including Japan, South Korea, Australia, Mongolia and others have been cooperative in this regard, although a jarring note is the objections raised by China in the Nuclear Supplies Group and other fora.

The key to the success of both the Look East and Act East policies, however, is the "land bridge" connectivity between India and the Southeast Asian region through Myanmar. Although this link is straddled with a number of problems such as relations with the military junta till recently, insurgency in the northeast region and others, India had approached this aspect through a multipronged approach. Today, India had evolved strategic partnerships with a number of countries in the Asia-pacific region, including with the ASEAN, Australia, Japan, Indonesia, Vietnam, Malaysia, South Korea and Singapore. At the multilateral level, besides intensifying relations with ASEAN Regional Forum and East Asian Summit, India had undertaken new initiatives such as Bay of Bengal Initiative for Multi-Sectoral Technical and Economic Cooperation (BIMSTEC), Asia Cooperation Dialogue, Mekong Ganga Cooperation and Indian Ocean Rim Association ASEAN Defence Ministers' Meeting and others.

In the recent times, the number of visits of high level leadership increased. Former PM Manmohan Singh visited Myanmar in 2012, followed by other visits by the military and the local leaders from northeast. PM Modi visited Naypyidaw in November 2014, reciprocated by State Councilor Aug San Suu Kyi in late 2016.[23]

While both India and China joined the ASEAN processes since the 1990s, and both agreed to the ASEAN being the "driver" in all the processes and agreed to the ASEAN conditions of Treaty of Amity and nuclear weapon free zone for the region, China of late has been asserting its position partly due to the sovereignty claims over the contiguous South China Sea as well as in the leadership contest with the US and Japan over the region. As India has no sovereignty claims in the region, and despite the ASEAN's initial negative responses towards Indian nuclear tests in 1998, many in the ASEAN see India as a balancer and provider of security to the region. After the July 12, 2016 Hague Tribunal ruling, China had also changed its position on ASEAN by increasing following a divide and rule policy in the grouping.[24]

VII. Economic Dimension

One of the most concrete results of the Look East policy is the tapping of economic and technological potential of the rising tiger economies in the East. Apart from Japan, Singapore had emerged as one of the largest investors in India and of late both exhibited increasing understanding on a number of issues.[25] India also considers that in the light of opposition, it was Singapore and Japan which supported India in entering the East Asian Summit process. Singapore's

---

23 For the recent visits of the President to Vietnam in September 2014, PM Modi to Naypyidaw in November 2014 and Foreign Minister Sushma Swaraj to Vietnam, Singapore and Myanmar see Manish Chand, "Act East: India's ASEAN Journey," November 14, 2014, http://mea.gov.in/in-focus-article.htm?24216/Act+East+Indias+ASEAN+Journey.

24 India's position on South China Sea, as articulated in various official statements, include that the sovereignty dispute need to be resolved in a bilateral fashion by the disputants, that to usher in peace and stability in the region crucial for free passage of goods and services, the concerned understandings including the Declaration of the Conduct since 2002 and the relevant multilateral institutions should be given full play, and international law including the 1982 UNCLOS be observed by all parties. After the July 2016 verdict, India suggested to all parties to the dispute that "utmost respect to the UNCLOS" provisions be observed in South China Sea. Given over 55 percent of Indian trade passing through the region, it is natural that India had supported freedom of navigation principle, specifically after the 2009 INS Airavat incident.

25 See Faizal bin Yahya, *Economic Cooperation between Singapore and India An alliance in the making?* (London: Routledge, 2008).

former Prime Minister Goh Chuk Tung once remarked that for the stability of the ASEAN region, India should play a balancer role like the wings of a jumbo jet. In 2003 Singapore and India signed a defence cooperation treaty.

India signed a free trade area with the ASEAN region in 2009, which came into effect in 2010.[26] Trade increased from $56 billion in 2011 to over $76 billion in 2015, constituting over 10 percent of Indian external trade and as the 4th largest trading partner for India. Both have a bilateral trade target of over $200 billion by 2012.[27] In comparison, China which has a free trade with the ASEAN has a bilateral trade of $365 billion by 2014, although China's investments in the ASEAN account for $9 billion by 2014 as compared to $38 billion from India to the region. India has also actively participated in the Regional Comprehensive Economic Partnership and by early 2016 eleven rounds of discussions took place in this format.

Apart from the institutional contacts with the ASEAN, for enhancing contacts, India has of late emphasizing on expanding relations with the immediate neighbours among the BIMSTEC countries. This is seen in making the BIMSTEC constituents as special invitees to the BRICS summit meeting held at Goa in October 2016. Also, India offered a credit line of over $1 billion for enhancing the infrastructure projects in the BIMSTEC countries.

VIII. Connectivity

For further economic integration and providing substance to the Act East policy, India had of late emphasizing on the connectivity with the ASEAN region through road, railway and maritime port links through Northeast region.[28]

---

26 ASEAN India, "ASEAN-India Relations," *ASEAN India*, September, 2017, http://mea.gov.in/aseanindia/20-years.htm.

27 See Ministry of External Affairs Government of India, "Address by Secretary (East) at the Inaugural Session of the International Relations Conference on 'India's Look East - Act East Policy: A Bridge to the Asian Neighbourhood'," December 13, 2014, http://mea.gov.in/Speeches-Statements.htm?dtl/24531/address+by+secretary+east+at+the+inaugural+session+of+the+international+relations+conference+on+indias+look+east++act+east+policy+a+bridge+to+the+asian+neighbourhood+pune+december+13+2014.

28 See Namrata Goswami, "Looking "East" Through India's North East - Identifying Policy "Challenges" and Outlining the "Responses," *IDSA Occasional Paper*, No. 2, June 2009.

The main projects include Kaladan Multi-modal Transit Transport Project, the India-Myanmar-Thailand Trilateral Highway Project and Rhi-Tiddim Road Project, Border Haats. The June 17, 2016 joint statement between India and Thailand noted the progress made in the India-Myanmar-Thailand Trilateral Highway connecting Moreh in Manipur to Mae Sot in Thailand through Myanmar.[29]

According to the estimates of the Asian Development Bank on the multi-infrastructure projects coming up in the Asia-Pacific region are numerous. It mentions about the 90-km Imphal-Moreh road constructed at a cost of $160 million, in addition to a railway project of over 95 km to cost $400 million. In addition, there are also the Chennai port elevated expressway and Sagar Island Deepwater port at Kolkata (at $1.3 billion). In Bangladesh, the new Sonadiya Deepwater Port is expected to cost over $1 billion. Myanmar has also embarked on several projects, including Tamu-Kalay (127 km rail project costing $98 million), Kalay-Mandalay (539 km rail project costing $162 million). In addition, the Myanmar to Thailand road projects involving the different lines of Endu-Kawkareik-Myawaddy-Yagyi-Kalewa-Tamu-Mae Sot project connects from India northeast to Thailand.[30]

IX. Military Dimension

The "action" oriented aspect of Act East policy includes the deployment of the armed forces in the Asia-Pacific region. This has been a significant change in the Indian postures given its traditional policy of independent foreign policy, non-alignment, no military troops or bases abroad. Today, India is actively engaging with the outside world both towards the west and the east, with specific emphasis on the Act East policy. PM Modi had initiated several programmes

---

29 Ministry of External Affairs Government of India,"India-Thailand Joint Statement during the visit of Prime Minister of Thailand to India," June 17, 2016, http://mea.gov.in/bilateral-documents.htm?dtl/26923/indiatha iland+joint+statement+during+the+visit+of+prime+minister+of+thailand+to+india.

30 This is based on Asian Development Bank report accessed at https://www.adb.org/sites/default/files/ publication/174393/regional-transport-infrastructure.pdf. See also the various infrastructure projects coming up in the Asia-Pacific region in the recent times at https://reconnectingasia.csis.org/map/.

including Sagarmala of enhancing the maritime infrastructure projects across the Indian coasts, "Project Mausam" of connecting to the Indian Ocean littorals in commercial and cultural contacts. These are also part of the overall Make in India, Digital India initiatives undertaken by the new government since 2014. At one time, India had been part of the counter-terrorism support operations at Ainee in Tajikistan. In addition to Haiphong & Nha Trang in Vietnam where Indian naval ships could dock, they also have access to Subic Bay & Clark Air Base in the Philippines. With PM Modi's "Indian Ocean diplomacy" in 2014, Indian ships have replenishment facilities at Agaléga Islands in Mauritius, and in northern Mozambique. Recently, as well with the construction of the Chabahar port in Iran, India intends to connect to the energy-rich Central Asian Republics. In 2001, the then Group of Ministers considered Indian focus from the Gulf of Aden to the Straits of Malacca's. Further, in 2007, the Indian armed forces have adopted a posture of Indian Ocean as their primary focus while South China Sea and beyond as of secondary importance. In 2015, this posture was acknowledged as the national policy. Act East policy is thus subservient to this overall national policy of considering Indian Ocean as of utmost importance to the country. India intends to usher in its "leading" role in this region.

Vietnam is considered to be the "central pillar" in India's Act East policy, specifically in defence ties.[31] Both countries embarked on a comprehensive mutual support with Vietnam providing urban counter-terrorism training to the Indian forces, while India assisting Vietnam in training in Kilo-class submarine operations, Su-30 aircraft and other platforms, in addition to toying the idea of supplying to Vietnam Brahmos cruise missiles and fast attack craft. Also, Vietnam granted the Indian naval forces permanent berthing facilities for visiting ships at Haiphong and Nha Trang in Vietnam. During the PM Modi's visit to Hanoi in late 2016, it was agreed that India will provide Vietnam with a credit line, expanded

---

31 On the eve of PM Modi's visit to Hanoi, just before the Hangzhou G-20 Summit meeting, in September 2016, Indian foreign ministry's Secretary (East) suggested that Vietnam forms the "central pillar". See Scroll Staff, "Vietnam a 'central pillar' of India's Act East Policy, government says as Narendra Modi begins visit" Scroll.in, September 2, 2016, http://scroll.in/latest/815579/vietnam-a-central-pillar-of-indias-act-east-policy-government-says-as-narendra-modi-begins-visit.

from the September 2014 figure of $100 million, to over $500 million. Vietnam also agreed to provide facilities for satellite link-up facilities for India. In addition to the energy sector cooperation where both the state-owned ONGC Videsh Limited and private sector ESSAR companies have entered drilling operations in South China Sea, bilateral cooperation fields are expanding substantially between India and Vietnam.

Indian armed forces conduct a variety of military engagements with a number of their counterparts in the Asia-Pacific region. These include the participation in Langkawi International Maritime Aerospace Exhibition (LIMA-) annual events in Malaysia; International Maritime Defence Expo (IMDEX) Asia at Singapore, ASEAN Regional Forum Disaster Relief Exercises in Malaysia, International Fleet Review at Sagami Bay, Japan; as a part of the Eastern Fleet Overseas Deployment, Indian naval ships have been deployed to South China Sea in May-June 2015 and participated in the SIMBEX events with Singapore, and PASSEX with their counterparts Indonesia, Australia, Thailand and Cambodia. Indian Navy also conducts coordinated patrols (CORPAT) with its counterparts in Myanmar, Thailand and Indonesia, in addition to being an "Observer Plus" in the Cobra Gold multilateral exercises.[32] With Australia, Indian navy conducts AUSINDEX exercises since 2015 for enhancing maritime security, cooperation and interoperability. Indian armed forces have been involved in pitching their hard power in the natural disaster relief measures in the Asia-Pacific region. These include dispatching 27 ships and 5,000 personnel for providing relief and evacuation of victims during the Tsunami of 2004-05 in Indonesia, Thailand, Sri Lanka, Maldives and other regions, involvement in "Operation Nargis" in Myanmar of relief measures, and the like.

The annual Malabar Exercises almost uninterruptedly conducted with the United States naval forces since the early 1990s, has become a trilateral event since 2015 with the participation of the Japanese Self Defence Maritime Forces.

---

32 Ministry of Defence Government of India, *Annual Report* 2016 (New Delhi: 2016) pp. 29-32, http://www.mod.nic.in/writereaddata/Annual2016.pdf.

In September 2007, all the three, in addition to Australia and Singapore joined the exercise in Bay of Bengal but were discontinued with China's demarches. It appears that such multilateral naval exercises could commence in near future.

On November 23, 2015, speaking at the Shangri-La, PM Modi observed that one of the main drivers of the recently enacted Act East policy include efforts to enforce freedom of navigation on the high seas. This observation comes in the wake of "territorial sea" proponents in the South China Sea that intends to create enclaves and possible restrictions on free trade passages in the region.[33] However, while the Obama-Modi joint vision statement of January 2015 called for consolidation of positions in the Indian Ocean and the South China Sea, the latter region is more complicated given the Chinese assertiveness in the region despite the July 12, 2016 judgement by the Permanent Court of Arbitration at The Hague quashing several sovereignty claims in the region. The United States want India to play a "lynchpin" role and as a "net security provider" in the Indi-Pacific region. While the above details suggest that the Indian armed forces have been active in the Act East policy, sensitivities were expressed on "joint patrolling" in the South China Sea region. This has been part of an ongoing dialogue among several stakeholders in the region.

Thus, India's Act East policy has re-invigorated its ties to the Indo-Pacific region. Diplomatic, political, commercial, defence people-to-people and cultural contacts increased between India and its eastern neighbours stretching from Japan and Fiji to Maldives. The number of high-level political visits increased recently bringing in its wake an increasing understanding on a number of issues of mutual concern. Given the lack of any territorial ambitions on the part of India, and showcasing its economic potential in the recent times with high growth rates had all attracted India to the East and Southeast Asian countries. Making available its hard power for disaster relief and other humanitarian measures further

---

33 Nilanjana Sengupta, "Modi lays out India's Act East policy in Singapore Lecture at the Shangri-La," *The Straits Times*, November 23, 2015, http://www.straitstimes.com/singapore/modi-lays-out-indias-act-east-policy-in-singapore-lecture-at-the-shangri-la.

solidified relations with these two regions. However, the unfolding of Act East policy has its own dissonances, mainly from China whose rise in the last three decades is altering the regional order. Thus, there is a potential for competition or even conflict in overlapping areas — such as South China Sea - between these two Asian giants, in addition to the role of Japan which is also retrieving its lost ground in the recent times. While India-China conflict is muted currently, competition for influence in the region is visible, although both have mutually beneficial economic relations in addition to conflict preventive confidence building measures. On the other hand, India and Japan (and Singapore) have been intensifying cooperation on a range of issues, including on how to counter an aggressive Chinese posture. They support a rule based order, including freedom of navigation in the maritime regions. Primarily intended to protect the burgeoning economic and trade links with the east, India's Act East policy is also acquiring teeth with the increasing engagement of the Indian armed forces with many of its counterparts in the region. When the Quad was formed in November 2017, Indian foreign ministry stated that :

> They [the Quad officials] agreed that a free, open, prosperous and inclusive Indo-Pacific region serves the long-term interests of all countries in the region and of the world at large. The officials also exchanged views on addressing common challenges of terrorism and proliferation linkages impacting the region as well as on enhancing connectivity.[34]

The above suggests that while India has expressed its interest in the Indo-Pacific idea and had committed several resources, such efforts are expressed through its own policies such as Act East policy, although the trilateral Malabar exercises can be termed as a concrete expression of this initiative.

---

34 Ministry of External Affairs Government of India, "India-Australia-Japan-U.S. Consultations on Indo-Pacific (November 12, 2017)," November 12, 2017, http://mea.gov.in/press-releases.htm?dtl/29110/IndiaAustraliaJapanUS_Consultations_on_IndoPacific_November_12_2017.

X. Observations

The idea of Indi-Pacific is of recent origin although discussions about it are more than a decade old. The success or failure of the Indo-Pacific will be based on the ability of the Quad to institutionalize its efforts and expand to co-opt the regional states in the ASEAN and others in the two oceans- an area increasingly seen as contested between China and other powers.[35] Since the Declaration of Conduct on the South China Sea in early 2000s, the ASEAN had for the first time exhibited serious weaknesses in not coming out with a consensus statement on the islands dispute. Laos, Cambodia, Myanmar — the non-claimants — are seen more close to the Chinese position and reportedly blocked the consensus. Other states which are claimants to the dispute either remained powerless or were divided by China's charm offensive. Philippines President Duterte's visit to Beijing and the latter's announcement of largesse of over $24 billion in proposed investments in Mindano and other areas have softened this country's position. Brunei, another claimant as well has been making positive statements towards Beijing. Malaysian purchases of military weapons from Beijing are creating conditions of friendship with China. This leaves Vietnam, which fought historical wars against the Middle Kingdom, to oppose China's incursions in the South China Sea. This became more evident after the deployment of Haiyang Shiyou rig by China in the Paracels since May 2014. A number of pitched battles both on the seas as well as in the civil society ensued. Vietnam has also diversified relations with the US, Japan and India in the hopes of balancing China.

Secondly, the idea of Indo-Pacific and the revised Quad is an answer to usher in security and stability in the region and to put China's unbridled assertiveness in its place. Of course the Quad is straddled with several problems-suggesting China could wade through.

The Quad has not been able to make a joint statement of purpose and an institutional mechanism but the US — India — Japan have organised

---

35 "The Indo-Pacific- Defining a Region," *Stratfor,* Nov. 15, 2017, https://worldview.stratfor.com/article/indo-pacific-defining-region.

"inter-operable" Malabar naval exercises since 2015 besides political coordination. The Quad could address "low politics" for the moment by tackling piracy, disasters, etc given China's active opposition and potential strengths. The economic side of the Quad is also weak with the Trans Pacific Partnership revoked. While Australian Prime Minister Turnbull in a recent visit to the US suggested to a joint regional economic block, this may take some time to fructify. However combined signalling — even in norms and symbolic measures — itself could make the Quad relevant for the evolving regional situation.

Moreover China could deter the Quad once again by sending different signals to different actors and coercive measures. While at the official level it had exhibited confidence or even suggesting that such initiatives could "destabilise" the region, think — tanks and others have raised the bogey of "containment" of China forgetting for the moment to the complex interdependence in the region. More mature Chinese analysts see the Quad at the most as a balancer to an unpredictable but rising China.

When Giants Vie: Great Power Rivalry, ASEAN, and the Regional Security Architecture in East Asia

See Seng Tan *

I. Introduction

East Asia's regional security architecture has come under strain in the light of growing Sino-American competition over the South China Sea (SCS). Increasingly, the actions of some of the participating countries in ASEAN-led multilateral arrangements — the ASEAN Regional Forum (ARF), the ASEAN Defence Ministers Meeting-Plus (ADMM+), among others — have threatened to turn those mechanisms into battlegrounds where participants spar over their respective national interests. In particular, perspectival and policy differences between China and the United States over the SCS disputes have threatened to turn ARF and the ADMM+ meetings into arenas of big-power sparring. Moreover, in advancing their own interests, those major powers have apparently sought to use their size and sway to shape the behaviours of the smaller and weaker ASEAN countries; more than any other great power, China has been singled out in this regard.[1] In a sense, arrangements like the ARF and the ADMM+ are being used as they were meant for, namely, as places of debate where states can argue and presumably work out their disagreements. However, the gridlocks that have ensued are leading many to dismiss those arrangements — and, crucially ASEAN given its central position in the regional architecture — as increasingly irrelevant to the security and stability of East Asia.

---

* Professor and Deputy Director, Institute of Defence and Strategic Studies; Rajaratnam School of International Studies, Nanyang Technological University

1 Frank Ching, "Beijing must not abuse new influence: The China Post," *The Straits Times*, January 6, 2016, http://www.straitstimes.com/asia/east-asia/beijing-must-not-abuse-new-influence-the-china-post.

The ability and will of the Association of Southeast Asian Nations (ASEAN) to effectively lead the region through building and maintaining a viable regional architecture that could deliver the peace and prosperity dividends sought after by all stakeholders has come into question.[2] Memorably, in 2008-2009, Australian Prime Minister Kevin Rudd proposed his 'Asia Pacific Community' (APC) as replacement for the discernibly moribund ARF.[3] Yet there is no question that Sino-US competition, as played out over the SCS disputes, has accentuated existing cleavages within ASEAN and created new ones,[4] thereby eroding further ASEAN's ability to manage its region. ASEAN's brand of multilateral diplomacy and institutionalism has grown less effective as regional strategic challenges have evolved over the last two decades. Most crucially, it appears ASEAN can no longer facilitate the negotiation among the great powers themselves of the *modus vivendi* needed to manage the region's stability.[5] In a sense, this is not particularly surprising since ASEAN's place and role in East Asian regionalism has largely confounded expectations — certainly realist expectations, which suppose small or weak states to be objects rather than subjects of great power attention.[6] However, measured against its past success, ASEAN, perhaps somewhat unfairly, has had to endure criticism for having 'failed' to manage big power relations.

Against this vexing backdrop for ASEAN, this paper proffers this modest

---

2 Granted, unhappiness with ASEAN and its leadership of the regional architecture has been around well before the present-day tensions over the SCS. Northeast Asian participants received unequal treatment at the hands of their ASEAN counterparts during the early days of the Forum. See, Michael Leifer, "The ASEAN Regional Forum: Extending ASEAN's Model of Regional Security," *Adelphi Paper*, Vol. 36, No. 302, 1996, p. 41.

3 Carlyle A. Thayer, "Kevin Rudd's multi-layered Asia Pacific Community initiative," *East Asia Forum*, June 22, 2009, http://www.eastasiaforum.org/2009/06/22/kevin-rudds-multi-layered-asia-pacific-community-initiative/.

4 See, for example: Bilahari Kausikan, "Consensus, centrality and relevance: Asean and the South China Sea," The Straits Times, August 6, 2016, http://www.straitstimes.com/opinion/consensus-centrality-and-relevance-asean-and-the-south-china-sea; Heng Sarith, "ASEAN: between China and America," *East Asia Forum*, July 17, 2013, http://www.eastasiaforum.org/2013/07/12/asean-between-china-and-america/.

5 Evelyn Goh, "ASEAN-led Multilateralism and Regional Order: The Great Power Bargain Deficit," *The Asan Forum* (Special Forum), May 23, 2014, http://www.theasanforum.org/asean-led-multilateralism-and-regional-order-the-great-power-bargain-deficit/.

6 Richard Stubbs, "ASEAN's Leadership in East Asian Region-Building: Strength in Weakness," *The Pacific Review*, Vol. 27, No. 4, 2014, pp. 523-541.

proposition: the obligation for East Asia's order and the regional architecture established in its service is not ASEAN's alone to bear, but one shared by all the stakeholders of that architecture, certainly China and the US. Far from normative, this contention is an empirical fact because, for all intents and purposes, a number of non-ASEAN countries have contributed significantly to defining the form and function of East Asia's multilateral architecture from its beginnings. The aforesaid developments have led some, fairly or otherwise, to formulate the following conventional wisdom about the grouping's ostensive place and role in East Asian regionalism. First and most fundamental is that of the primacy of ASEAN in the conceptualization, formation and maintenance of regional architecture. ASEAN, it is often said, sits in the 'driver's seat' of the regionalism vehicle and works hard to maintain its centrality in the architecture.[7] At the ASEAN Ministerial Meeting in Hanoi in July 2010, ASEAN dialogue partner countries unequivocally reaffirmed their support for 'ASEAN Centrality' as well as their declared hope that 'ASEAN would continue to play a central role in the emerging regional architecture'.[8] The second and consequent assumption is the apparent passivity of the non-ASEAN partners in regionalism in the light of their volitional concession to ASEAN's regional leadership. In short, they are perceived as mere followers or norm-takers, rather than leaders or norm-makers, in East Asian regionalism.[9] Thirdly, the logical conclusion therefore is that non-ASEAN participants in East Asian regionalism, having acknowledged ASEAN as the rightful occupant of the driver's seat in the regionalist enterprise, are thereby absolved of responsibility and accountability for the perceived inefficiencies and ineffectiveness of regional architecture, because blame rests principally with ASEAN and the perceived flaws of its institutional design and model of diplomacy cum security.[10]

---

7 For instance, Article 1 of the ASEAN Charter stresses the need to 'maintain the centrality and proactive role of ASEAN as the primary driving force in its relations and cooperation with its external partners in a regional architecture that is open, transparent and inclusive'.

8 ASEAN, "ASEAN Centrality and EAS tops AMM agenda," July 20, 2010. http://www.asean.org/news/item/aseancentrality-and-eas-tops-amm-agenda-ha-noi-20-july-2010#.

9 Amitav Acharya, "Ideas, identity, and institution-building: From the 'ASEAN way' to the 'Asia-Pacific way'?," *The Pacific Review*, Vol. 10, No. 3, 1997, pp. 319-346.

10 Allan Gyngell, "Design Faults: The Asia Pacific's Regional Architecture," *Lowy Institute Policy Brief*, July 18, 2007; Ralf Emmers and See Seng Tan, "The ASEAN Regional Forum and Preventive Diplomacy: Built

These three assumptions are rather commonplace in the academic debate on East Asian regionalism.[11] The reality, however, is that non-ASEAN partners, variously promoting and protecting their respective interests, have actively defined and molded the substance and style of East Asia's regional security architecture. As mentioned above, ASEAN's ambition to define a proprietary role for itself in the ARF was seen by some as an attempt at unwarranted exclusivism; for instance, the Japanese argued that 'non-ASEAN countries should also actively get involved in the management of the ARF in the light of equal participation'.[12] If anything, their affirmations of ASEAN's centrality in the regional architecture indirectly reflect their corporate participation — uneven and unequal, to be sure — together with their ASEAN counterparts in the norm-making process.[13] And they are still doing it, albeit in increasingly less salubrious ways as evidenced by a number of testy exchanges between Chinese and US representatives at recent ARF and ADMM+ meetings.

II. China, the US, and the Regional Security Architecture in East Asia

Notwithstanding legitimate concerns regarding ASEAN's increasingly troubled trusteeship of the regional security architecture it helped to establish, many observers are agreed that the vital contribution of ASEAN to regional order was its success in persuading the great powers to commit to a structure multilateral confidence-building at a time of strategic uncertainty following the ending of the Cold War. Understandably, middle powers welcomed the move for the opportunities it would supposedly afford them to shape the region in ways they would otherwise not have been able to do on their own; for example,

---

to Fail?" *Asian Security*, Vol. 7, No. 1, 2011, pp. 44-60.

11 See the essays in See Seng Tan, ed., *Regionalism in Asia (Vol. III): Regional Order and Architecture in Asia* (Abingdon, Oxford: Routledge, 2009).

12 Japanese Foreign Minister Kono Yohei, cited in Takashi Yuzawa, *Japan's Security Policy and the ASEAN Regional Forum: The Search for Multilateral Security in the Asia-Pacific* (Abingdon: Routledge, 2007), p. 63.

13 Hiro Katsumata, "Establishment of the ASEAN Regional Forum: constructing a 'talking shop' or a 'norm brewery'?" *The Pacific Review*, Vol. 19, No. 2, 2006, pp. 181-198.

Australia and Japan led in the establishment of the Asia-Pacific Economic Cooperation (APEC).[14] On the other hand, great powers are arguably less likely to support and participate in multilateral arrangements not of their own making,[15] not least because of the institutional constraints placed on their ability to act unilaterally — in theory.[16] It bears reminding that both China and the US, for different reasons, viewed post-Cold War proposals to build the architecture for multilateral security dialogue in East Asia with great suspicion. According to J.D. Yuan, 'China is strongly opposed to establishing any institutionalised mechanisms for dealing with regional security issues since the countries in the region are vastly different in terms of history, culture, political and social systems, and different visions of national security and priorities'.[17] Likewise, in the view of a noted analyst of US foreign policy, the US 'has always been ambivalent about multilateralism'.[18] Yet ASEAN, working alongside other regional advocates, succeeded in persuading both the Chinese and the Americans to invest in the region's emerging multilateral venture.

---

14 See Andrew Carr, "Is Australia a Middle Power: The Systemic Impact approach," *Australian Journal of International Affairs*, Vol. 68, No. 1, 2014, pp. 70-84; Mark Beeson and Richard Higgott, "The changing architecture of politics in the Asia-Pacific: Australia's middle power moment?," *International Relations of the Asia-Pacific*, Vol. 14, No. 2, 2014, pp. 215-237; Yoshihide Soeya, "Prospects for Japan as a middle power," *East Asia Forum*, July 29, 2013, http://www.eastasiaforum.org/2013/07/29/prospects-for-japan-as-a-middle-power/. Also see the chapters on Australia and Japan in See Seng Tan, *Multilateral Asian Security Architecture: Non-ASEAN Stakeholders* (Abingdon, Oxon: Routledge, 2015), pp. 41-62, 63-86.

15 Crucially, Washington's first mover role in Europe's post-war multilateral framework saw the United States taking the lead to build formal multilateral arrangements – the United Nations (UN), the North Atlantic Treaty Organization (NATO), and the Bretton Woods monetary system undergirded by the International Monetary Fund (IMF) and the World Bank – that would contribute to not only Europe's reconstruction (in conjunction with the Marshall Plan) but indeed global governance itself. See, G. John Ikenberry, *After Victory: Institutions, Strategic Restraint, and the Rebuilding of Order after Major Wars* (Princeton, N.J.: Princeton University Press, 2001). On NATO, see, Frank Schimmelfennig, "Multilateralism in Post-Cold War NATO: Functional Form, Identity-Driven Cooperation," paper for AUEB International Conference on "Assessing Multilateralism in the Security Domain," Delphi, Greece, June 3-5, 2005, pp. 2-46.

16 David M. Malone and Yuen Foong Khong, eds., *Unilateralism and U.S. Foreign Policy: International Perspectives* (Boulder, CO: Lynne Rienner, 2003).

17 Jing-dong Yuan, *Asia-Pacific Security: China's Conditional Multilateralism and Great Power Entente* (Carlisle, PA: Strategic Studies Institute, US Army War College, 2000), p. 7.

18 Michael Mastanduno, "After Bush: A Return to Multilateralism in U. S. Foreign Policy?" *Nanzan Review of American Studies*, Vol. 30, 2008, pp. 33-46, see p. 34.

China has evolved from a wary neophyte at multilateral diplomacy to a self-assured connoisseur and convenor of the practice.[19] According to David Shambaugh, 'China's perception of [multilateral] organizations [has] evolved from suspicion, to uncertainty, to supportiveness'.[20] China's involvement in East Asian multilateralism has not only provided it with a platform from which to promote if not protect its economic and security interests, but also, or so hoped, to assuage regional concerns and reassure neighbours on how China would deploy its rising power and influence.[21] For much of the 1990s and the 2000s, China advanced, with relative success, the idea that its rise in power and influence was an essentially 'peaceful' development that did not threaten others. The idea that China's rise as having been relatively peaceful, status quo-oriented, and a key pillar for a stable regional order has also received strong backing by noted scholars.[22] It patiently endured and benefited from US primacy and leadership in Asia, but of late has demonstrated an increasing tendency to question that given the perceived decline in US power and influence in contrast to its own rising power and influence. Rather than 'backyards', 'buffer zones' and 'spheres of influence', China's preference has been to emphasize a regional 'community-building' approach.[23] Echoing previous Chinese leaders from

19 See, for example: Guoguang Wu, "Multiple Levels of Multilateralism: The Rising China in the Turbulent World," in Guoguang Wu and Helen Landsdowne, eds., *China Turns to Multilateralism: Foreign Policy and Regional Security* (Abingdon, Oxon: Routledge, 2008), pp. 267-289; Hongying Wang and Erik French, "China's Participation in Global Governance from a Comparative Perspective," *Asia Policy*, No. 15, 2013, pp. 89-114.

20 David Shambaugh, "China Engages Asia: Reshaping the Regional Order," *International Security*, Vol. 29, No. 3 (Winter 2004/05), pp. 64-99, see p. 69.

21 Chien-peng Chung, *China's Multilateral Cooperation in Asia and the Pacific: Institutionalising Beijing's 'Good Neighbour Policy'* (Abingdon, Oxon: Routledge, 2010).

22 See, David C. Kang, "Getting Asia Wrong: The Need for New Analytical Frameworks," *International Security*, Vol. 27, No. 4, 2003, pp. 57-85; Alastair Iain Johnston, "Is China a Status Quo Power?," *International Security*, Vol. 27, No. 4, 2003, pp. 5-56; Alastair Iain Johnston, *Social States: China in International Relations, 1980-2000* (Princeton N.J.: Princeton University Press, 2008). But not every Asian country is persuaded that a China-centric order in Asia is a good thing. See, Amitav Acharya, "Will Asia's Past Be Its Future?," *International Security*, Vol. 28, No. 3, April, 2003, pp. 149-164.

23 See, Yugang Chen, "Community-Building or Rebalancing? China and the United States in Southeast Asia," in Mingjiang Li and Dongmin Lee, eds., *China and East Asian Strategic Dynamics: The Shaping of a New Regional Order* (Lanham, MD: Lexington Books, 2011), pp. 3-18; Zhang Xiaoming, "The Rise of China and Community Building in East Asia," *Asian Perspective*, Vol. 30, No. 3, 2006, pp. 129-148.

Deng Xiaoping on, Chinese President Xi Jinping has claimed that China would never seek 'hegemony or expansion' in the Asia-Pacific — the 'Indo-Pacific' seems to be the preferred term these days — even as its efforts to strengthen its diplomatic and military footprint in the region have raised concerns about Chinese intentions.[24] In the international economic arena, it has supported the World Trade Organization (WTO) and promoted unilateral and multilateral trade liberalization.[25] On the other hand, if it is true that great powers as a rule do not make good multilateral actors,[26] then China's involvement in East Asia's multilateral arrangements, as recent years particularly under Xi's leadership have shown, has also exposed the limits of socialization to institutional norms particularly as China has gone from strength to strength.[27]

China's gradual involvement in East Asian multilateral diplomacy can be attributed in part to ASEAN's efforts to integrate it into the emerging post-Cold War regional security architecture. During the Cold War period, ASEAN went from an anti-Chinese, anti-communist alliance in its formative years, at least in the eyes of Beijing, to a partner of sorts as a consequence of common aversions.[28] Even then, it warily viewed East Asian multilateralism — particularly if it were to take after European institutions and practices — as a prospective mechanism with which the West would employ to conscribe China's rise.[29] However, other

---

24 Jeremy Blum, "Former foreign minister says 'China will never seek to become a hegemonic power'," *South China Morning Post*, September 18, 2013, http://www.scmp.com/news/china-insider/article/1312346/former-foreign-minister-says-china-will-never-seek-become; Patrick Donahue and Brian Parkin, "Xi Says China's Military Expansion Not Aimed at Asian Hegemony," *Bloomberg News*, March 29, 2014, http://www.bloomberg.com/news/2014-03-28/xi-says-china-s-military-expansion-not-aimed-at-asian-hegemony.html.

25 Razeen Sally, "Free Trade Agreements and the Prospects for Regional Integration in East Asia," *Asian Economic Policy Review*, Vol. 1, No. 2, 2006, pp. 306-321.

26 Steven Holloway, "US Unilateralism at the UN: Why Great Powers Do Not Make Great Multilateralists," *Global Governance*, Vol. 6, No. 3, 2000, pp. 361-381.

27 Hongying Wang, "Multilateralism in Chinese Foreign Policy: The Limits of Socialization," *Asian Survey*, Vol. 40, No. 3, 2000, pp. 475-491.

28 China needed ASEAN's diplomatic backing against China's main Cold War adversaries, Vietnam and the Soviet Union, whereas ASEAN needed China's support in its diplomatic effort to prevent non-communist Southeast Asia from falling into Vietnamese hands. Tan, Multilateral Asian Security Architecture: Non-ASEAN Stakeholders, p. 91.

29 According to J.D. Yuan, 'An OSCE-type institutional arrangement not only will not be able to deal with the complexity of issues but also likely falls under the control of certain powers. Indeed, Chinese analysts

participating countries, including the ASEAN states, envisioned the APEC and ARF as regional platforms to deeply engage post-revolutionary China and help socialise its evolution towards becoming a 'normal' and 'responsible' power.[30] Part of that arguably involved the others conceding to China's unease with the third and final phase of a proposed three-stage roadmap for security cooperation in the ARF. In particular, conflict resolution, conceived in the ARF Concept Paper as the culmination of confidence-building and preventive diplomacy, was viewed by the Chinese as problematic for the implications it held for the possibility for intrusion and intervention by other countries in China's domestic affairs. At the second ARF Senior Officials Meeting (SOM) in Bandar Seri Begawan in August 1995, the Chinese succeeded in getting the offending stage replaced with 'elaboration of approaches to conflict'.[31] During the mid-1990s through much of the 2000s, China also used the regional architecture — at the ARF in March 1997, for the first time — to disseminate its 'new security concept' which, on the surface, read like a repackaging of its time-honoured 'five principles of peaceful coexistence' while advancing the notion that military alliances are an outmoded relic of the Cold War or should be seen as such.[32] With ASEAN's help, China adroitly employed the regional architecture — including 'Track 2' multilateral platforms such as the Council for Security Cooperation in the Asia-Pacific (CSCAP), the Pacific Economic Cooperation Council (PECC) and the Network of Northeast Asian Think Tanks (NEAT), inter alia — to promote its rise or

---

assert that a direct transplant of the CSCE model to the Asia-Pacific region is impractical and may even be counterproductive'. Yuan, Asia-Pacific Security: China's Conditional Multilateralism and Great Power Entente, p. 7.

30 Jusuf Wanandi, "ASEAN's China Strategy: Towards Deeper Engagement," *Survival*, Vol. 38, No. 2, 1996, pp. 117-128; Amitav Acharya, "ASEAN and Conditional Engagement," in James Shinn, ed., *Weaving the Net: Conditional Engagement with China* (New York: Council on Foreign Relations, 1996, pp. 220-248. There are, to be sure, limits to socialization; see, Wang, "Multilateralism in Chinese Foreign Policy: The Limits of Socialization"; Alice D. Ba, "Who's Socializing Whom? Complex Engagement in China-ASEAN Relations," *The Pacific Review*, Vol. 19, No. 2, 2006, pp. 157-179.

31 Michael Leifer, "China in Southeast Asia: Interdependence and Accommodation," in David Goodman and Gerald Segal, eds., *China Rising: Nationalism and Interdependence* (Abingdon: Routledge, 2013), pp. 156-171, see p. 166.

32 David M. Finkelstein, "China's "New Concept of Security"," in Stephen J. Flanagan and Michael E. Marti, eds., *The People's Liberation Army and China in Transition* (Washington, DC: National Defense University Press, 2003), pp. 197-209.

development as inherently 'peaceful'.[33] Indeed, throughout the 2000s or at least the period referred to as China's so-called 'charm offensive' years[34] — in 2002, China and ASEAN established a free trade agreement as well as the Declaration on the Conduct by Parties in the South China Sea (DOC); in 2003, China became the first major power to sign ASEAN's Treaty of Amity and Cooperation (TAC) and it inked a strategic partnership with ASEAN — many regional countries regarded China for the most part as 'a good neighbour, a constructive partner, a careful listener, and a non-threatening regional power'[35] — adjectives that might seem at odds with today's perceptibly more 'hegemonic' China.[36]

Following the Asian financial crisis (AFC) of 1997, China's reputation as a regional stakeholder and putative provider of regional public goods benefited indirectly from the perceived highhandedness of the United States and the International Monetary Fund (IMF), which not only imposed upon ailing East Asian economies bitter structural readjustments but effectively killed Japan's proposal for an Asian Monetary Fund (AMF).[37] Together with Japanese and South Korea support, Chinese backing was key to ASEAN's formation of the ASEAN+3 in 1997 and the Chiang Mai Initiative (CMI), a currency swap arrangement, in 2000. China's perceived dominance in the ASEAN+3 presumably led other East Asian neighbours to seek ways to 'dilute' that through expanding the EAS — first by including Australia, India and New Zealand when the EAS inaugurated in 2005, and subsequently Russia and the US in 2011. On the other hand, China's recent emphasis of its other multilateral ventures, such as the Shanghai Cooperation Organization (SCO) and the Conference on Interaction and Confidence-Building Measures in Asia (CICA) — the latter a platform used

---

33   On the use of Track 2 networks to promote specific regional visions, see, See Seng Tan, *The Making of the Asia-Pacific: Knowledge Brokers and the Politics of Representation* (Amsterdam: Amsterdam University Press, 2013).

34   Joshua Kurlantzick, *Charm Offensive: How China's Soft Power Is Transforming the World* (New York: New Republic Books, 2008).

35   Shambaugh, "China Engages Asia: Reshaping the Regional Order," p. 64.

36   Mushahid Ali, "China's Hegemonic Trajectory: Intimidating ASEAN?," *RSIS Commentaries*, No. CO16245, October 4, 2016, https://www.rsis.edu.sg/wp-content/uploads/2016/10/CO16245.pdf.

37   Phillip Y. Lipscy, "Japan's Asian Monetary Fund Proposal," Stanford Journal of East Asian Affairs, Vol. 3, No. 1, 2003, pp. 93-104.

by Xi in 2014 to promote his exclusivist idea that Asia's security should be the sole prerogative of Asians alone (i.e., Xi's so-called 'Asian security concept'),[38] and in 2016 to propose 'a new regional security cooperation structure that would "suit Asia's unique traits" and also a new crisis-control mechanism to prevent regional issues from flaring up'.[39] Furthermore, at the APEC summit in Manila in 2015 and presumably as a counterproposal to the Trans-Pacific Partnership (TPP), Xi called for Asia-Pacific leaders to align their development plans and establish the so-called 'Free Trade Area of the Asia Pacific'.[40] Together with other Chinese initiatives such as the Asian Infrastructure Investment Bank (AIIB) and the Belt and Road Initiative (BRI), the string of new multilateral ventures advanced by China have understandably worried ASEAN advocates, who view them as evidence of increasing Chinese impatience with ASEAN and the ASEAN-based regional architecture as well as a Chinese exercise in 'unilateral multilateralism' apart from ASEAN.[41]

On its part, the US consistently rejected proposals and discouraged initiatives for multilateral institution-building in Cold War Asia. In the 1980s, the United States rejected Soviet leader Mikhail Gorbachev's proposal for an Asia-wide security forum along the lines of the Conference for Security Cooperation in Europe (CSCE) — an idea at the core of Gorbachev's comprehensive 'Asia-Pacific agenda' — because it revived US suspicions reserved for Leonid Brezhnev's proposal in 1969 for a system of collective security in Asia.[42] As the

---

38 Zhang Yu, "Xi defines new Asian security vision at CICA," *The Global Times*, May 22, 2014, http://www.globaltimes.cn/content/861573.shtml.

39 Kor Kian Beng, "Xi Jinping calls for security structure that 'suits Asia'," *The Straits Times*, April 29, 2016, http://www.straitstimes.com/asia/east-asia/xi-jinping-calls-for-security-structure-that-suits-asia.

40 Ned Levin, "China's Xi Urges New Free-Trade Pact at APEC," *The Wall Street Journal*, November 18, 2015, http://www.wsj.com/articles/chinas-leader-urges-new-free-trade-pact-at-apec-1447827988.

41 This author's communications with ASEAN-based analysts and practitioners. In fairness, China's 'new multilateralism' is in fact global, and not just East Asian, in ambition and orientation. See, Alice Ekman, "China's multilateralism: higher ambitions," *Issue Alert*, No. 2 (Paris: European Union Institute for Security Studies, 2016).

42 Leszek Buszynski, "Russia and the Asia-Pacific Region," *Pacific Affairs*, Vol. 65, No. 4, Winter, 1992-1993, pp. 486-509; Tom J. Farer, "The Role of Regional Collective Security Arrangements," in Thomas G. Weiss, ed., *Collective Security in a Changing World* (Boulder, CO: Lynne Rienner, 1993), pp. 153-189; Arnold L. Horelick, "The Soviet Union's Asian Collective Security Proposal: A Club in Search of Members," *Pacific Affairs*, Vol. 47, No. 3, Autumn, 1974, pp. 269-285.

'hub' of the post-WWII San Francisco system of bilateral alliances with a number of Asia-Pacific countries — such as Australia, Japan, the Philippines, South Korea, Taiwan and Thailand — and a key partner of the arrangements like the Australia, New Zealand, US Security Treaty (ANZUS) as well as the now defunct Southeast Asia Treaty Organization (SEATO), the United States worried over whether advocacy for security multilateralism meant the concomitant rejection of its alliances in the region. In this regard, the assurance the Americans got from ASEAN was the ARF was neither a replacement for nor alternative to the system of US-Asian alliances and the balance of power, but a complement to it, whether as an adjunct or an equal.[43]

Importantly, the initial US reluctance to participate in the regional security architecture did not connote an absolute rejection of multilateralism as a whole.44 It was President Bill Clinton who initiated the upgrading of APEC from a gathering of economic ministers to a summit by inviting heads of government to the APEC meeting in Seattle in 1993.45 And despite the emphasis of his administration on the war on terrorism and the antipathy towards multilateralism evident among neoconservatives advising him on foreign policy, George W. Bush firmly supported APEC. Unlike the ARF, EAS and ADMM+, the APEC is not an

---

43 The 'ARF-as-adjunct' argument is from Michael Leifer, The ASEAN Regional Forum: Extending ASEAN's Model of Regional Security, *Adelphi Paper* No. 302 (London: Oxford University Press for International Institute for Strategic Studies, 1996), whereas the 'ARF-as-equal' argument – where the ARF is understood as a 'mechanism for defusing the conflictual by – products of power balancing practices' – is from Yuen Foong Khong, "Review article: Making bricks without straw in the Asia Pacific?," *The Pacific Review*, Vol. 10, No. 2, 1997, pp. 289-300, see p. 296. Also see, David Dickens, Lessening the Desire for War: The ASEAN Regional Forum and Making of Asia Pacific Security, *Working Paper* 11/98 (Wellington: Centre for Strategic Studies, Victoria University of Wellington, 1998), p. 2.

44 For that matter, Washington's initial scepticism did not prevent it from complaining about its perceived exclusion from such processes. For example, James Baker, as Secretary of State for George H. W. Bush, reportedly reacted angrily to Australia's initial proposal for the APEC, which did not include the United States. See, Michael Wesley, "The dog that didn't bark: the Bush administration and East Asian regionalism," in Mark Beeson, ed., *Bush and Asia: America's Evolving Relations with East Asia* (London: Routledge, 2006), p. 69.

45 'We have to develop new institutional arrangements that support our national economic and security interests internationally," as Clinton noted in 1993. 'We're working to build a prosperous and peaceful Asia Pacific region through our work here in APEC'. Cited in Frank Langdon and Brian L. Job, APEC Beyond Economics: The Politics of APEC, *Working Paper* No. 243 (Notre Dame, IN: The Helen Kellogg Institute for International Studies, October 1997), p. 3.

ASEAN initiative; however, ASEAN's influence on the final form and function of APEC is evident.46 Together with his counterparts from the ASEAN states, President Bush established the ASEAN-US Enhanced Partnership (2006-2011) and the ASEAN-US Leaders' Meeting. It was also under Bush that the United States became the first dialogue partner of ASEAN to appoint an ambassador to ASEAN. While much has been made of his Secretary of State Condoleeza Rice's absences at the 2005 and 2007 ARF meetings, it bears noting that two Democrat presidents also missed meetings — Clinton vis-à-vis APEC in 1995 and Barack Obama vis-à-vis APEC and EAS in 2013 — as a result of federal shutdowns in the US.

With his avowed advocacy of multilateralism and his rebalance to Asia, President Obama has built upon the foundation laid by his predecessor and has robustly engaged the region through America's accession to the Treaty of Amity and Cooperation (TAC) in 2009 and its subsequent entry into the EAS in 2011. American participation in the EAS is important as the summit brings together all the great powers of the Asia Pacific within a single setting — India, currently

---

46 Before APEC could be established, its architects had to assure ASEAN that it would not be overshadowed by APEC, which a key concession: the APEC meeting would be held in an ASEAN country every second year – beginning with Singapore in 1990 – and the APEC secretariat would be based in an ASEAN country, Singapore. With this guarantee in hand, ASEAN threw its support behind the APEC with the Kuching Consensus of 1990, which urged for increased sensitivity in the APEC process to the wide variances in economic development and socio-political systems among the member economies in the region, as well as the need for consultation on economic issues rather than the adoption of mandatory directives that all members must implement. With the agenda of APEC driven each year by the host country, the arrangement has not only given the ASEAN countries considerable experience in the conduct of large-scale multilateral intergovernmental meetings but also in influencing the direction and substance of regional trade discussions. See, Man-jung Mignonne Chan, "APEC's Eye on the Prize: Participants, Modality, and Confidence-Building," in K. Kesavapany and Hank Lim, eds., *APEC at 20: Recall, Reflect, Remake* (Singapore: Institute of Southeast Asian Studies, 2009), pp. 41-54; Robert Gilpin, "APEC in a New International Order," in Donald C. Hellmann, Kenneth B. Pyle, eds., *From APEC to Xanadu: Creating a Viable Community in the Post-War Pacific* (Armonk, N.Y.: ME Sharpe, 1997), pp. 34-35; Jeff Loder, Jean Michel Montsion, and Richard Stubbs, "East Asian Regionalism and the European Experience," in Alex Warleigh-Lack, Nick Robinson, and Ben Rosamond, eds., *New Regionalism and the European Union: Dialogues*, Comparisons and New Research Directions (Abingdon, Oxon: Routledge, 2011), pp. 80-96; John McKay, "APEC: Successes, Weaknesses, and Future Prospects," in Daljit Singh and Anthony L. Smith, eds., *Southeast Asian Affairs 2002* (Singapore: Institute of Southeast Asian Affairs, 2002), pp. 42-53; Naoko Munakata, *Transforming East Asia: The Evolution of Regional Economic Integration* (Washington, D.C.: The Brookings Institution, 2006).

not a part of the APEC, is represented in the EAS — potentially strengthening the 'ASEAN+8' configuration — also reflected in the ADMM+ process — as 'a crucial pattern for regional cooperation'.[47] Under Obama, the ASEAN-US leaders' meeting was upgraded to a summit-level gathering — the most recent of which was held in Sunnylands, California in early 2016 — and an ASEAN-US strategic partnership was established in 2015. These developments underscored the Obama administration's regard for ASEAN as a 'fulcrum for the region's emerging regional architecture'.[48] While this might have partially assuaged regional angst over whether Washington would continue its support of ASEAN centrality in East Asian regionalism,[49] it also raised expectations regarding the effectiveness of the regional architecture to 'produce results', as Hillary Clinton had emphasized in her speech during her tenure as US Secretary of State.[50] For a brand of multilateralism that has principally favoured process over progress, the emphasis of the Americans — indeed, of many if not most non-ASEAN stakeholders – on results constitutes a litmus test, especially for an ASEAN increasingly given to crisis. Needless to say, the presidency of Donald Trump, who has persistently emphasized bilateralism and a transactional approach to foreign policy, has thrown the entire issue of US support for multilateralism wide open.[51] But while President Trump took the US out of TPP in his first week in office — although it would now seem, amid escalating trade tensions

---

47 David Capie and Amitav Acharya, "The United States and the East Asia Summit: A New Beginning?" *PacNet*, No. 64, November 14, 2011, https://csis-prod.s3.amazonaws.com/s3fs-public/legacy_files/files/publication/pac1164.pdf.

48 US Secretary of State Hillary Clinton in October 2010, cited in Aaron Sirila, "Clinton: 'Renewed American leadership in Asia'," *Asia Matters for America*, 4 November 2010, http://asiamattersforamerica.org/asia/clinton-renewed-american-leadership-in-asia.

49 Indeed, the way ASEAN advocates keep inadvertently misquoting Mrs Clinton's reference to ASEAN as 'a fulcrum' with 'the fulcrum' perhaps reflects their anxiety and subsequent relief over US of endorsement and support for ASEAN. See Seng Tan, "Rethinking "Asean Centrality" in the Regional Governance of East Asia," *Singapore Economic Review*, Vol. 63, No. 1, 2018, p. 20.

50 As Hillary Clinton has argued, 'It's more important to have organizations that produce results, rather than simply producing new organizations'. Cited in See Seng Tan, "Competing Visions: EAS in the regional architecture debate," *East Asia Forum*, November 15, 2011, http://www.eastasiaforum.org/2011/11/15/competing-visions-eas-in-the-regional-architecture-debate/.

51 Leon Hader, "The Limits of Trump's Transactional Foreign Policy," *The National Interest*, January 2, 2017, http://nationalinterest.org/feature/the-limits-trumps-transactional-foreign-policy-18898.

with China, that Trump is prepared to reconsider re-entering negotiations on the rechristened Comprehensive and Progressive Agreement for Trans-Pacific Partnership (CPTPP)[52] — the US remains involved in security multilateralism in the region. While not an avowed fan of multilateralism as his predecessor claimed to be, Trump has mostly behaved as any 'forum shopper' might, picking and choosing the institution most appropriate to the US's perceived needs,[53] such as the ADMM+ which Washington continues to strongly support.

III. China-US Competition: Implications for the Regional Security Architecture

As the foregoing analysis has shown, the Chinese and Americans have clearly evolved from being initially suspicious of multilateralism — China more generally, the US specifically over previous attempts by Imperial Japan and the Soviet Union within East Asia — to powers that see utility in the enterprise, but presumably as à la carte multilateral actors rather than wholehearted devotees of the form and function.[54] And while these two big powers and other non-ASEAN stakeholders might have increasing cause to question ASEAN's stewardship of the regional architecture, it is the SCS disputes and ASEAN's evident inability to influence them in any positive way that has underscored for many the increasing irrelevance of both ASEAN — notwithstanding ASEAN's standard disclaimers about its political neutrality and that it is neither an adjudicating body nor a conflict resolution mechanism and was never meant to function as such – and the regional architecture to the region's security. Realist analysts have long predicted the rise of strategic competition among the great powers in post-Cold War East Asia and the likelihood of the region becoming a 'cockpit' of big power rivalry.[55]

---

52 Jack Crowe, "Trump Considers Changing Mind on TPP," *National Review*, April 12, 2018, https://www.nationalreview.com/news/trump-tpp-us-considers-re-entry/.

53 Jürgen Rüland, "The rise of "diminished multilateralism": East Asian and European forum shopping in global governance," *Asia Europe Journal*, Vol. 9, No. 2-4, 2012, pp. 255-270.

54 Stewart Patrick, Multilateralism à la Carte: The New World of Global Governance, *Valdai Papers*, No. 22 (Moscow: Valdai Discussion Club, July 2015).

55 See, Richard K. Betts, "East Asia and the United States after the cold war," *International Security*, Vol. 18, No. 3, 1993-1994, pp. 34-77; Barry Buzan and Gerald Segal, "Rethinking East Asian security," *Survival*,

Needless to say, there have been important dissenting opinions,[56] although some of their authors have themselves begun to doubt the region's ability to keep the peace in the light of growing tensions in the SCS.[57] Big power tensions have also crept into their interactions at the ARF and more recently the ADMM+. Remarkably, their 2016 editions were relatively free of acrimony and discord, or at least nothing that was serious enough to debilitate the meetings as had previously happened. 'ASEAN and China worked hard to avoid a showdown, which would have left all parties red-faced if a fallout were to occur during the Commemorative Summit to mark the silver jubilee of ASEAN-China Dialogue Relations', as one observer explained. 'ASEAN was also eager not to allow the South China Sea issue clout and dominate bilateral ties'.[58] But this was certainly not the case at earlier gatherings.

The 2010 ARF at Hanoi is remembered for the vehemence with which then Chinese Foreign Minister Yang Jiechi reacted to comments made by his US counterpart Hillary Clinton as amounting to 'an attack on China',[59] which the Chinese interpreted as American interference in the SCS, a 'core interest' for Beijing. It was instructive, for example, to learn from Clinton, as revealed in her memoir, that her reference to the peaceful resolution of competing sovereignty

Vol. 36, No. 2, 1994, pp. 3-21; Aaron L. Friedberg, "Ripe for Rivalry: Prospects for Peace in a Multipolar Asia," *International Security*, Vol. 18, No. 3, Winter, 1993-1994, pp. 5-33. For a more recent example, see, John J. Mearsheimer, "The Gathering Storm: China's Challenge to US Power in Asia," *Chinese Journal of International Politics*, Vol. 3, No. 4, 2010, pp. 381-396.

56 Richard A. Bitzinger and Barry Desker, "Why war is unlikely in Asia: facing the challenge from China," *Survival*, Vol. 50, No. 6, 2008, pp. 105-128; Muthiah Alagappa, ed., *Asian Security Order: Instrumental and Normative Features* (Stanford, CA: Stanford University Press, 2003).

57 As Richard Bitzinger, who was played down the likelihood of major war in East Asia, recently warned, '[China's] 'continental militarization' of the SCS not only diminishes the 'open order' of the Southeast Asian maritime sphere. It also greatly raises the likelihood that the South China Sea will become a flashpoint for escalating conflict. China is not only militarizing the SCS, it is making it too important for Beijing to lose. This, in turn, raises the premium of a first strike by China on its regional rivals, in order to shore up its claims. More and more, China is playing a risky game of chicken, and at the same time, it does not seem to appreciate the grave potential consequences of its actions'. Richard A. Bitzinger, "China's militarization of the South China Sea: Building a strategic strait?" *Asia Times*, June 21, 2016, http://atimes.com/2016/06/chinas-militarization-of-the-south-china-sea-building-a-strategic-strait/.

58 Tang Siew Mun, "Six takeaways from Asean summits," *The Straits Times*, September 15, 2016, http://www.straitstimes.com/opinion/six-takeaways-from-asean-summits.

59 Kai He, *China's Crisis Behaviour: Political Survival and Foreign Policy After the Cold War* (Cambridge: Cambridge University Press, 2016), p. 102.

claims to the SCS is a US 'national interest' was apparently made in response to the felt anxieties of some of the regional countries toward China's growing assertiveness.[60] If anything, consecutive allegations by the ASEAN states about Chinese 'bullying' apparently reinforced her own sense, garnered from her involvement at the Strategic and Economic Dialogue (S&ED) in Beijing in May 2010, that the Chinese might have overreached with their identification of the South China Sea as a 'core interest' alongside Taiwan and Tibet, and their inflexible insistence that (in Clinton's words) 'China would not tolerate outside interference' in the SCS as such.[61] One view has it that the Chinese felt they had been unfairly waylaid by the Americans, which was in a sense understandable since, before Clinton's comments, the Obama administration had largely eschewed confrontation with the Chinese.[62] Yang also infamously reminded the ASEAN countries of the indubitable 'fact' that China is a 'big country' and that they were 'small countries'.[63] Furthermore, the Chinese apparently vented their fury against ASEAN states —— Singapore in particular —— deemed to be aligned closely with the US.[64]

The 2014 ARF held in *Naypyidaw* was equally notable for the starkly different positions adopted by Beijing and Washington. In the light of China's controversial decision to station an oil rig in Vietnamese waters in May 2014 —— which provoked anti-Chinese riots in Vietnam —— US Secretary of State John Kerry proposed a moratorium on 'provocative acts' in the SCS.[65] Kerry's proposal

---

60 Hillary Rodham Clinton, *Hard Choices* (New York: Simon and Schuster, 2014), p. 79.

61 Cited in Geoff Dyer and Tom Mitchell, "Hillary Clinton: The China hawk," *Financial Times*, September 6, 2016, https://www.ft.com/content/92b23c8e-7349-11e6-bf48-b372cdb1043a.

62 Gordon C. Chang, "Hillary Clinton Changes America's China Policy," *Forbes*, July 28, 2010, http://www.forbes.com/2010/07/28/china-beijing-asia-hillary-clinton-opinions-columnists-gordon-g-chang.html.

63 John Pomfret, "U.S. takes a tougher tone with China," *The Washington Post*, July 30, 2010, http://www.washingtonpost.com/wp-dyn/content/article/2010/07/29/AR2010072906416.html.

64 As one observer has described of the abundant pyrotechnics at the 2010 ARF, 'Yang [Jiechi] stared down the foreign minister of Singapore, a country known in the region as one of America's staunchest friends. The Singaporean foreign minister, a normally placid man named George Yeo, stared right back'. Joshua Kurlantzick, "The Belligerents," *New Republic*, January 27, 2011, https://newrepublic.com/article/82211/china-foreign-policy.

65 Anne Gearan, "U.S., China tussle over sea claims," *The Washington Post*, August 10, 2014, https://www.washingtonpost.com/world/us-china-tussle-over-sea-claims/2014/08/10/2f613504-2085-11e4-8b10-7db129976abb_story.html.

was roundly rejected by his Chinese counterpart Wang Yi.[66] The secretary general of ASEAN, Le Luong Minh, claimed that the ASEAN leaders did not discuss the US proposal, which in his view was not the critical issue. Rather, it was up to the ASEAN 'to encourage China to achieve a serious and effective implementation' of the 2002 DOC, which committed signatories to 'self-restraint' on actions such as land reclamation and building on disputed islands and reefs.[67] Since then both parties have produced a 'framework' Code of Conduct (COC) that presumably lays the groundwork for an actual COC. But the lack of reference to the COC as binding has raised doubts about its relevance to SCS security even if the parties succeed in delivering a code.[68]

When the ADMM+ met for only its third time in Kuala Lumpur in November 2015, it was forced to scrap a planned joint statement on the SCS. Unlike ASEAN meetings, however, joint declarations are not mandatory for the ADMM+. The initial plan to issue one was abandoned following disagreements among the non-ASEAN countries over whether to include mention of the SCS disputes in the statement. It had been widely — and wrongly — reported by the international press that the failure of the ADMM+ to issue a joint statement was reminiscent of ASEAN's disunity in Phnom Penh in July 2012.[69] Moreover, the failure to agree on a statement fosters the impression that the ADMM+, for all the encouraging progress it has hitherto achieved in defence diplomacy — achievements that are more functional than strategic in nature, it should be said[70]

---

66 Paul Mooney and Lesley Wroughton, "U.S. call for South China Sea 'freeze' gets cool response from China," *Reuters*, August 9, 2014, http://www.reuters.com/article/us-asean-southchinasea-idUSKBN0G904O20140809.

67 "Southeast Asia ministers didn't discuss US plan on sea claims, diplomat says," *The Straits Times*, August 9, 2014, http://www.straitstimes.com/asia/se-asia/southeast-asia-ministers-didnt-discuss-us-plan-on-sea-claims-diplomat-says.

68 Ian Storey, "Assessing the ASEAN-China Framework for the Code of Conduct for the South China Sea," *ISEAS Perspective*, August 8, 2017, <https://www.iseas.edu.sg/images/pdf/ISEAS_Perspective_2017_62.pdf>.

69 See Seng Tan, "Claims of Asean disunity at summit unfounded," *The Straits Times*, November 22, 2015, http://www.straitstimes.com/opinion/claims-of-asean-disunity-at-summit-unfounded.

70 According to a 2016 study, the utility of defence diplomacy – as ADMM-Plus activities have been labelled – as an instrument of statecraft to defuse tensions, reduce hostility, and shape the behaviour of states toward each other has been overstated. Instead, their research shows that defence diplomacy is considerably more useful as an approach for dealing with precise and immediate security issues rather than in shaping major

— may go the way of the ARF. However, the key difference at the ADMM+ meeting in Kuala Lumpur was that all ten ASEAN member countries — including the four SCS claimants — stood firmly against the inclusion of the SCS in the proposed joint declaration, while ensuring its mention in the chairman's statement issued by Malaysia in its role as ASEAN chair for 2015.[71] Reportedly, the collective position adopted by the ASEAN countries was in anticipation of the likely inability of the ADMM+ members to achieve the consensus needed for a joint declaration owing to their differences over the SCS.[72] But it is also noteworthy that the Chinese and Americans, alongside their 16 fellow partner countries, have actively participated in military-to-military exercises in the six areas of defence collaboration under ADMM+ auspices, helping the arrangement to achieve a level of multilateral security cooperation never before experienced in the post-Cold War Asia-Pacific region.[73]

IV. Whither ASEAN and Regional Security Architecture?

In some respects, the aforementioned stresses placed upon the regional architecture constitute a sort of 'new normal' in East Asian regionalism. Doubts over the durability of the architecture's institutional design and the effectiveness of its diplomatic-security conventions in managing if not resolving those stresses are usually placed at ASEAN's feet, and that is not incorrect to an extent. However, as we have seen, the reality is that the perceptibly flawed regional architecture and its diplomatic-security conventions are equally the outcome of the choices of and contributions by non-ASEAN partners. In short, the so-called ASEAN Way is not ASEAN's alone, but it is 'Everybody's Way'. It could

---

strategic outcomes. See, Daniel Baldino and Andrew Carr, "Defence Diplomacy and the Australian Defence Force: Smokescreen or Strategy?" *Australian Journal of International Affairs*, Vol. 70, No. 2, 2016, pp. 139-158.

71 See Seng Tan, "The ADMM-Plus: Regionalism That Works?" *Asia Policy*, No. 22, July, 2016, pp. 70-75, see p. 73.

72 This author's communication with Singapore defence officials in November 2015.

73 For a comparison of the achievements of the ADMM+ and the ARF, see, See Seng Tan, "A Tale of Two Institutions: The ARF, ADMM-Plus and Security Regionalism in the Asia Pacific," *Contemporary Southeast Asia*, Vol. 39, No. 2, August, 2017, pp. 259-264.

therefore be said that the regional architecture, arguably and ironically, works best when flawed. Or, to put it another way, what are often viewed as design flaws of the regional architecture are deliberately built in to accommodate the respective national interests of all partners and stakeholders.[74] As such, given that both ASEAN and non-ASEAN countries are equally accountable for the present condition of the regional architecture, any prospect for an improvement to the state of regionalism in East Asia necessarily entails the collective responsibility of all concerned.

Firstly, the argument for a functioning and effective EAS makes good intuitive sense, but the goal to improve institutional efficacy has to take into account the political cum strategic factors that have shaped and shoved the regional architecture. The debate over the perceived flaws of the regional architecture and how to fix it is well known and will not be rehearsed in detail here. To date, attention has largely focused on the EAS, a leaders-led forum, as the best hope for a viable multilateralism in East Asia. When member nations of the EAS agreed in Hanoi in April 2010 to enlarge its membership from 16 to 18 countries by including Russia and the US, it was suggested by some that EAS expansion is the culmination of the regional debate over Kevin Rudd's APC proposal.[75] That being said, it is tempting nonetheless to see the EAS as the logical overarching institution under which the ARF, ADMM+, and other ministerial processes could come. For instance, at the 2015 EAS meeting, Japanese Prime Minister Shinzo Abe argued that an enhanced EAS ought to be the 'premier forum' for regional security in East Asia.[76] Notwithstanding reservations with the APC proposal, the idea that the EAS should be empowered

---

74 The power of veto given only to the Permanent 5 members of the UN Security Council is just such a 'design flaw' of global multilateralism, which P5 powers have resolutely refused to give up because it benefits their respective interests.

75 Thom Woodroofe, "Is the East Asia Summit Rudd's gift to the world?," *Australian Policy Outline*, January12, 2012, http://apo.org.au/commentary/east-asia-summit-rudd%E2%80%99s-gift-world. On the other hand, this suggestion has been robustly refuted by many others. See, Aaron Connolly, "Canberra's Clouseau Strategy," *Lowy Interpreter*, January 5, 2011, http://www.lowyinterpreter.org/post/2011/01/05/Canberras-Clouseau-strategy.aspx.

76 "The 10th East Asia Summit Meeting," *Ministry of Foreign Affairs of Japan*, November 22, 2015, http://www.mofa.go.jp/a_o/rp/page3e_000426.html.

with the capacity to 'steer' the various regional arrangements has also garnered growing support among a number of ASEAN-based intellectuals. It has been argued that the EAS has the potential to become the 'apex summit' for strategic dialogue in East Asia.[77] Elsewhere, mindful of the anger the APC idea and its 'concert of powers' corollary evoked in some ASEAN quarters,[78] Rizal Sukma, former head of the Jakarta-based Centre for Strategic and International Studies and current Indonesian ambassador to the UK, argued that the EAS 'should function as a sort of steering committee for the Asia-Pacific region in two interrelated ways'.[79]

The key challenge, however, remains whether regional consensus over a negotiated *modus vivendi* among the major powers, especially China and the US — importantly, an ASEAN-backed and facilitated *modus vivendi* — is possible. A key failing of the concert of powers idea, or at least the version advanced by the Australians, was not only in its neglect to include ASEAN within that concert (neither as a power nor facilitator), but in the failure of the Rudd government to consult ASEAN and obtain its support. But it also assumed that Beijing would back the idea. Such an assumption constituted a big leap of faith in the light of possible Chinese suspicions over Western containment of its rise via the EAS, whose very establishment in late 2005 (with the inclusion of Australia, India and New Zealand beyond the ASEAN+3 member countries, much to China's chagrin) was framed by some as an exercise in power-balancing – unhelpfully misinterpreted by the Chinese on occasion, intentionally or inadvertently, as

---

77 Singapore Institute of International Affairs, *Rethinking the East Asia Summit: Purpose, Processes and Agenda, Policy Brief* (Singapore: Singapore Institute of International Affairs, September 2014), p. 2.

78 Amitav Acharya, "Asia-Pacific Security: Community, Concert or What?" *PacNet*, No. 11, March 12, 2010, https://csis-prod.s3.amazonaws.com/s3fs-public/legacy_files/files/publication/pac1011.pdf.

79 As Sukma has argued, 'First, it should be allowed to function as a steering committee for coordinating various regional institutions in the region such as the ASEAN Plus Three (APT), the ASEAN Regional Forum (ARF), the ASEAN Defence Ministers Meeting Plus (ADMM Plus), and the Asia-Pacific Economic Cooperation (APEC). Second, there is also the need for the EAS members of the G20 to form an informal caucus to coordinate their policies and interests at the global level'. Rizal Sukma, "Insight: East Asia needs a steering committee," *The Jakarta Post*, October 4, 2012, http://www.thejakartapost.com/news/2012/09/04/insight-east-asia-needs-a-steering-committee.html.

containment — as much as in multilateral institution-building.[80] Tellingly, the late Lee Kuan Yew once lamented the apparent predilection of the Chinese media to erroneously conflate the notions of balancing and containment, as evidenced by the latter's mistranslations of his use of the term 'balance' in English as 'conscribe' (or contain) in Chinese.[81] The enlargement of the EAS to include a rebalancing US — after all, the Obama administration publicly presented its interest and membership in the EAS as part of its rebalance strategy[82] — likely exacerbated rather than mitigated those suspicions. According to Mark Beeson and Diane Stone:

> *The EAS has become surprisingly prominent largely as a consequence of the United States seeking to reengage or 'pivot' back to the region. Whatever one thinks of the merits of or necessity for such a policy, that these institutions are seen by the Obama administration as potentially important ways of exerting influence is significant and in marked contrast to the Bush era. The principal attractions of the EAS for the United States are: first, it is a member and so are key allies like Australia and, second, it potentially offers a way of curbing China's influence.[83]*

That the Chinese have understood this logic of having its influence curbed all too well — and have presumably rejected it, if only implicitly — is perhaps seen in its recent attempts to start new multilateral initiatives (AIIB, CICA, BRI, etc.) and to invest in extant ones (SCO) that ensure a predominant place and role for China in such — the very thing China was unable to do with the EAS.

---

80 For example, Singapore ostensibly wanted India in the EAS because, in Lee Kuan Yew's words, 'it would be a useful balance to China's heft'. Cited in Amitav Acharya, *Singapore's Foreign Policy: The Search for Regional Order* (Singapore: World Scientific, 2008), p. 99.

81 See Seng Tan and Oleg Korovin, "Seeking Stability in Turbulent Times: Southeast Asia's New Normal?" in Daljit Singh, ed., *Southeast Asian Affairs 2015* (Singapore: ISEAS Yusof Ishak Institute, 2015), pp. 3-24, see p. 10.

82 Jeffrey A. Bader, "U.S. Policy: Balancing in Asia, and Rebalancing to Asia," *Brookings India-US Policy Memo*, September 23, 2014, https://www.brookings.edu/opinions/u-s-policy-balancing-in-asia-and-rebalancing-to-asia/.

83 Mark Beeson and Diane Stone, "Patterns of Leadership in the Asia-Pacific: A Symposium," *The Pacific Review*, Vol. 27, No. 4, 2014, pp. 505-522, see p. 516.

The buy in by one and all cannot be ensured until everyone is assured they will neither be unfairly targeted nor marginalized by the rest.

Secondly, all stakeholders of the regional architecture have to do what they can to reduce tensions amongst themselves, which is an easier-said-than-done proposition in the light of the massive trust deficit that presently exists between and among East Asian countries.[84] All multilateral architectures, whether at the global or regional level, work only insofar as their members are prepared to give and take. In an intriguing riposte to Michael Leifer's contention that the 'prerequisite' for a successful ARF is likely to be 'the prior existence of a stable balance-of-power',[85] Yuen Foong Khong has proposed, in contrast to Leifer, that the ARF is not just 'a valuable adjunct' to the workings of the balance of power, but 'a mechanism for defusing the conflictual by-products of power balancing practices'.[86] If so, the fact that interstate war has not broken out over China-US rivalry over the SCS could perhaps be explained by attributing it to the success of the regional architecture as a venue where contending parties parley and defuse their mutual tensions. On the other hand, it is equally evident that the regional architecture is not necessarily an alternate mechanism to the balance of power, as conceived by Khong, because regional countries have in effectively turned the regional architecture and its component parts into arenas for power-balancing. Writing at the end of the Cold War, Inis Claude had this to say about the importance of strategic moderation to the success of the balancing model adopted by the Concert of Europe:

> *That this moderation is viewed as the essential foundation for the functioning of the balance-of-power system rather than as a consequence of its functioning is evidenced by the fact that the fading and ultimate collapse of the efficacy of that [Concert of Europe] system is customarily*

---

84 Bilahari Kausikan, "The roots of strategic distrust: the US, China, Japan and Asean in East Asia," *The Straits Times*, November 18, 2016, http://www.straitstimes.com/opinion/the-roots-of-strategic-distrust-the-us-china-japan-and-asean-in-east-asia; J.M. Norton, "The Sources of US-China Strategic Mistrust," The Diplomat, April 24, 2014, http://thediplomat.com/2014/04/the-sources-of-us-china-strategic-mistrust/.

85 Leifer, *The ASEAN Regional Forum: Extending ASEAN's Model of Regional Security*, pp. 57-58.

86 Khong, "Review article: Making bricks without straw in the Asia Pacific?," p. 296.

attributed to the decline of those factors that sustained moderation.[87]

In most if not all historical multilateral-institutional examples, it has required major powers to willingly exercise strategic restraint and self-moderation. The US did it (or so claimed) in the early postwar period to facilitate global arrangements like the United Nations and the IMF, France and Germany did it in the 1950s to facilitate European integration and the make the EU work, Indonesia did it in 1967 to facilitate the formation and maintenance of ASEAN, and major powers like China and the US have been doing it since the 1990s — unevenly and unequally, as the foregoing analysis has suggested — to facilitate the regional architecture. Granted, the national interest comes first for all countries, and rightly so. While mutual strategic restraint and cooperation might not be the obvious policy choice for rising and rebalancing powers and the rising expectations of their peoples, it is fundamentally the right choice for the region as a whole.

Finally, ASEAN has to do considerably better as an institution in order to guarantee its centrality in East Asian regionalism. Here again there are constraints that are not easily done away with. The inability of ASEAN to preserve cohesion within its own ranks has unsettled the organization. Conflicting security perspectives and political tensions have long persisted between ASEAN countries, but the grouping has in the past been able to manage its members' differences and prevented such from harming their common interests.[88] In this regard, much has been made of China's so-called 'divide and rule' strategy aimed at creating divisions among ASEAN states and undermining their institution's ability to adopt a collective position on the SCS which China deems inimical to its interests.[89] The failure of ASEAN to issue a joint statement at its ministerial

---

87 Inis L. Claude, "The Balance-of-Power Revisited," *Review of International Studies*, Vol.15, No.2, April, 1989, pp. 77-85, see p. 80.

88 Explaining why ASEAN has not been particularly cohesive, one analyst noted that the ASEAN member states 'have different interests and self-identity, all of which bleeds into policy. In this kind of environment, it is hard to agree on matters and get deals done, and when they are sealed, they are usually reactive in nature and scope'. See, Brad Nelson, "Can Indonesia Lead ASEAN?" *The Diplomat*, December 5, 2013, http://thediplomat.com/2013/12/can-indonesia-lead-asean/.

89 Michael Thim, "China's 'divide and rule' attitude in Southeast Asia is good for no one, including itself," *South China Morning Post*, August 8, 2016, http://www.scmp.com/comment/insight-opinion/

meeting in Phnom Penh in July 2012 — the first time in its history — as a result of actions by Cambodia in its tenure as the chair of ASEAN for that year, or China's 'four-point consensus' on the SCS which it cobbled together with Brunei, Cambodia and Laos in April 2016[90] — without the knowledge, much less support, of the other ASEAN members — have become prime examples of perceived Chinese interference in ASEAN's affairs. Likewise, while ASEAN countries as a whole have welcomed the US rebalance — or, for that matter, Japan's own 'pivot' to the region[91] — and the material gains that come with it, they have also worried over the prospect of being forced by external powers to take sides amid the growing strategic competition.[92] On the other hand, the danger to ASEAN unity is not simply a matter of external influences and interference. The apparent reluctance by some ASEAN countries to make good on their commitments to the grouping's express vision and timeframe for regional integration — namely, the ASEAN Community with its economic, political-security and socio-cultural 'pillars' — has likewise fuelled doubts about the sincerity and determination of member countries regarding their collective goals.[93] Without question, these developments have eroded international confidence in ASEAN's credibility as a regional leader in Southeast Asia, much less East Asia.

Tellingly, it is not only non-ASEAN countries that have questioned ASEAN's commitment and capacity to and for deeper integration. As a former foreign minister of Indonesia has warned, 'ASEAN's centrality, or place in

article/2000792/chinas-divide-and-rule-attitude-southeast-asia-good-no-one.

90 Nirmal Ghosh, "South China Sea consensus "shows up Asean fault lines," *The Straits Times*, April 25, 2016, http://www.straitstimes.com/asia/se-asia/south-china-sea-consensus-shows-up-asean-fault-lines.

91 Koh Swee Lean Collin, "Japan and the 'Maritime Pivot' to Southeast Asia," *The Diplomat*, December 14, 2015, http://thediplomat.com/2015/12/japan-and-the-maritime-pivot-to-southeast-asia/; Céline Pajon, "Japan's "Smart" Strategic Engagement in Southeast Asia," *Asan Open Forum*, 6 December 2013, http://www.theasanforum.org/japans-smart-strategic-engagement-in-southeast-asia/.

92 Robert Sutter, "China's Rise, Southeast Asia, and the United States: Is a China-centred Order Marginalizing the United States?" in Evelyn Goh and Sheldon W. Simon, eds., *China, the United States, and South-East Asia: Contending Perspectives on Politics, Security, and Economics* (Abingdon, Oxon: Routledge, 2008), pp. 91-106, see p. 96.

93 Lee Jones, "Explaining the failure of the ASEAN economic community: the primacy of domestic political economy," *The Pacific Review*, 2016, pp. 1-24, DOI, http://dx.doi.org/10.1080/09512748.2015.1022593.

the driving seat, is not a given'.[94] Likewise, while noting that ASEAN has done well at providing a 'centrality of goodwill', the late Surin Pitsuwan, the secretary-general of ASEAN from 2008 to 2012, has however argued that what the association needs to furnish is a 'centrality of substance' as it faces more demanding tests to its solidarity that arise from on-going shifts in the power dynamics and trade patterns of East Asia.[95] That said, ASEAN officials past and present differ on what is possible with ASEAN On the one hand, a former deputy secretary-general of ASEAN made the following appeal in the run-up to the 2015 deadline for the realization of the AEC:

> *While 'process-based regionalism' — the series of meetings, dialogues, consultations and engagements that ASEAN has put in place for internal economic integration and relations with its major trading partners — has produced spectacular results for ASEAN as convener of regional meetings and pace-setter for ASEAN+1, ASEAN+3, and East Asia Summit these institutional processes alone may not be adequate to maintain ASEAN's centrality and to achieve the AEC by 2015. To meet its goals, ASEAN must ensure the substantive implementation of its economic agreements, declarations, plans and programs. I call this form of regionalism supported by concrete results and outcomes based on a structured and rules-based regime 'results-based regionalism'... To maintain ASEAN's centrality in the region and to achieve the goal of AEC by 2015 it is imperative that ASEAN shifts aggressively towards 'result-based regionalism'. We must act now.[96]*

---

94 Marty Natalegawa, "Aggressively waging peace: ASEAN and the Asia-Pacific," *Strategic Review: The Indonesian Journal of Leadership, Policy and World Affairs*, Vol. 1, No. 2, 2011, pp. 40-46, see p. 40.

95 "Speech By Outgoing Secretary-General (2008–2012) H.E. Surin Pitsuwan Ceremony for the Transfer of Office of the Secretary-General of ASEAN," January 1, 2013, http://www.asean.org/resources/2012-02-10-08-47-56/speeches-statements-of-the-former-secretaries-general-of-asean/item/speech-by-outgoing-secretary-general-2008-2012-he-surin-pitsuwan-ceremony-for-the-transfer-of-office-of-the-secretary-general-of-asean-3.

96 S. Pushpanathan, "Opinion: No place for passive regionalism in ASEAN," *The Jakarta Post*, April 7, 2010, http://www.thejakartapost.com/news/2010/04/07/no-place-passive-regionalism-asean.html.

On the other hand, another former secretary-general of ASEAN has offered a more cautious take:

> *ASEAN is still far from being economically integrated as a region. And there is little prospect that it will be fully integrated, as envisioned, in the near future, much less by 2015. But whether or not the AEC is achieved by 2015 should not be held against the literal rendering of the specific measures to realise ASEAN economic integration, as provided for in the Strategic Schedule appended to the AEC Blueprint. Rather, the plan to realise the AEC by 2015 should be looked at as a re-affirmation of the ASEAN leaders' aspiration for, and commitment to, efficiency in trading, market openness and links with the international community. The year 2015 should be considered not as a hard-and-fast target, in which ASEAN, its objectives and the way it conducts business are suddenly transformed. Rather, it should be regarded as a benchmark to help measure ASEAN's progress toward regional economic integration.*[97]

A related concern to an enhanced ASEAN regionalism as intrinsic to the proper exercise of its centrality in the regional architecture is that of leadership *within* ASEAN. Fiascos like the 2012 ASEAN ministerial meeting in Phnom Penh have fostered the impression of an aimless and rudderless organization. Writing in 2016, one analyst has bemoaned the alarming absence of decisive and substantive leadership at the most recent suite of ASEAN-based regional meetings, which has rendered ASEAN susceptible to outside influence.[98] In this regard, the role and place of Indonesia as the *de facto* leader of ASEAN can no longer be assumed. Much has been made of Indonesia's apparent sense of entitlement as

---

97 Rodolfo C. Severino, "Let's be honest about what ASEAN can and cannot do," *East Asia Forum*, January 31, 2014, http://www.eastasiaforum.org/2014/01/31/lets-be-honest-about-what-asean-can-and-cannot-do/.

98 As Termsak Chalermpalanupap has observed, 'Now there is no known core group of steady hands in the AMM. Neither is there a potential 'quarterback," like ex-Indonesian Foreign Minister Dr. Marty Natalegawa, to assist the ASEAN chairman and rally the troops or mend fences in time of crisis or political debacle, like what happened at the 45th AMM in Phnom Penh in July 2012. Under these debilitating circumstances, ASEAN is vulnerable to interference and manipulation by external parties'. Termsak Chalermpalanupap, "No ASEAN Consensus on the South China Sea," *The Diplomat*, July 21, 2016, http://thediplomat.com/2016/07/no-asean-consensus-on-the-south-china-sea/.

ASEAN's leader.[99] On the other hand, its leadership has been deemed by some as 'incomplete'. Although Indonesia has sought to establish a stable and autonomous security environment, conducted conflict meditation efforts in the Cambodian conflict and the SCS disputes, and developed institutional mechanisms to promote security, democracy and human rights, its leadership in ASEAN has been deemed incomplete owing not only to the 'sectorial' leadership practised by Indonesia but also resistance from other ASEAN members to Jakarta's preference for an autonomous regional order and in recent years its democratization.[100]

In recent times, a litany of frustrations over not being able to successfully exercise its leadership —— the inability to persuade the more conservative newer members of ASEAN to support the development of a more robust charter in 2007, the failure to convince Cambodia against succumbing to Chinese political leverage, among others[101] —— has led to calls from prominent Indonesian strategic thinkers for a 'post-ASEAN foreign policy' for Indonesia.[102] Whether by accident or design, that wish has in a sense been fulfilled, ironically so, with the emergence of Joko Widodo as Indonesian president in 2014 because of his relative disinterest in foreign policy in general and in ASEAN in particular, and his apparent preference for strategic bilateral partnerships than multilateralism.[103] Crucially,

---

99 Pattharapong Rattanasevee, "Leadership in ASEAN: The Role of Indonesia Reconsidered," *Asian Journal of Political Science*, Vol. 22, No. 2, 2014, pp. 113-127.

100 Ralf Emmers, "Indonesia's role in ASEAN: A case of incomplete and sectorial leadership," *The Pacific Review*, Vol. 27, No. 4, 2014, pp. 543-562.

101 As has been observed by one commentator: 'Beijing's non too subtle exercise of political leverage over Phnom Penh recently with regard to the Code of Conduct has exposed divisions within ASEAN and frustrations among Jakarta's foreign policy elite. But more seriously for Jakarta, machinations behind the April ASEAN Summit meeting illustrate the risks of a foreign policy predicated upon the cohesion and on-going integration of ASEAN as a bulwark against major powers'. Greta Nabbs-Keller, "Beijing's divide and rule strategy exposes Jakarta," *The Lowy Interpreter*, April 18, 2012, http://www.lowyinterpreter.org/post/2012/04/18/ASEAN-Beijings-divide-rule-and-strategy-exposes-Jakarta.aspx.

102 Rizal Sukma, "Indonesia needs a post-ASEAN foreign policy," *The Jakarta Post*, June 30, 2009, http://www.thejakartapost.com/news/2009/06/30/indonesia-needs-a-postasean-foreign-policy.html; Rizal Sukma, "A post-ASEAN foreign policy for a post-G8 world," *The Jakarta Post*, October 5, 2009, http://www.thejakartapost.com/news/2009/10/05/a-postasean-foreign-policy-a-postg8-world.html; Jusuf Wanandi, "Insight: RI's foreign policy and the meaning of ASEAN," *The Jakarta Post*, May 6, 2008, http://www.thejakartapost.com/news/2008/05/05/insight-ri039s-foreign-policy-and-meaning-asean.html.

103 Malcolm Cook, "Duterte, Jokowi and ASEAN," ISEAS - Yusof Ishak Institute, May 10, 2016, https://www.iseas.edu.sg/medias/commentaries/item/3072-duterte-jokowi-and-asean.

the Indonesian leader has apparently deprioritized the AEC — whilst urging Indonesians to ready themselves for the potentially aversive effects wrought by the implementation of the AEC by end 2015[104] — in favour of economic partnerships with China and Japan, Indonesia's largest and second largest export markets, respectively. President Widodo has also eschewed the idea of Indonesia playing a mediatory role in the SCS disputes by insisting that 'that is a problem for other countries'.[105] All these examples reflect an Indonesia neither interested in leading ASEAN nor seeking to enhance the regional security architecture.

At the ASEAN-China Defence Ministers' Informal Meeting (ACDMIM) that took place in February 2018 in Singapore, both sides took away something. For the Chinese, the agreement to conduct the first joint maritime exercise scheduled for the later part of 2018 was the realization of a goal to boost defence ties with the ASEAN states.[106] For the ASEAN and its member countries, it marked the deepening of their security relations with major powers since the ACDMIM was the third such initiative with a big power following equivalent sessions ASEAN conducted with the US and with Japan beginning in 2014.[107] More crucially, from the perspective of ASEAN, these '+1' engagements with great powers reflect the consolidation of ASEAN centrality in the wake of the damage done to it by great power rivalry and the impact that has had on intra-ASEAN cohesion and unity. Yet as this paper has sought to show, the pride and place of ASEAN in the regional security architecture in East Asia are by no means assured, but are contingent on the consent and goodwill of the big powers and ASEAN's own ability to deliver effectively on its promises.

---

104 "Indonesia must be ready for ASEAN Economic Community: President Jokowi," *Antara News*, December 26, 2015, http://www.antaranews.com/en/news/102207/indonesia-must-be-ready-for-asean-economic-community-president-jokowi.

105 Cited in Avery Poole, "Is Jokowi Turning His Back on ASEAN?," *The Diplomat*, September 7, 2015, http://thediplomat.com/2015/09/is-jokowi-turning-his-back-on-asean/.

106 Danson Cheong, "Asean defence chiefs to increase cooperation against terrorism, hold maritime exercise with China," *The Straits Times*, February 6, 2018, http://www.straitstimes.com/asia/se-asia/asean-defence-chiefs-to-increase-cooperation-against-terrorism-hold-maritime-exercise.

107 Prashanth Parameswaran, "China Reveals New Proposal to Boost Defense Ties With ASEAN," *The Diplomat*, October 17, 2015, https://thediplomat.com/2015/10/china-reveals-new-proposal-to-boost-defense-ties-with-asean/.

Maintaining Liberal Regional Order: Japan's Approaches to ASEAN and Taiwan in Free and Open Indo-Pacific Strategy

Sachiko Hirakawa *

I. Introduction

In August 27, 2016, at the opening session of the Sixth Tokyo International Conference on African Development (TICAD VI) held in Kenya, Japanese Prime Minister Abe Shinzo first publicly proposed the concept of Indo-Pacific strategy. In the keynote speech, he argued that Japan "bears the responsibility of fostering the confluence of the Pacific and Indian Oceans and of Asia and Africa into a place that values freedom, the rule of law, and the market economy, free from force and coercion, and making it prosperous." Abe laid out his vision of Japan working together with Africa "in order to make the seas that connect the two continents into peaceful seas that are governed by the rule of law."[1]

A year later when the US President Donald Trump visited Japan on November 6, 2017, Abe persuaded Trump that a free and open maritime order based on the rule of law is a cornerstone for peace and prosperity of the international community. The two leaders affirmed that Japan and the US will work to promote common regional strategy by developing the Indo-Pacific concept with "free and open" value emphasis. They directed relevant ministers and institutions to flesh out detailed cooperation.[2] A few days later, at the APEC CEO Summit at Da Nang, Vietnam, Trump highlighted the importance of "Free and Open Indo-Pacific" as the shared regional future vision.[3] Then the word of

---

* Associate Professor, Center for International Education, Waseda University

1 Address by Prime Minister Abe Shinzo at the opening session of TICAD VI, August 27, 2016, https://www.mofa.go.jp/afr/af2/page4e_000496.html.

2 Japan-U.S. Working Lunch and Japan-U.S. Summit Meeting, November 6, 2017, http://www.mofa.go.jp/na/na1/us/page4e_000699.html.

3 He stated, "...I've had the honor of sharing our vision for a free and open Indo-Pacific — a place where sovereign and independent nations, with diverse cultures and many different dreams, can all prosper side-

"Free and Open Indo-Pacific Strategy (FOIPS)" received much wider attention from the international community. The FOIPS is originally Japan's advocacy and initiative, which the US followed.

It is naturally understood that the strategy was reactively scripted to China's "Belt and Road Initiative (BRI)," which covers whole Eurasian continent extending European and even African continents by both land and sea routes. Even though Japan has formally kept the stance that it welcomes China's peaceful rise, Japan has been gradually seen China's strengthening of military forces and unilateral attempts to change the status-quo as serious concerns and threats to Japan and the regional order. China's recent active regional initiatives such as BRI focus on mainly socio-economic development but little attention to be paid to human rights, democracy, and transparency of governance. This fundamentally conflicts with Japan's regional vision and fundamental values. For ensuring stable and predictable international environment, Abe government has pursued its foreign policy of "Proactive Contribution to Peace" based on the principle of international cooperation. In this context, FOIPS is regarded as a fruited sequence of Japan's recent proactive diplomacy.

This paper discusses Japan's approach to ASEAN and Taiwan, both of which are critically important to architect Japan's favorable regional order through such FOIPS. It first examines the origin and development of FOIPS. Then it focuses on Japan's approach to ASEAN from security dimension and subsequently Japan's approach to Taiwan from socio-economic dimension. Finally, the author suggests Japan's FOIPS should connect the two approaches in a comprehensive and integrated regional strategy. Japan-initiated regional strategy should by nature take more non-military, socio-economic, and soft power approaches to Asian partners. This has a magnetic effect with ASEAN and Taiwan, which are not state sovereignty-oriented but people-oriented liberal actors in the region. Therefore, Japan's initiative has potential to be an alternative to hierarchical hegemonic stability.

---

by-side, and thrive in freedom and in peace." *White House*, "Remarks by President Trump at APEC CEO Summit, Da Nang, Vietnam. https://www.whitehouse.gov/briefings-statements/remarks-president-trump-apec-ceo-summit-da-nang-vietnam/.

II. Origin and Development of FOIPS

Postwar Japan, a pacifist trading nation, has consistently sought for constructing a liberal regional order. By the constitutional restriction, Japan's means to increase its national power is limited to non-military methods such as diplomacy, economic activities, and "soft power" which can be achieved not by coercion but by attraction. Meanwhile, Japan's physical conditions requires overseas natural resources and markets. Therefore, the free and open regional conditions are essentially fundamentals to Japan's national survival and prosperity. Therefore, Japanese economic diplomacy essentially aims to confirm the free "space" for private sectors' to pursue overseas activities smoothly. The role of government is to coordinate the good environment for such activities.[4]

In the 1980's, Japan started to organize regional economic cooperation frameworks of the Asia-Pacific such as PECC (Pacific Economic Cooperation Council), which finally led to the formation of APEC in 1989. After the 1997 Asia Financial Crisis, regional cooperation even without US participation developed and Japan actively supported ASEAN's initiatives to coordinate broader regional frameworks such as ASEAN Plus Three (APT). However, at the turn of the century, Japan's regional policy was strategically re-examined with growing China's presence. Regarding the membership of the proposed EAS (East Asia Summit), while China argued the EAS membership should limit within APT countries, Japan insisted to bring additional three countries, namely Australia, New Zealand, and India into the "East Asia" summit. These three Japan's "like-minded" countries were invited in the sprits of "open regionalism." The formation of ASEAN Plus Three Plus Three (APTT, then EAS) of 2005 resulted in Japan's diplomatic win over China. Abe was in fact then Chief Cabinet Secretary and the real person who actively worked on it with Prime Minister Koizumi Junichiro.

During the first Abe administration (September 2006 -August 2007,) his awareness of Indo-Pacific was already well outlined. For example, in May 2007,

---

4 More details are discussed in Sachiko Hirakawa, "Japan: living in and with Asia," in Lee Lai To and Zarina Othman eds, *Regional Community Building in East Asia: Countries in Focus* (Routeledge, 2016).

a quadrilateral sub-cabinet-level dialogue was implemented together with the US, Australia and India. When he delivered a speech in Indian parliament in August 2007, Abe stated that a region that could be described "broader Asia" was emerging through the "dynamic coupling" brought about by the Pacific Ocean and Indian Ocean.[5] He pointed out that the strategic global partnership between Japan and India was a vital element of the concept of an "arc of freedom and prosperity," which means outer rim of the Eurasian continent including "the successfully budding democracies."[6] This idea was previously introduced as "Value oriented diplomacy" in 2006 by then Foreign Minister Aso Taro. Aso's foreign policy emphasized "universal values" such as democracy, freedom, human rights, rule of law, and the market economy. Thus, the combination of Abe and Aso, who are Deputy Prime Minister and Financial Minister today, is an original planner of the Indo-Pacific strategy and they are working partner today.

After the period of the government of the Democratic Party of Japan (DPJ), Abe assumed the second administration in Dec 2012. He immediately announced the similar policy in updated version. In his planned speech[7] titled "The Bounty of the Open Seas: Five New Principles for Japanese Diplomacy," Abe declared that "Japan's national interest lies eternally in keeping Asia's seas unequivocally open, free, and peaceful…in maintaining them as the commons for all the people of the world, where the rule of law is fully realized." Then he cited the US focus to the confluence of the two oceans, the Indian and the Pacific, and emphasized the significant role of Japan-US alliance. This speech was done in "the very region where we stand by" in time of the fortieth year of relations between Japan and ASEAN.

The five new principles clearly show Japan's consistent beliefs. The first is protecting freedom of thought, expressions, and speech in this region where two

---

5 Ministry of Foreign Affairs, "Confluence of the Two Seas," August 22, 2007.

6 Speech by Mr. Taro Aso, Minister for Foreign Affairs, "Arc of Freedom and Prosperity: Japan's Expanding Diplomatic Horizons" Nov 30, 2006, http://www.mofa.go.jp/announce/fm/aso/speech0611.html.

7 Prime Minister Shinzo Abe's policy speech, which was supposed to be delivered during his stay in Jakarta, but actually was not done because of some unavoidable changes in his itinerary to deal with the Algerian hostage crisis, https://japan.kantei.go.jp/96_abe/statement/201301/18speech_e.html.

oceans meet. These are universal values that humanity has gained. The second is ensuring that the seas, which are the most vital common to us all, are governed by laws and rules, not by might. The third is pursuing free, open, interconnected economies as part of Japan's diplomacy. The forth principle is bringing about ever more fruitful intercultural ties among peoples between Japan and this region. The fifth is promoting exchange among the younger generation for the future.

Thus, Japan's regional concept has consistently from the Asia-Pacific to Indo-Pacific with the same aspiration.

Meanwhile, the second Abe administration actively modified Japan's security policies to react the increasingly severe security environment surrounding Japan such as China's maritime expansions and North Korean offensive nuclear development. Especially, the year 2015 was historical turning point in terms of Japan's national peace and security legislation.

First, Japan-US Defense Cooperation were updated for the first time in eighteen years since 1997, which promoted cooperation not only for defense of Japan but also for the Asia-Pacific regional security in the name of "the areas surrounding Japan." The new 2015 Guideline represents an expansion of alliance cooperation both geographically and across domains. For regional and global peace and security, it outlines activities such as peacekeeping operations, international humanitarian assistance and disaster relief, maritime security, partner capacity building, noncombatant evacuation operations, intelligence, surveillance and reconnaissance, training and exercises, and logistic support, as well as the promotion and strengthening of trilateral and multilateral security and defense cooperation. In addition, the new Guideline established an Alliance Coordination Mechanism (ACM), a general framework to ensure the effectiveness of Japan-US cooperation.

Secondly, the legislation for peace and security was decide by the Diet after totally over 200 hours debate in the two Houses. The bill has two objectives; First is strengthening Japan's national security including gray zone situations. Second is strengthening of cooperation with the international community to ensure international peace and stability. To these ends, ten laws, including the Self-Defense Forces Act, were amended and a new law, the International Peace Support Act, which has been described as a permanent and general law regarding international peace cooperation, was enacted. The updated legislation was based on the Cabinet Decision of July 1, 2014, which permitted limited exercises of collective self-defense rights in a form that maintain logical consistency with the interpretation of Constitution. Thus, these updated Guidelines and relevant legislative improvements underpin the formation and pursuit of FOIPS.

III. Japan's Approach to ASEAN: Launching Defense Cooperation

1. Vientiane Vision to ASEAN

Taking full advantages of the above described security policy updates, Japan started active approach to ASEAN. Vientiane Vision is perhaps the first proactive "defense diplomacy" or "defense cooperation initiative" launched in postwar Japan's history. In November 2016, Japan proposed the Vientiane Vision which is a guiding principle for Japan's defense cooperation with ASEAN. It was announced as Japan's own initiative by Defense Minister Inada Tomomi at the second ASEAN-Japan Defense Ministers' Informal Meeting held in Vientiane, Lao PDR. The "vision" for the first time shows, in a transparent manner, the full picture of the future direction of defense cooperation with ASEAN as a unitary actor.

The vision focused on the following three points.

(1)To consolidate the order based on the principles of international law governing peaceful conduct among states, Japan supports ASEAN's efforts to uphold principles of international law, especially in the field of maritime and air space.

(2)To promote maritime security which is a foundation for the regional peace and prosperity, Japan supports ASEAN's efforts to build up capabilities for

Intelligence, Surveillance and Reconnaissance (ISR) and Search and Rescue (SAR) at sea and air space.

(3) To cope with increasingly diversifying and complex security issues, Japan supports ASEAN efforts to build up capabilities in various fields.

Practical defense cooperation is conducted by effectively combining the following diverse measures:

(1) Sharing understanding and experience regarding international law, especially in the field of maritime security, through i.e. conducting researches and sponsoring seminars, etc., with a view to its effective implementation.

(2) Conducting capacity building cooperation in various fields such as HA/DR, PKO, landmine and UXO (Unexploded Ordnance) clearance, cybersecurity, defense buildup planning (sharing know-how), etc.

(3) Transferring equipment and technology, developing human resources regarding defense equipment and technology co-operation, holding seminars on defense industries, etc.

(4) Continued participation in multilateral joint training and exercises, inviting ASEAN observers to Self-Defense Forces' training, etc.

(5) Inviting Opinion Leaders from ASEAN, etc.

These listed implementations are to receive annual follow-up through Japan-ASEAN Defense vice-ministerial forum. Meanwhile, the Japan-US Security Consultative Committee (SCC) held in August 2017 decided to jointly strengthen Japan's such efforts in Southeast Asia. The joint statement wrote, "Regarding cooperation with Southeast Asian nations, the Ministers affirmed their intention to further enhance capacity building programs and defense equipment and technology transfers in areas including maritime security, defense institution building, and humanitarian assistance and disaster relief (HA/DR)."[8] Furthermore, in the Japan-US summit in November, capacity building on maritime law enforcement was pointed out one of the three core pillars of

---

8 Joint Statement of the Ecurity Consultative Committee, August 17, 2017, http://www.mod.go.jp/e/d_act/us/index.html.

FOIPS.[9] Overall, organizing such soft and broader security network may have an effect on demanding China to increase cost for paying more attention to regional situations.

2. Three Aspects of the Strategy

Japan's security approach to countries in the Indo-Pacific region has three aspects.[10] The first is strengthening of Japan's presence and leadership in "maritime Asia" in the region. The Ministry of Defense has initiated the strengthening of coordination in maritime spheres through port calls by the SDF and joint exercises with the coastal nations of Southeast Asia as well as Australia and India. In February 2016, following a visit to Vietnam in May of the previous year, P-3C patrol aircraft of the Maritime Self-Defense Forces (MSDF) were dispatched to Da Nang in central Vietnam, where they conducted their first-ever search and rescue joint table-top exercise based on a scenario requiring coordinated action between the P-3C aircraft and the Vietnamese navy. In March 2016, the MSDF minesweeper tender Uraga and minesweeper Takashima docked at Port Klang in Malaysia for the first time in three years, and in April the destroyers Ariake and Setogiri made their first port calls at Cam Ranh Bay in Vietnam. The Uraga and Takashima also docked at Cam Ranh Bay in May.

In April 2016, an MSDF Overseas Training Cruise Unit including the Ariake, the Setogiri, and the submarine Oyashio, docked at Subic Bay on the Philippine island of Luzon for the first time in fifteen years, and subsequently MSDF destroyers and other vessels made several calls at the same port. The MSDF has also pursued active operation in the Southeast Asian seas region, conducting goodwill exercises with the Malaysian navy in April 2016, the Indonesian navy in August, the Philippines navy in September. With these opportunities, bilateral meetings and conferences with defensive authorities of countries has also been increasing every year.

---

9 Japan-U.S. Working Lunch and Japan-U.S. Summit Meeting, Nov 6, 2017, http://www.mofa.go.jp/na/na1/us/page4e_000699.html.

10 The National Institute for Defense Studies, Japan, East Asian Strategic Review 2017, 244. The information in this section is largely cited from the same book of 2017 and 2018.

From May to July 2017, the MSDF dispatched the biggest destroyer *Izumo* to Southeast Asian and Indian waters and pursued defense exchange events. In May, Izumo, together with its comrade *Sazanami*, participated in the international review fleet in Singapore and visited Cam Ranh Bay ports where they held a seminar on search and rescue activities. In June, Izumo had a port call at Subic Bay to welcome President Duterte and held the Japan-ASEAN Ship Rider Cooperation Program on board *Izumo*, as an initiative of Japan's "Vientiane Vision."

Also, the MSDF has taken an active part in multilateral maritime security efforts, including the Komodo 2016 exercises organized by the Indonesian navy in April and Indonesia's international fleet review. In May 2016, the MSDF dispatched one destroyer and 360 personnel, the third largest number among the participating countries, to take part in the Maritime Security Field Training Exercise of the expanded ASEAN Defense Ministers' Meeting (ADMM-Plus), participating in on-site inspection, fleet escort, and search and rescue exercises in the sea areas from Brunei to Singapore. The MSDF has also dispatched transport ships and destroyers to take part in regularly held multilateral maritime exercises, such as Pacific Partnership 2016 held from June to August by the US armed Forces, the Rim of the Pacific Exercises (RIMPAC) 2016 held by the US armed Forces in coordination with its allies during the same period and Kakadu 16 held by Australian Army in September. During Pacific Partnership 2016, a joint cruise in the South China Sea by a US navy hospital ship and an MSDF transport ship was conducted, and during RIMPAC 2016, cruise training was conducted in a voyage from Japan to Hawaii by Japan, the US, India, Indonesia, and Singapore.

The second aspect is enhancing the capabilities of partner countries. Following the replacement in April 2014 of the Three Principles od Arms Exports by the Three Principles on Transfer of Defense Equipment and Technology, Japan concluded an agreement with the Philippines for the transfer of defense equipment and technology in February 2016. It was Japan's first agreement of this kind with an ASEAN member. According to this agreement, Japan will transfer to the Philippine navy up to five naval TC-90 training aircraft, assist in the education and training of Philippine navy personnel, and provide support in the field of maintenance. It is expected that they can be used for humanitarian aid and disaster relief (HA/DR) and for monitoring the situation in the seas around the Philippines. Meanwhile, as part of its capacity building technical assistance project, the Ministry of Defense held a seminar on HA/DR together with the United Kingdom in January 2016 for participants from the ASEAN, and in July it provided its first capacity building support for the maintenance of ship diesel engines. Likewise, in response to the first 2 + 2 meetings of foreign and defense ministers of Japan and Indonesia held in December 2015, capacity building support for the creation of nautical charts was provided to the Indonesian Navy in March 2016.

This approach toward ASEAN's maritime nations is a response to the situation in the South China Sea. China has been aiming to change the current situation using fishing boats and China coastguard vessels, of which number is exceeding those owned by the coast guards of the concerned countries in the region. Capacity building of maritime law enforcement agencies such as means by providing patrol boats is critically important to prevent the escalation of the event of contingencies. In addition, Japan's approach to continental ASEAN countries are also active. For example, in February 2016, the Ministry of Defense held a disaster response seminar as its first capacity building support project for the Lao People's Armed Forces (LPAF), and from July to August 2016 it provided on-site instruction on search, rescue and relief in the HA/DR field. For Thai Ministry of Defense personnel, In April and May 2016, the Ministry of Defense provided capacity building support regarding international aviation law and aviation safety, respectively. At the Japan-Thailand Defense Minister's Meeting in June, two countries confirmed that they would further strengthen bilateral defense cooperation through initiatives such as participation in exercises and the establishment of the staff talks between the Royal Thai Army and Japan's Ground Self-Defense Force (GSDF).

The third aspect is advocating and sharing of norms and principles. After the award on July 12, 2016, of an arbitral tribunal in the Hague in the Netherlands between the Philippines and Chin in the South China Sea, the Ministry of Foreign Affairs announced its view that the award was final and legally binding on the disputing countries and urged China and the Philippines to abide by the Court's award.[11] The Ministry of Defense also have actively played in promoting the sharing of norms and principles with other countries in the region. In his speech at the Shangri -La-Dialogue in May 2015, Defense Minister Nakatani proposed the Shangri-La Dialogue Initiative (SDI), which included the "wider promotion of common rules and laws at sea and in the air in the region." He again emphasized the SDI at the 2016 Shangri-La Dialogue. In addition to stating his hope for the early conclusion of the Declaration of the Conduct of Parties in the South China Sea (DOC), he expressed support for the US freedom of navigation operations in the South China Sea and the peaceful resolution of problems based on international law, including arbitration.

The Ministry of Defense is also holding seminars on international aviation law as one of its capacity building support projects underlined in the 2016 Vientiane Vision. In addition to the seminars in Thailand mentioned earlier, the Ministry has already implemented such seminars in Indonesia, the Philippines, Malaysia, and Vietnam. Another important initiative is the promotion of the Code for Unplanned Encounters at Sea (CUES) adopted at the Western Pacific Naval Symposium (WPNS). The MSDF adopted the CUES in the training and exercises it conducted with the Philippines navy in May and June 2015, and with the Malaysian navy in August 2015 and April 2016. After the conclusion of the above mentioned Komodo 2016 exercise, on which the WPNS 2016 Ship Rider Program was held on board *Izumo*. In this program, seminars were held at sea and the participants discussed the topic such as the rule of law and maritime security. Use of the CUES has gradually been expanding as an international maritime standard.

---

11 Arbitration between the Republic of the Philippines and the People' s Republic of China regarding the South China Sea (Final Award by the Arbitral Tribunal) (Statement by Foreign Minister Fumio Kishida), July 12, 2016, http://www.mofa.go.jp/press/release/press4e_001204.html.

It was applied in the ADMM-Plus Maritime Security Field Training Exercises held in May 2016 and its application was discussed at the Indian Ocean Naval Symposium (IONS) held by India. In September 2016, China and the ASEAN reached an agreement to apply the CUES in the South China Sea. Furthermore, China and the US have conducted exercises using the CUES in order to avoid the crisis at sea, and it is possible that similar training by the MSDF and Chinese navy may be implemented in the future.

Thus, the promotion of international norms and principles has become one of the important tasks in the defense exchange and cooperation with other countries being undertaken by both the Ministry of Foreign Affairs and the Ministry of Defense. This indicate that "defense diplomacy," the use in diplomacy of defense assets in peacetime, has become rapidly more important in recent international relations.

IV. Japan's Approach to Taiwan: Updating Non-governmental Relations

1. Updating Relations Between Two People-oriented Governments

Even though Japan maintains One China policy and only pursue private and practical relations with Taiwan in accordance with 1972 Japan-China Joint Communique, Japan's recent governmental approach to Taiwan is upwardly flexible because Japan's recent economic diplomacy is becoming more strategic with security and political consideration. In fact, in 2017 Foreign Ministry's Economic Affairs Bureau newly published the year book titled "Economic Diplomacy of Our Country（我が国の経済外交）." According to the book, in 2016, Japanese government advanced economic diplomacy from the three aspects: (1) rulemaking to strengthen a free and open international economic system, (2) supporting Japanese companies' overseas business expansion by promoting public-private cooperation, and (3) promoting resource diplomacy and attracting investment and tourists.

All aspects are useful and advantageous to solidify a new strategic foundation of Japan-Taiwan relations. It is essentially possible because Japan and Taiwan share basic values and their people to people relations are so deep and firm enough to require upgraded coordination between the two governments. When Tsai Ing-wen was elected in January 2016, Abe government unusually responded with its congratulatory expressions. Foreign Minister Kishida Fumio issued his statement as below.[12]

(1) Ms. Tsai Ing-wen was elected in today's presidential election in Taiwan. The Government of Japan congratulates Ms. Tsai on her victory and praises the smooth implementation of the election which demonstrates that democracy in Taiwan has deeply taken root.

(2) Taiwan is an important partner and a precious friend of Japan. We share basic values and enjoy close economic relationship and people to people exchange. The Government of Japan will work toward further deepening cooperation and exchanges between Japan and Taiwan, based on the existing position to maintain Japan-Taiwan relations as working relationship on a non-governmental basis.

(3) We expect that the issue surrounding Taiwan will be resolved peacefully by direct dialogue between the concerned parties and that it will contribute to the peace and stability of the region.

Perhaps many Japanese people accepted the comment naturally without any question. Taiwan is always Japanese peoples' favorite interests and travel destination. They feel very close to Taiwan's culture and society through a lot of information by commercial magazines and TV shooting locations. For instance, by then three translated biographies of Tsai Ing-wen already became available at bookstores in Japan.[13]

---

12 The Result of the Presidential Election in Taiwan, (Statement by Foreign Minister Fumio Kishida), 16 January, 2016, http://www.mofa.go.jp/press/release/press3e_000053.html.

13 The biography of Chen Sui-bian was published in 2000, but no biography of Ma Ying-jeou was translated. The books related to Lee Teng-hui have been outstandingly popular in Japan.

Even though such many Japanese did not know, indeed this was the very first official comment by Japanese government on Taiwan's domestic politics since 1972. Two days later, in answering to the question prepared by the ruling party member, even Prime Minister Abe himself stated, "Taiwan has been our long-time friend. That presidential election is regarded as the evidence of Taiwan's freedom and democracy. I will wholeheartedly congratulate on Ms. Tsai Ing-wen's victory. I expect cooperation and peoples' exchanges between Japan and Taiwan will hereafter advance."[14] Tsai administration also demonstrated its importance on Taiwan-Japan relations by deciding to put big names to top positions of Taiwan's contact windows with Japan; ex-Premier of the Republic of China Hsieh Chang-ting to the head of Taipei Economic and Cultural Representative Office in Japan, ex Vice Premier of the Republic of China Chiou I-jen for Head of the East Asia Relations Commission in Taipei. The result was significantly productive. On January 2017, the Interchange Association, Japan's contact agency with Taiwan, was renamed into the Japan-Taiwan Exchange Association.

The old name of institution suggests a political compromise to consider China's intolerance of exposure of Taiwan's name to the outside world. However, the 45 years' steadily accumulated achievements of economic-social relations between Taiwanese and Japanese have found the good practical reason to change the name of institution: The 2016 survey showed that only 14% of Taiwanese recognized the name of the Interchange Association despite more than six-millions of peoples' mutual exchanges, therefore, adding clear identification of "Japan-Taiwan" to the old name was necessary for Taiwanese and Japanese.[15] As the Association's mission is to facilitate practical relations, this reason sounds appropriately logical for renaming the institution within the existing 1972 framework. It cannot be denied that the reason logically represents the characteristics of two governments such as democratic values, people-oriented society, and bottom-up political processes. In May 17, 2017, the Association of

14 January 18, 2018, At the budget committee of the House of Councilors.
15 Speech by Numata Mikio at the opening ceremony of Japan-Taiwan Exchange Association, January 3, 2018, https://www.koryu.or.jp/news/?ItemId=176&dispmid=4259%20.

East Asian Relations, a counterpart of the Japan-Taiwan Exchange Association was renamed into the Taiwan-Japan Relations Association. The disgraced "name" issue between Japan and Taiwan was now overcome in the form for China not to stop forcefully despite its official complains.

Abe cabinet also tested further move. In March 25, 2017 Japanese State Minister for Internal Affairs and Communications Akama Jiro made a one-day trip to Taiwan and participated in the event of promoting Japan's agricultural products and sightseeing tourism organized by the Japan-Taiwan Exchange Association. It was the first official visit to Taiwan by Japanese high-ranking official since 1972. Akama stated, "we acknowledge that China-Japan relations remains the most important, at the same time Japan also keep non-governmental practical relations with Taiwan."[16] Expectedly China strongly complained, however, Japan's Chief Cabinet Secretary Suga Yoshihide evaluated the visit as "meaningful" to deepen economic ties and people exchanges between Japan and Taiwan. Foreign Minister Kishida also explained that the visit did not conflict with Japan's basic principle. Thus, Abe cabinet's assertive but cautious approach to Taiwan is clearly found. This is a remarkable change in the Abe-Tsai era.

2. Taiwan's Possibilities in Regional FTA

For Japan, TPP (Trans-Pacific Partnership) was not only trade liberalization framework, but strategically important because it was expected to be the mechanism of deepening economic interdependence among Asia-Pacific countries which shared the basic values such as freedom, democracy, human rights, and rule of law. In addition to liberalizations of trade and investment, TPP did cover various fields such as intellectual property, E-commerce, government procurement, environment, and so on. This 21st type rule-based integration is expected to lead the free and fair economic sphere and solidify global value chains. In this sense, TPP could be essential to Japan's comprehensive security and regional stability. With this belief, Japan joined the TPP negotiation in 2013

---

16 Sankei Shimbun, March 26, 2017. On March 30, Akama visited and reported to Abe. Abe said, "Otukaresama (Xinku-la)." Nikkei Shimbun, March 30, 2017.

and played the major roles to lead the tough negotiations until reaching the signing of agreement by twelve delegations including the US in February 2016.

Therefore, even after the withdrawal of the US under Trump's administration later in the same year, Japan consistently coordinated and led the process of negotiations among remaining eleven countries and finally confirmed the Comprehensive and Progressive Agreement for Trans-Pacific Partnership CPTPP in Da Nang, Vietnam in November 2017 and the signing was done in Auckland in March 2018. CPTPP (TPP-11) is a new arrangement to enact the substantial contents of TPP even without US participation. It basically incorporated original TPP provisions with 20 suspensions and therefore remains a high-standard agreement.

From this perspective, Japan reasonably hoped Taiwan to join CPTPP. In fact, the CPTPP Agreement Article 5 stipulates: "any State or separate customs territory may accede to this Agreement, subject to such terms and conditions as may be agreed between the Parties and that State or separate customs territory."[17] Taiwan is surely qualified to CPTPP membership. Given the US is not a Party of CPTPP, Japan is a key actor to decide new members including Taiwan. In the daily press conference of June 26, 2017, Chief Cabinet Suga, answering to the question by a reporter with Nikkei Shimbun regarding Taiwan's will to join CPTPP, stated that the TPP-11 is open-ended and inclusive, therefore Japan welcomes participations of Taiwan and other economies which show wills of joining this high-standard framework, and provide necessary information to be shared with them.[18]

During President Ma's period, Taiwan concluded two bilateral comprehensive FTAs, in Taiwan called ECAs (Economic Cooperation Agreement) in 2013 without causing China's oppositions: ANZTEC (Agreement between New Zealand and the Separate Customs Territory of Taiwan, Penghu, Kinmen and Matsu on Economic Cooperation) with New Zealand and ASTEP

17　Text of the CPTPP is found at http://www.tpp.mfat.govt.nz/text.
18　Prime Minister of Japan and His Cabinet, https://www.kantei.go.jp/jp/tyoukanpress/201706/26_p.html.

(Agreement between Singapore and the Separate Customs Territory of Taiwan, Penghu, Kinmen, and Matsu on Economic Cooperation.) The levels of two ECAs are high enough to match the TPP standard. Meaningfully, these two countries are also members of both TPP and RCEP (Regional Comprehensive Economic Partnership). However, at the same time it should be pointed out they had been already China's FTA partners and well behaved with consultation to China when negotiating with Taiwan. Therefore, gaining CPTPP members' support without causing China's negative interruption requires new strategic thinking and thought out preparation with like-minded partners collectively.

The possibility of Taiwan-Japan EPA can be alternatively examined. Unlike New Zealand and Singapore, Japan does not have a comprehensive FTA with Taiwan. However, Japan and Taiwan accumulated functional agreements piece by piece. In September 2011, Japan and Taiwan remarkably concluded an investment agreement, which mutually guaranteed nondiscriminatory treatment. This was the first high-level agreement since 1972. Furthermore, the concluded date preceded the similar investment agreement with China signed in August 2012. This exceptional case suggests that with prudential diplomatic skills, Japan and Taiwan still have some possibilities to make a substantial agreement. However, Taiwan's negative attitude on importing the agricultural products from the district which suffered from nuclear disaster after 2011 East Japan earthquake remains a major obstacle for Japan to advance any trade negotiations.

V. Concluding Remarks

This paper found that Japan's "Free and Open Indo-Pacific Strategy" necessarily put emphasis on relations with ASEAN and Taiwan to accomplish its regional vision. It is not only because of their geographical importance or priority, but essentially because ASEAN and Taiwan are both Japan's "value" partners with which Japan can actively promote favorable regional order.

ASEAN and Taiwan can be arguably seen as liberal "non-state" actors who overcame sovereignty-centric ideology in this region. Even though each member of ASEAN has political and economic differences, ASEAN as one collective unit clearly already stands as "People's ASEAN" by the ASEAN Charter. The guiding principles for action stipulated by the ASEAN Charter does include liberal values such as adherence to rule of law, good governance, democracy, human rights, and trade rules while the principles non-interference is also addressed.[19] In fact, APSC (ASEAN Political Security Community) security cooperation mainly focuses on non-traditional security areas. In addition, because of Japan's unique security policy and SDF's characteristics, Japan-ASEAN security and defense cooperation largely take the forms of human security perspectives, which primarily concerns individuals free from the state sovereignty.

This basic nature also gives the chance to involve Taiwan to participate in the cooperation and the author believes that Japan should find the way to connect ASEAN and Taiwan into human security initiatives to complete Japan's favorable regional vision under FOIPS. Although Taiwan's diplomatic sovereignty is not recognized by Japan and other regional states, Taiwan is an equal member of regional community such as APEC, and in socio-economic fields Taiwanese people have strong ties with people in the region. While ASEAN is a good example of regional integration, Taiwan is a good evidence of transborder networking of peoples or civil societies. The challenges and experiences of ASEAN and Taiwan are giving us many inspirations for the future global society. In conclusion, both ASEAN and Taiwan are essential to a liberal regional order, which Japan should defend and support in the Free and Open Indo-Pacific strategy.[20]

---

19 See ASEAN Charter, Article 2, especially Section (h), (i), (n).
20 All web site information was accessed finally on May 6, 2018.

川普政府亞洲政策分析：印太戰略與意涵

高佩珊 *

壹、前言

雖然「印太」字眼在歐巴馬政府時期就曾被提出；例如，希拉蕊擔任國務卿時曾經於 2010 年 10 月在夏威夷演講時拋出「印太」這個概念，[1] 但當時無論是有關美國外交政策的文件和重要政府官員的演講仍維持使用「亞太」（Asia-Pacific）這個名詞。川普上任後開始積極提倡印太概念，隨著國務卿與總統及高級官員多次的公開談話，皆將過去習慣使用的亞太改成印太，意即將印度納入美國在亞洲的權力布局棋盤之中，將美國亞洲外交政策重點擴展至印度洋。無論是美國亞太政策的升級版或被許多學者認為是圍堵中國「一帶一路」（One Belt, One Road）計畫的對抗版，美國的印太戰略都值得吾人仔細推敲爬梳。

因此本文將從印太名詞之起源、印太概念到具體成型為戰略、戰術之運用，以及美國能否成功操作此戰略等，做一研究與整理。藉由認識美國亞洲新政策，亦能瞭解印太新戰略下關鍵組成國家包含日本、澳洲與印度等國對於美國亞洲權力布局各自的看法，對於自身在印太戰略架構下所扮演的角色與可發揮之影響的想法。最後評論此戰略對於美國與中國關係未來發展可能造成的影響，以及在大國權力競爭下，台灣未來可能的選項做出結論與建議。

貳、研究理論與文獻分析

國際關係學界當中現實主義（Realism）學派強調以「權力」（power）和「國家利益」（national interest）兩個核心概念；認為國際政治就是國家之間權力的相爭，國家利益惟有透過權力方能實現。現實主義自 1930 年代末發展至 1960 年代初，其中雖曾遭受來自其他學派的挑戰與論辯，經由新現實主義（Neorealism）學派在 1960、1970 年代現實主義理論再度發揚光大，獨領風騷

---

* 中央警察大學國境警察學系暨研究所副教授

1 Hillary Rodham Clinton, "America's Engagement in Asia-Pacific," US Department of State, October 28, 2010, https://2009-2017.state.gov/secretary/20092013clinton/rm/2010/10/150141.htm.

直到建構主義成為另一顯學。然而，時至今日，在世界政治版圖上，此刻現實主義學派觀點顯然再度被主要強權國家奉為圭臬，用以制定外交政策。

若以攻勢現實主義（defensive realism）為例，此派學者認為國際體系結構導致國家必須追求「相對權力」（relative power）的最大化，而且國家之間的權力鬥爭是不可避免的。權力是國家及國際政治運作的手段，惟有擁有最大的權力才能保障國家最大的安全。因此，國家必須時刻注意、比較自身與對手國的相對權力，注意國際權力的分配與變動，並努力使自身成為國際體系當中最強大的國家，以降低對手國家對其權力的挑戰；因此，惟有當一個國家在一個區域取得霸權地位後，[2] 它才能成為一個維持甚至是滿足於現狀的國家（status quo state）。[3]

按照此邏輯，美國長年存在於亞洲區域，並以自身政治、經濟與軍事等優勢維護其區域霸權地位，若無挑戰者出現，它便會是一個維持現狀的國家。但眾所皆知在中國經濟實力大幅提升，甚至於 2010 年超越日本成為世界第二大經濟體時，不只在亞洲區域，甚至在整個國際權力格局中，都在累積挑戰美國所需的經濟與軍事力量。按照攻勢現實主義的邏輯，在中國尚未取得區域霸權地位之前，中國大陸勢必不斷積累實力以達成成為區域霸權的目標；如此一來，無論願意與否，它便成為美國在亞洲權力布局的挑戰者、國際權力格局的修正者（revisionist state），因此中國成為區域霸權的目標勢必會遭逢美國的反對。

攻勢現實主義大師米爾斯海默（John J. Mearsheimer）認為美國是現代歷史上唯一取得區域霸權地位的國家，在未來的權力與安全競賽中將會阻止另一個實力強大的霸權出現。[4] 吾人若將現實主義觀點用以對照川普上台後提出的亞

---

2 與米爾斯海默（John J. Mearsheimer）看法相同，吉爾平（Robert Gilpin）也認為歷史上沒有任何國家能夠完全支配整個國際體系，美國充其量也只是區域霸權（regional hegemony），而非全球霸權（global hegemony）。霸權是指在許多主要的權力範疇擁有優勢地位與主導權威，能夠引領或影響國際上議題的設定，從而制定與其理念接近的國際關係遊戲規則，並能以實力確保這些遊戲規則的履行。見 John J. Mearsheimer, "Through the Realist Lens: Conversation with," Interview by Harry Kriesler, Conversations with History, April 8, 2002, http://globetrotter.berkeley.edu/people2/Mearsheimer/mearsheimer-con0.html；包宗和主編，《國際關係辭典》（台北：五南，民 103 年），頁 54-55。

3 周湘華、董致麟主編，《國際關係理論與應用》（新北：新文京，民 103 年），頁 13。

4 米爾斯海默（John J. Mearsheimer）認為區域霸權國的出現將不利於美國，因為當某些國家成為區域霸權意指它們將能干涉、介入周遭國家事務，如同現今美國涉足於世界各地事務一樣。他指出美國在歐洲並無潛在對手，因此美國現今最重要的任務便是維持在西半球的優勢，防止中國在亞洲

洲新政策，便不難理解美國在亞太提倡印太戰略的原因與動機。

由於美國與中國兩大國在亞洲區域與國際議題的長期競爭，國際關係學界與國際政治經濟領域、甚至企業界皆有許多學者長期投注心力，研究美中關係以了解雙方往來存在之問題。因本文旨在探究川普印太戰略之成形，並了解此戰略對美中關係發展可能造成之衝擊，故在文獻方面先探討幾本對於美中外交政策與對外關係研究之中、英文書籍及期刊著作，無論是從美國與亞太國家面相研究印太戰略，或是從美中雙邊關係進行的分析，皆能提供研究參考方向。

例如，大陸學者朱清秀在其文章〈日本的印太戰略能否成功？〉就從日本對「印太」的認知，[5]日本對此將採取的行動和應對戰略進行分析。該篇文章解析日本強化西南防衛、介入南海爭端、聯合美國與澳洲及印度牽制和制衡中國的理由。朱清秀認為日本對於印太的認知與理解具有明顯的海洋中心傾向，因此所採取的行動和戰略皆具有牽制中國的內涵。此篇文章架構從印太概念的內涵、日本的印太概念認知、日本的印太戰略，及日本印太戰略面臨的挑戰做出分析，提供本文關於印太戰略起源具有價值之參考，惟此文章發表於美國印太戰略具體成形之前，無法對於美國戰略思考進行分析，本文將加強此部份論述。

另一篇由中國社會科學院世界經濟與政治研究所國際戰略室主任薛力所寫的〈印太戰略對一帶一路影響幾何？〉文章，[6]則從「戰略文化」與「力量平衡」論述美國印太戰略的成形，並質疑它對中國一帶一路建設可能造成的影響。薛力認為川普於 2017 年 11 月至亞洲五國訪問後，帶動印太戰略在中國的研究熱潮，證明美國仍是對中國外交影響最大的國家。他認為西方戰略文化的一大特點便是，先確定對手後再制定相應對的戰略，對於川普政府而言，「修正主義國家」（中國與俄羅斯）與「流氓國家」（如伊朗與北韓）便是美國此時期的對手；因為中國的崛起不僅打破亞洲地區的平衡力量，同時也對美國的價值觀與利益形成挑戰。薛力指出，美國國內推動印太戰略的主要力量是美國軍方與

成為區域霸主。見 John J. Mearsheimer, "We Asked John Mearsheimer: What Should Be the Purpose of American Power?" The National Interest, August 21, 2015, http://nationalinterest.org/feature/we-asked-john-mearsheimer-what-should-be-the-purpose-13642.

5 朱清秀，〈日本的印太戰略能否成功？〉，《東北亞論壇》，第 3 期，2016 年 8 月 3 日，頁 103-115。

6 薛力，〈印太戰略對一帶一路影響幾何？〉，《世界知識》，第 3 期，2018 年 2 月 1 日，頁 73。

共和黨建制派；此外，美國的盟友如澳洲、日本與印度對於此戰略也十分積極主動，但印太戰略能否成功還受到中美關係的特殊性與一帶一路本身的特點。薛力認為美中關係的複雜性使得美國很難定位中國，且作為中國外交頂層設計的一帶一路能讓中國與周遭國家共同獲益，使得印太戰略難以有效影響一帶一路建設。大陸學者們明顯都對印太戰略的成功與否提出質疑。

由外國學者 Satish Chandra 與 Baladas Ghoshal 共同主編，集合 13 位學者的英文專書《The Indo-Pacific Axis: Peace and Prosperity or Conflict?》《印太核心：和平、繁榮或衝突？》，[7] 為近期最新出版主要從印度及亞太主要國家面向分析論述印太區域安全的專書。書中多篇文章分別從印度角度出發探討印太戰略、印度與澳洲的戰略關係、美國勢力涉入印度洋和美國轉向亞太政策對印度的意涵、日本與中國的印度洋戰略、文化經濟海洋路線的擴展等議題。其中比較特別的是 Shankari Sundararaman 發表的 "Understanding the Indo-Pacific: Why Indonesia will be Critical?" 〈瞭解印太：為何印尼相當重要？〉一文，[8] 試圖從印太戰略內主要組成的四個國家以外的角度單獨分析印尼的角色。作者從印尼的角度分析印太概念的意涵，並指出印尼在東南亞區域內、南海內所佔據的地緣政治重要性，但印尼國內出現不滿身為東南亞國協（Association of Southeast Asian Nations）區域內最大國的國家，卻長期為東協成員國所忽略。因此，藉由論述重視海洋安全、海上航行通道自由的的印太概念，作者提出印尼在此戰略中可扮演的角色與價值。

由於本文將探討美國印太戰略對於美中關係的影響，以及台灣該如何面對中美兩大強國與區域國家所提出的重要外交戰略，因此亦參考關於美中台關係的著作。新加坡學者 Sheng Lijun 所撰寫的 China's Dilemma: The Taiwan Issue，[9] 從中國的角度出發，探討中國與美國以及台海兩岸的關係。該書認為台灣的國際地位與身分自冷戰結束以來一直是中國所面臨的最複雜的政治問題之一；對於中國而言，台灣涵蓋在其國家整體發展策略內，因此美國與中國雙

7　Satish Chandra and Baladas Ghoshal, eds., *The Indo-Pacific Axis: Peace and Prosperity or Conflict?* (New York: Routledge, 2018).

8　Shankari Sundararaman, "Understanding the Indo-Pacific: Why Indonesia will be Critical?," in *The Indo-Pacific Axis: Peace and Prosperity or Conflict?* (New York: Routledge, 2018), Chapter 12.

9　Sheng Lijun, *China's Dilemma: The Taiwan Issue* (London: I.B.Tauris, 2001).

方在台灣問題上將持續進行角力，未來兩岸可能發生衝突。以上幾本文獻都能提供本文研究許多參考方向。

參、川普印太戰略意涵

一、印太戰略概念緣起

印太戰略在川普政府上台後取代美國過去外交政策長期使用的亞太戰略一詞，國際關係學界學者因此紛紛對印太戰略的詞彙所隱含之意涵作出分析，有些學者認為它只是一個新名詞但具體意涵不變；有些人認為印太顧名思義就是包含印度洋和印度，認為川普企圖將亞太升級為印太並將印度拉入美國新的亞洲政策當中，利用印度制衡中國力量的增長。

在分析印太戰略目標、戰略意涵和具體戰術運用之前，本文先解釋印太名詞概念起源。學界大致同意川普所提倡的印太戰略概念，起源於日本首相安倍晉三於 2006 年初次上台時所推行的從東南亞經由中亞到中歐再到東歐的「自由與繁榮之弧」（Arc of Freedom and Prosperity）價值觀外交（Value Oriented Diplomacy）。[10] 也就是將此弧線內與日本擁有相同價值觀，即政治民主與經濟自由的歐亞國家連結一起，此外交政策當時被視為圍堵中國的設計。

安倍在 2007 年 8 月出訪印尼、印度與馬來西亞多國時，便提出以日本、印度、美國和澳洲等國組成此自由與繁榮之弧的構想；[11] 惟當時安倍內閣執政時間僅僅短暫一年（2006 年 9 月至 2007 年 9 月），此構想未能有效推展。2012 年底再度回鍋擔任首相的安倍將過去的價值觀外交擴大升級為一個民主鑽石同盟。安倍執政下的日本外交重心顯然一直是以全力布局亞洲放眼世界之大戰略為主；他主張日本除與夏威夷、澳洲與印度連結，從印度洋到西太平洋組成一個保護形同鑽石的網絡維護海洋權益外，也應擴大與英國、馬來西亞、新加坡、紐西蘭和大溪地的法國太平洋海軍合作。[12] 與「自由與繁榮之

---

10 原野城治，〈俯瞰地球的安倍外交—專訪內閣官房參事谷內正太郎（1）〉，日本網，2013 年 8 月 20 日，https://www.nippon.com/hk/currents/d00089/。

11 〈自由與繁榮之弧需要台灣〉，自由時報，2007 年 9 月 6 日，http://talk.ltn.com.tw/article/paper/152693。

12 〈聯合美、澳、印度 安倍組對中包圍網〉，自由時報，2013 年 1 月 29 日，http://news.ltn.com.tw/

弧」的構想相比，升級版的「亞洲民主安全之鑽」（Asia's Democratic Security Diamond）更加具體也更具有企圖心。以上兩項外交設計皆已將印度與印度洋區域涵蓋在內。

2013 年 2 月安倍在美國國際戰略研究中心（Center for Strategic and International Studies）演講以「日本回來了」（Japan is Back）為題，[13] 首次使用「印太」一詞，指出日本在印太地區的主要任務為做為規則的提倡者、全球共同利益的捍衛者，主張日本應該與美國、印度、韓國和澳洲等民主的海洋國家，共同維護海洋安全。但當時日本的官方文件包含「國家安全保障戰略」（National Security Strategy），[14] 皆尚未使用印太一詞，直至 2016 年 8 月安倍在第六屆「東京非洲發展會議」提出印太構想。[15] 安倍提出日本希望與非洲一起將連接亞洲與非洲兩大陸的海洋建設成為以法維護的和平大海，將太平洋與印度洋、亞洲與非洲的交流構建成重視自由、法治和市場經濟的區域。[16]

自此，「自由開放的印太戰略」成為日本外交新重心。緊接著同年 9 月，安倍與印度總理莫迪會晤時表示，[17] 印度是連接亞洲和非洲最重要的國家，希望印度與日本緊密攜手推動印太戰略構想的具體化。安倍積極爭取印度對此戰略之支持，並認為可以與印度自莫迪上台以來採取的「東進政策」（Act East Policy）做一對接，[18] 主導印太地區事務。11 月莫迪訪問日本期間，表示日印雙方觀點逐漸匯集（convergence of views），強健的日印關係有助於雙方在亞

---

news/world/paper/650457。

13 趙全勝主編，〈日本安倍晉三政府的亞太區域政經發展策略研析〉，《日本外交研究與中日關係》（台北市：五南，民 104 年），頁 406。

14 日本為取代 1957 年制定的「國防基本方針」，於 2013 年國家安全保障會議（National Security Council）創立時，首次制定「國家安全保障戰略」。關於此戰略主要內容參見 Ministry of Foreign Affairs of Japan, "National Security Strategy", April 6, 2016, http://www.mofa.go.jp/fp/nsp/page1we_000081.html.

15 見吳懷中，〈安倍政府印太戰略及中國的應對〉，《現代國際關係》，第 1 期，2018 年 1 月 30 日，頁 13-21。

16 日本國首相官邸，〈總理演說〉，2013 年 1 月 29 日，https://www.kantei.go.jp/cn/97_abe/statement/201608/1218963_11159.html。

17 〈莫迪訪日與南亞格局〉，中時電子報，2016 年 11 月 14 日，http://www.chinatimes.com/newspapers/20161114000352-260109。

18 印度自 1991 年拉奧總理上台後大力推動「東望政策」（Look East Policy）經濟改革計畫，使經濟成長率平均維持在 6%；2014 年莫迪上台後更加積極地施行「東進政策」，則被認為具有強權外交戰略意涵。見張棋炘，〈莫迪外交 - 從東望邁向東進〉，南亞觀察，2016 年 9 月 3 日，https://southasiawatch.tw。

洲與世界扮演穩定的角色。[19] 相當有趣的是，日本與印度的緊密合作正是因為雙方對於中國勢力崛起的共同擔憂與川普凡事以美國第一，可能降低美國對與亞洲事務的介入；但日本與印度的合作最後推動川普對於印太戰略的接受與積極推動。川普 2017 年 11 月初至亞洲訪問時，便提出「印太戰略」為美國新亞洲政策，主張美國要和日本、印度與澳洲等民主國家共同維護太平和印度洋的自由與穩定。

倘若此戰略組成國家的先決條件為「民主國家」，自然將中國排除在外，因而造成中國學界對此戰略之負面看法。值得注意的是，無論哪一個大國所主張的外交構想，都突顯印度現今在亞洲所能扮演之角色，在全球政治與經濟格局中所處的重要位置。印度自然以國家利益為第一考量，選擇性的分別與中國、美國與日本合作，巧妙運用印度之戰略價值。[20]

二、川普印太戰略意涵

印太戰略從名詞的出現、概念的源起，到逐漸成型為一具體戰略，其具體意涵被不同國家做出不同的解讀。例如，中國學者普遍認為印太戰略意在圍堵中國、抗衡中國，根本目的在於牽制中國在亞洲地區的影響。[21] 東南亞國家也則希望藉由積極拉攏美、印、日等強權，能達到平衡中國勢力成長所帶來的可能威脅。至於日本與澳洲，則早在川普印太戰略推出之前，便各自構建以自己國家為戰略中心的印太概念，因此對美國印太戰略作出積極回應。各國皆自自身國家安全戰略角度，仔細評估美國印太戰略實行的可行性與有效性，做出戰略選擇。至於美國施行印太戰略之戰略目標與意涵，則可從川普上台後對於中國的重新定位與至今政府公布的三份重要文件一窺究竟：

---

19 Elaine Lies, "Strong Japan-India Ties can help stabilize the world, says Modi," Reuters, November 11, 2016, https://www.reuters.com/article/us-india-japan-idUSKBN1360CX。

20 無論是 2016 年 3 月美印日在菲律賓北部鄰近南海的海域與 6 月在日本舉行的的「馬拉巴爾」三國聯合軍演，亦或是同年 4 月中俄印國外交部長共同於莫斯科就南海問題發表的聯合公報，皆突顯印度巧妙運用自身戰略價值為國家謀取最大利益。相關報導參見〈美日印在非北聯合軍演〉，大紀元，2016 年 3 月 3 日，http://www.google.com.tw/amp/www.epochtimes.com/b5/16/3/3/n4653287.htm/amp；〈俄羅斯印度外長公開聲明在南海問題上支持中方立場〉，觀察者，2016 年 4 月 20 日，http://m.guancha.cn/military-affairs/2016_04_20_357641.shtml。

21 朱清秀，〈日本的印太戰略能否成功？〉，頁 106。

（一）國家安全戰略報告（National Security Strategy）

根據 2017 年 12 月 18 日公布的美國「國家安全戰略報告」，[22] 美國國家利益涵蓋「四大支柱」：保衛國土、美國人民和美國生活方式；促進美國繁榮；以實力維護和平；增進美國的影響。此份文件指出美國世界地位遭遇的重大挑戰和趨勢，包含中國和俄羅斯這種「修正型大國」（revisionist powers）採取技術、宣傳和脅迫等方式，試圖塑造一個與美國利益和價值觀對立的世界。在與對外戰略有關的第三支柱「以實力維護和平」，該份文件表明美國將保障和平並威懾敵對勢力，重建美國軍事力量。

此外，美國將運用戰略競爭新時代的一切治國方略手段，包括外交、資訊、軍事和經濟等手段保護美國利益並加強太空和網路能力。美國的盟國和夥伴使美國的實力更加強大並保護共同的利益。美國期待盟邦能承擔更大責任，應對共同威脅。最重要的是美國將確保世界關鍵地區如印度 - 太平洋、歐洲以及中東的實力平衡繼續有利於美國。

美國在此份文件中提及中國 33 次，並將中國定位為「戰略競爭者」（strategic competitor）、「對手國」（rival power）、「敵對國」（adversary）等等。美國認為其與中國和俄羅斯之間的戰略競爭是現今主要挑戰，反恐已經不再是國家安全主要威脅；中國與俄羅斯才是「美國安全與繁榮的主要挑戰」。[23] 此份報告還指出，中國正在歐洲以不公平貿易行為與透過對重點行業、敏感技術和基礎設施進行投資以獲得戰略位置；在非洲，中國則通過賄賂精英、主導採礦業並把非洲許多國家綑綁於大批債務與承諾，破壞非洲國家的長期發展。[24] 因此，美國國家安全戰略的重點便是幫助維持區域力量的平衡，維護合作的國際經濟環境。

與 2010 年 5 月歐巴馬時期公布的國家安全戰略相比，[25] 當時歐巴馬改變小

---

22 關於此報告綱要可見美國在台協會，〈美國國家安全戰略綱要〉，2017 年 12 月 18 日，https://www.ait.org.tw/zhtw/white-house-fact-sheet-national-security-strategy-zh/。

23 報告全文參見 The White House, "National Security Strategy", December 18, 2017, https://www.whitehouse.gov/wp-content/uploads/2017/12/NSS-Final-12-18-2017-0905.pdf.

24 理乍得魏茨，〈美國新國家安全戰略報告與中國〉，中美聚焦，2018 年 1 月 8 日，http://zh.chinausfocus.com/peace-security/20180108/24869.html。

25 亞洲世紀和中國崛起，〈2010 年奧巴馬政府美國國家安全戰略報告全文〉，2010 年 5 月 27 日，

布希「單邊主義」、「先發制人」的做法，強調合作與對話的「多邊外交」、「結盟」的概念，並表示美國合作對象將從傳統盟友擴大到中國和印度等崛起中的新興大國，和中國在雙方關切的共同問題上合作；意即與中國維持「戰略穩定」的關係。2015 年 2 月的國家安全戰略，同樣以正面論述與中國關係，美國表示歡迎中國的和平崛起和穩定繁榮，將致力加強與中國的建設性關係，降低產生誤會的風險，共同促進亞洲和世界的安全與繁榮。[26] 當時美國應對的核心問題以打擊極端組織、反對恐怖主義，且網路安全與核擴散問題才是美國最大的安全威脅。

（二）國防戰略報告（National Defense Strategy）

與國家安全戰略報告相同，2018 年 1 月 19 日出版的美國「國防戰略報告」也將中國與俄羅斯認定為「美國的繁榮和安全的主要核心挑戰」。[27] 此份文件提到中國 12 次，並指出現階段恐怖主義並非美國國家安全主要關注點，國家間的戰略競爭才是。國防部長馬蒂斯（James Mattis）表示美國「越來越多得面對來自中國、俄羅斯這些要求改變現狀力量的威脅，這些國家試圖打造與其威權模式一致的世界秩序」，[28] 但美國的軍事優勢卻在逐漸喪失。

根據該份報告，中國是美國的「戰略競爭對手」，中國使用掠奪性的經濟手段恐嚇鄰國，並在南海進行軍事化行動；中國正在運用自己的軍事現代化、影響戰和掠奪式的經濟活動來逼迫鄰國重整印度洋—太平洋地區的秩序，使之有利於中國，目標在尋求於印太地區獲取霸權，並在未來取代美國以取得全球龍頭地位。因此，美國將通過強化前沿軍力部署與提升同盟合作的方式，維持在核心區域的軍事優勢，且將更加重視利用盟國和夥伴國的力量，以印太、歐洲、中東地區和西半球為重點區域，建構相互尊重、互相合作與分擔責任區域

---

http://risechina.blogspot.tw/2012/12/2010.html/。

26 〈美國發布 2015 年國家安全戰略報告〉，央視網，2015 年 2 月 7 日，http://m.news.cntv.cn/2015/02/07/ARTI1423272867811939.shtml。

27 報告全文參見 The US Department of Defense, "Summary of the 2018 National Defense Strategy", January 19, 2018, https://www.defense.gov/Portals/1/Documents/pubs/2018-National-Defense-Strategy-Summary.pdf.

28 〈美國防戰略報告：將中俄視為頭號威脅〉，BBC 中文網，2018 年 1 月 19 日，http://www.bbc.com/zhongwen/trad/world-42755369。

盟友夥伴關係體系。與 2008 年公布的國防戰略相比，[29] 當時美國雖然將中國視為潛在的競爭對手，但仍尋求以長期、多面向的和平時期的交往方式。川普政府對於中國角色的定位、應對中國的崛起，顯然更為強硬更具對抗色彩。

（三）核態勢評估報告（Nuclear Posture Review）

2018 年 2 月 2 日公布的 100 頁「核態勢評估報告」，[30] 指出美國雖然不想將中國與俄羅斯視為對手國（adversary），但卻明確指出中俄戰略政策與核武能力等國對美國國家安全帶來的挑戰；因此，美國現在需要深具靈活的、彈性的和適應力強的核能力以保護美國和美國的盟友與夥伴國家，以促進戰略穩定。

該報告明白指出如同中國國家主席習近平在中國共產黨第十九次全國代表大會的發言，中國將在 2050 年前「建設世界一流軍隊」，持續提高核武能力。中國宣示的政策和主義並沒有改變，但其核武現代化的計畫長期缺乏透明度，令外界難以明白中國的意圖。無論是中國開發的裝載在洲際彈道導彈（ICBM）、潛射彈道導彈的核彈頭，或射程能到達美國本土、海外領土、美國盟友和美軍海外基地的戰區彈道導彈，中國同時也在擴展非核武器的軍事能力，包括太空戰和網絡戰能力，但外界都難以得知相關細節。無論是中國的軍事現代化或中國追求在亞洲區域的主導地位，皆成為美國利益在亞洲面臨的主要挑戰。美國採取的戰略應對為，防止北京錯誤的認為可以通過有限使用戰區核能力確保優勢、有限地使用核武器是可以被接受的。

此份報告明白將中國、俄羅斯和北韓視為主要核威脅，指出美國為敵人量身製造核震懾策略，在「極端狀況」下美國將考慮使用核武器，以保護美國、其盟國和夥伴的關鍵利益。至於甚麼情況為極端狀況，則處於戰略模糊以維護美國利益。美國將維持可信賴的軍事能力，讓中國在衝突升高時權衡得失，了解使用核武器將得不償失。該份文件指出美國已經做好回應中國的核武侵犯和

---

29 The US Department of Defense, "National Defense Strategy", August 5, 2008, https://www.defense.gov/Portals/1/Documents/pubs/2008NationalDefenseStrategy.pdf.

30 報告全文參見 The US Department of Defense, "Nuclear Posture Review", February 2, 2018, https://www.defense.gov/News/SpecialReports/2018NuclearPostureReview.aspx.

非核侵犯行為，美軍在亞太地區的演習已經明白展示美國的準備。[31] 與 2010 年報告相比，[32] 當時的報告雖然要求提高美中兩國核力量的透明度，也指出美國有必要與中國及俄羅斯保持戰略穩定。相較於歐巴馬時期的核態勢評估報告，川普政府明顯改變對中國及俄羅斯的看法。

　　以上三份重要官方文件皆顯示印太戰略提出背後的戰略意涵來自於美國對中國的重新定位，認為中國企圖在印太地區限制美國的存在並試圖掌控地區主導權。因此針對中國積極主動的國際行為，美國必須做出反應。因此，不可否認美國的印太戰略的確包含應對中國的崛起，平衡或制衡中國在亞洲區域持續上升的的影響力和各種軟、硬實力之戰略意涵。

肆、印太戰略戰術運用

　　既然川普政府確立「自由開放的印太戰略」為美國亞洲新外交戰略架構，除其所隱含之具體意涵外，亦須確立實行此戰略之戰術工具之運用。除以實行民主政治制度作為印太戰略組成國家之先決條件外，意即在外交上組成具有相同民主價值觀交往的陣線，在軍事上亦須透過提升軍事預算與共同軍事演習、軍事力量展示和軍售武器等方式有效實行此戰略，以平衡中國力量之增長，確保美國在亞洲地區的領導地位。

　　2017 年 11 月美國國會參眾兩院壓倒性通過 2018 年度近 7000 億美元的國防支出法案，[33] 12 月川普簽署「國防授權法案」（National Defense Authorization Act）時表明強軍構想；此預算將用以購買戰艦、90 架 F-35 隱形戰機、24 架 F／A-18 超級大黃蜂噴氣式戰鬥機以及 3 艘沿海作戰艦艇，現役部隊人數也將增加 1 萬 6000 人以上。[34] 美國國防部長馬提斯（James Mattis）說

---

31 〈美國核態勢報告：“極端情況”下將使用核武器〉，美國之音，2018 年 2 月 5 日，https://www.voacantonese.com/a/pentagon-nuclear-report-20180202/4238460.html。

32 〈中國與核態勢評估報告：從戰略互信到戰略穩定〉，卡內基國際和平基金會，2010 年 7 月 22 日，https://carnegieendowment.org/2010/07/22/zh-event-3013。

33 美國國防預算 2017 年度與 2018 年度分別為 6190、6920 億美元。見〈美國 2018 年度國防預算近 7000 億美元 年增 12%〉，上報，2017 年 9 月 19 日，https://www.upmedia.mg/news_info.php?SerialNo=25096；〈力挺川普振軍力 美國會敲定 7000 億美元年度國防預算〉，自由時報，2017 年 11 月 9 日，http://news.ltn.com.tw/news/world/breakingnews/2247877。

34 〈推動強軍和反恐川普簽署國防授權法案〉，大紀元，2017 年 12 月 13 日，http://www.epochtimes.com/b5/17/12/12/n9951117.htm。

明這些預算將用於召募更多軍隊、訓練士兵以及購買船艦等等。[35] 2018 年 2 月
川普政府提出的 2019 年度 4.4 兆美元預算案中，國防預算仍高達 6861 億美元。[36]

除大幅提升國防軍事預算外，對外美國也持續透過與印太戰略組成國家
或區域國家簽訂軍事同盟條約、[37] 舉行聯合軍事演習、軍售武器或軍事援助相
關盟國、建立和保持國防軍事層級人員交流與對話機制等方式維持美國在印太
區域的戰略價值。例如，每年固定舉行的韓美聯合兩棲登陸演習、美日印聯合
軍演，以及美日、美澳、美菲共同軍演等。美國航空母艦「羅斯福號」（USS
Theodore Roosevelt CVN-71）、[38]「卡爾文森號」（USS Carl Vinson），[39] 持續
在南海區域及重要國際水域進行的「例行演練」或「自由航行行動」（Freedom
of Navigation Operation），[40] 包括派遣軍艦、軍機進入其他國家試圖管制的區域，
並就行動總結美軍的任務與相關活動；目的在展現國際社會不接受這種限制。
透過軍事演習與美軍的例行性演練與航行任務，皆能展示美國軍事力量，表明
美國有能力實現其戰略目標。

---

35 〈川普跟金正恩拚了國防預算創史上最高紀錄〉，中時電子報，2018 年 2 月 11 日，http://www.
chinatimes.com/realtimenews/20180211002165-260417。
36 〈川普 2019 年預算〉，風傳媒，2018 年 2 月 13 日，http://www.storm.mg/article/399467。
37 美國與亞洲盟國多簽有安全同盟條約，例如：《美菲共同防禦條約》、《美日安保條約》、《美韓
共同防禦條約》、《美澳紐安全條約》、《美泰軍事同盟協議》等。
38 「羅斯福號」核子動力航空母艦打擊群今年 4 月在南海執行「例行演練」後，便前往菲律賓停靠並
邀菲律賓將領登艦。見〈中美航艦南海尬軍力美軍公布軍演照片〉，聯合新聞網，2018 年 4 月 11 日，
https://udn.com/news/story/7331/3080114。
39 母港位於聖地牙哥的「卡爾文森號」為多用途航空母艦，1983 年至今多次至南海巡航，在福特級
核子動力航空母艦正式服役前為全球最大軍艦，每 20 秒可彈射出一架作戰飛機；曾於 2011 年時海
葬賓拉登。「卡爾文森號」戰鬥群包括柏克級驅逐艦邁爾號（USS Wayne E. Myer）以及航艦機隊。
2017 年 2 月美國派遣隸屬於海軍第三艦隊的「卡爾文森號」鄰近南海進行空軍演訓。
40 「自由航行行動」由卡特總統於 1979 年提出，當某些國家對於航行自由有越權的主張或限制
（excessive claims or restrictions）時，美國便會派遣軍艦或軍機至該區域巡航，迫使世界各國遵守「聯
合國海洋法公約」成文化的國際習慣和國家實踐。美國表示此舉不僅是為了全球利益（global public
good），也是為了保障美國安全，確保世界各國能自由航行。此行動計畫已為美國海軍遍及全球的
例行性任務。見劉瑋哲，〈什麼是美海軍自由航行計畫（FONOP）〉，洞見國際事務評論網，2016
年 2 月 1 日，http://www.insight-post.tw/insight-knowledge/20160201/14430。

伍、結論

　　美國國家戰略報告反映出美國對於全球及區域局勢改變的看法，提出美國須採取的應對方式及國家安全政策的制定方向。因此，印太戰略的提出便是美國對亞洲格局變動的關切，包含美國對中國的重新定位，認為中國在印太地區限制美國的存在並試圖掌握區域事務主導權，且在此區域中國展現的戰略企圖心與積極主動性的國際行為，致使美國必須做出反應。因此，印太戰略背後的意涵便是應對中國的崛起，平衡（balance）或制衡（contain）中國在亞洲區域持續上升的的影響力和各種軟、硬實力的提升。

　　川普提倡之印太戰略由概念（concept）逐漸具體成型為戰略（strategy），搭配戰略手段與工具的使用，自然造成各國對此戰略之積極回應或重視。雖然印太戰略部分區域與中國提出的一帶一路計畫重疊，致使外界易將此戰略是為圍堵或抑制中國勢力之外交戰略設計，然此戰略未來成功與否仍需取決於亞洲各國對自身國家利益之計算與考量。

　　由於各國在印太地區面對不同的挑戰，勢必從自身國家安全角度評估美國印太戰略的可行性與有效性，巧妙在美國與中國之間保持平衡。對於台灣而言，無論是日本需經由台灣與東南亞才能到達印度，日本海上石油運輸線需經過台灣，亦或是從地緣戰略角度來看，台灣在印太地區佔據的重要位置，在美國的戰略布局中皆具價值性。然而，如何能夠維持兩岸關係穩定的同時，還能巧妙運用自身實力與大國交往，在在考驗我政府智慧。

G-0 或 G-2？未來國際體系的轉變

桂　斌 *

壹、前言

　　美國一手打造的國際體系，在過去七十年間創造人類歷史上最長的承平時期與經濟繁榮。蘇聯於 1991 年解體後，美國成為全球唯一超級強權。進入廿一世紀，長期反恐戰爭與金融風暴使美國國力與國際聲望亦受影響。時至今日，美國雖然仍然是包括軍事、經濟、金融在內的世界超級大國，但其地位也已經與蘇聯瓦解時有所不同，開始面對中國此一新崛起者的挑戰。

一、後冷戰時期國際格局的演變

　　1990 年代是美國獨霸的年代。受到經濟、科技與全球化的影響，美國作為唯一的全球性強權，在軍事上，美國是全球軍事投資最多、預算最多也最先進的國家；經濟上，美國作為全球最大的經濟體以及全球商品最大的最終市場，對全球經濟抱持舉足輕重的影響力，並隨著柯林頓總統時期「新經濟」（New Economic）的興起，使得美國成為全球金融與投資的主要標的；文化上，美國文化隨好萊塢與流行產業擴散至全球，也是接收最多留學生的國家，在軟實力的排名上也是全球首屈一指。[1]

　　進入 21 世紀，911 事件使得整體國際格局產生轉變。美國在單邊主義下發動全球反恐戰爭，引發許多國家的不滿，但美國也被迫與其他國家進行合作與協調，單邊主義被削弱，形成「一超多強」的國際體系。在反恐戰爭結束後，美國國力相對下滑，特別在 2008 年金融風暴後，中共在國際上之話語權日漸增加，新的結構逐步成形。

---

* 淡江大學國際事務與戰略研究所博士候選人
1 依據英國波特蘭公關公司聯合美國南加州大學共同發布之《2017 The Soft Power 30》，2017 年法國取代美國成為全球軟實力排名第一的國家，https://softpower30.com/。

二、中國的崛起與美國的應對

中國大陸在 21 世紀扮演一個快速成長的新興強權。憑藉每年 8% 以上的經濟成長率，2011 年已取代日本成為全球第二大經濟體，甚至若依按照人均 GDP 數據及「購買力平價」做調整排名，中國已超越美國成為世界最大經濟體。[2] 隨著經濟成長，中國也不斷擴大其影響力，除試圖成為東亞的最主要權力行為者，逐步排除美國的影響力之外，也在非洲、拉美等第三世界區域擴大影響力，取得自然資源維繫經濟發展，運用投資從需要幫助的國家和政府那裡獲得想要的標的。

習近平上台後，中共一改「韜光養晦」之傳統外交政策方針轉向「有所作為」[3]，在中國共產黨第十九屆黨代表大會中習近平以總書記的身分提出中國要走自己的道路，堅持「四個自信」（道路自信、理論自信、制度自信、文化自信）[4]；在杭州與漢堡 G-20 峰會中，習近平以國家主席的身分提出「中國方案」，強調「給世界上那些既希望加快發展又希望保持自身獨立性的國家和民族提供了全新選擇，為解決人類問題貢獻了中國智慧和中國方案。」[5] 目前，中共之核心外交政策是：「要堅持獨立自主的和平外交方針，堅持把國家和民族發展放在自己力量的基點上，堅定不移走自己的路，走和平發展道路，同時決不能放棄我們的正當權益，決不能犧牲國家核心利益。要堅持不干涉別國內政原則，堅持尊重各國人民自主選擇的發展道路和社會制度，堅持通過對話協商以和平方式解決國家間的分歧和爭端，反對動輒訴諸武力或以武力相威脅。」[6]

---

2 郭匡超，〈超越美國 中國已成世界最大經濟體〉，中時電子報，2014 年 9 月 28 日，http://www.chinatimes.com/realtimenews/20140928001392-260408。

3 黎盟，〈專家：習近平重塑"韜光養晦有所作為"〉，人民網，2015 年 8 月 15 日，http://politics.people.com.cn/BIG5/n/2015/0815/c1001-27466723.html。

4 〈習近平：決勝全面建成小康社會 奪取新時代中國特色社會主義偉大勝利—在中國共產黨第十九次全國代表大會上的報告〉，新華社，2017 年 10 月 27 日，http://www.gov.cn/zhuanti/2017-10/27/content_5234876.htm。

5 任琳，〈G20 杭州峰會：中國參與全球經濟治理主場外交〉，人民網，2016 年 10 月 17 日，http://world.people.com.cn/n1/2016/1017/c1002-28784810.html；韓潔、安蓓、陳煒偉，〈推動世界經濟繁榮發展的"中國方案"—解讀習近平主席在 G20 漢堡峰會發言時的四點主張〉，新華社，2017 年 7 月 8 日，http://www.xinhuanet.com/world/2017-07/08/c_1121286903.htm。

6 黎盟，〈專家：習近平重塑"韜光養晦有所作為"〉。

　　在政治上，中國在東海與南海區域積極以行動伸張主權，包括東海防空識別區與南海大規模填海造陸。在經濟上則提出「一帶一路」之新戰略，並搭配設立「亞洲基礎設施投資銀行」作為配套措施，試圖將經濟力轉化為政治影響力。最明顯的例子在巴基斯坦，「中巴經濟走廊」正在促進巴基斯坦的經濟增長，創造就業機會。作為回報，中國正在建設的瓜達爾港（Gwadar Port）將使中國在印度洋獲得更大的影響力。在東南亞，菲律賓總統杜特蒂（Rodrigo Roa Duterte）不喜美歐的批評，中國則承諾會協助菲律賓改善落後的基礎設施，使其不再強烈反對中國在南海的主張，還在東盟展現親中國的立場。但另一方面，這種投資也使許多國家債台高築，例如斯里蘭卡即因高額債務被迫向中國移交南部漢班托塔港（Hambantota Port）的資產和經營權。[7]

　　另一方面，主打「美國優先」原則贏得總統大選的美國總統川普（Donald J. Trump）在正式執政後，其「美國優先」政策美國似乎帶領著美國朝向更單邊主義、孤立主義的方向前進，並準備放棄過去所承擔的國際領導責任，讓世界陷入權力真空。川普在上任不到一年的時間裡已經採取了一系列使美國退出國際領導地位的舉動，包括退出《跨太平洋夥伴關係協議》、《巴黎氣候變化協議》和聯合國教科文組織。川普決策執著於自己的直覺、即期利益與單邊獲利，在當今棘手的安全、經貿、環境三大議題上，把華府改造成一個「擴權卸責」（more strength with less responsibility）的領導。[8]

　　這類此消彼漲的訊息並非意味美中兩國會進行一種自願的、和平的權力轉移。相反的，隨著中國近年來諸多作為呈現與美抗衡的態勢、中國朝向自由民主方向轉型希望日益渺茫以及美中貿易逆差問題，使得美國的民意與政界的立場逐漸轉變。2008 年歐巴馬總統提出「重返亞洲」，但這並非是要與中國直接的對抗；2010 年 11 月，首爾 G-20 高峰會中的中美兩國首腦會談可被視為潮流轉向的徵兆，全球經濟、貿易、北韓和伊朗問題，以及人權問題都是這次會談議題，[9] 歐巴馬提出警告，如果中國不採取更多行動來遏制北韓的好戰行為，

7　林庭瑤，〈斯里蘭卡漢班托塔港經營權 移交中共〉，聯合新聞網，2017 年 12 月 9 日，https://udn.com/news/story/7331/2866305。

8　張登及，〈川普政策風格對兩岸關係影響評估與展望〉，財團法人國家政策研究基金會，2017 年 5 月 18 日，https://www.npf.org.tw/1/16956。

9　管淑平，〈G20 峰會 歐胡敲定雙邊會〉，自由時報，2010 年 10 月 29 日，http://news.ltn.com.tw/

他將採取措施來保護美國免受北韓核彈威脅。中國議題，涵蓋安全和經濟兩方面的擔憂，也是歐巴馬第二任總統競選焦點之一，共和黨候選人羅姆尼（Mitt Romney）指責歐巴馬在面對中國領導層時態度不夠強硬；歐巴馬政府自身也在與中國打交道的過程中發現中國在氣候變化標準上對美國的主張不屑一顧，在向伊朗施壓上一再拖延，還開始在南海領土主張問題上強硬實行自己的主張，使其決心改變作法，但美國戰略東移也讓中國對美國的動機產生深切懷疑。[10]

在 2016 年美國總統大選中，川普與希拉蕊•柯林頓對中國的政策其實大同小異，川普指控中國大陸刻意壓低匯率，主張大陸不停止人民幣貶值，就對陸製產品提高關稅；希拉蕊則指中國大陸藉由在市場以低於成本價格傾銷貨品，傷害美國的就業與企業，並公開反對給予中國大陸市場經濟體地位，兩人也都主張要施壓中國更進一步阻止北韓的核武計畫。[11] 這顯示出美國內部對於中國的看法已經出現改變轉向強硬，且此種改變開始逐漸成為主流民意。

世界各國也開始擔憂傳統權力平衡趨向瓦解，使得區域局勢複雜化、緊張化。在歐洲，也有一些國家與社會輿論開始擔心中國的鉅額投資以及控制重要領域關鍵工業可能造成的長期風險，也對北京在一些歐盟成員國或是即將加入歐盟的國家中建立戰略立足點感到擔憂；[12] 東協內部部分國家諸如越南、新加坡也開始警覺中國可能的威脅。

三、G-0 觀念的出現

全球最大的政治風險諮詢公司歐亞集團（Eurasia Group）主席布雷默（Ian Bremmer）提出「零極」（G-0）的概念。在 2011 年，歐亞集團提出該年度的

news/world/paper/439261。

10 Mark Landler, "Obama's Evolution to a Tougher Line on China", New York Times, September 9, 2012, https://www.nytimes.com/2012/09/21/us/politics/obamas-evolution-to-a-tougher-line-on-china.html

11 廖漢原，〈美國史上最分歧總統大選 政見比一比〉，中央社，2016 年 7 月 31 日，http://www.cna.com.tw/news/firstnews/201607310022-1.aspx。

12 Helena Legarda and Michael Fuchs, "As Trump Withdraws America from the World, Xi's China Takes Advantage", Center for American Progress, November 29, 2017, https://www.americanprogress.org/issues/security/reports/2017/11/29/443383/trump-withdraws-america-world-xis-china-takes-advantage/。

最大風險是「G-0 世界」[13]。G-0，意指沒有一個國家或國家集團具有推動真正的國際議程的政治和經濟槓桿的能力或意願，各國將政策著眼點放在內部事務之上，使提供全球治理的重要國際機構變成國家間相互對抗而非合作的競技場，造成國際建制的癱瘓甚至瓦解。[14]G-0 的背景是西方影響力的下降和發展中國家政府將焦點置於國內所造成的國際政治的權力真空。過去，美國利用其軍事和經濟力量來推動全球合作，但現在缺乏資源繼續作為全球公共產品的主要提供者，而中國則沒有興趣接受國際領導帶來的負擔。[15] 美國總統川普在2017 年 6 月宣佈退出應對全球氣候變暖的國際框架《巴黎協定》正是一例。川普認為這一協定「懲罰」了美國，讓美國失去了數百萬的工作崗位；中國外交部呼籲各方應共同珍惜和維護這一來之不易的成果，並表示中方願與有關各方加強合作，共同推動《巴黎協定》實施細則的後續談判和有效落實，推動全球綠色、低碳、可持續發展。[16] 不過值得注意的是，美國認定在協定中中國爭取到極有利的條件，按照該協定，中國將得以在未來 13 年間增加溫室氣體排放量，並獲准額外建造大量煤電廠，這也是川普決議退出的原因之一。[17]

　　造成 G-0 出現的原因有二。第一，西方自由世界兩大核心美國與歐盟各自面臨諸多難題，迫使歐美政策制訂者更聚焦國內事務；第二，「異者」的崛起。國際政治舞台上開始重新出現不認同美國與西方政治價值觀的強權與集團，例如中國、俄羅斯、金磚四國等。雖然他們目前仍缺乏設定國際日程的能力，但是它們有足夠的資源和能力阻撓美國的計畫，使得全球共識日趨瓦解。[18]

13　"Top Risks for 2011: G-Zero tops the list", Eurasia Group, January 4, 2011, https://www.eurasiagroup.net/media/top-risks-for-2011-g-zero-tops-the-list.

14　Ian Bremmer and Nouriel Roubini, "Explain why a "G-Zero World" means conflict, protectionism, and trade wars", *Eurasia Group*, February 7, 2011, https://www.eurasiagroup.net/media/ian-bremmer-and-nouri-el-roubini-exlain-why-a-g-zero-world-means-conflict-protectionism-and-trade-wars.

15　Ian Bremmer and Nouriel Roubini, "A G-Zero World," *Foreign Affairs*, March/April 2011, https://www.foreignaffairs.com/articles/2011-01-31/g-zero-world.

16　〈特朗普宣佈退出巴黎協定 國際社會大表失望〉，BBC 中文網，2017 年 6 月 2 日，http://www.bbc.com/zhongwen/trad/world-40129266。

17　The White House, "Statement by President Trump on the Paris Climate Accord", June 1, 2017, https://www.whitehouse.gov/briefings-statements/statement-president-trump-paris-climate-accord/.

18　Harry Kazianis, Coping with a G-Zero World. Ian Bremmer speaks with The Diplomat about a "G-Zero World," China's rise and why no single nation can fill the global power vacuum, The Diplomat. June 20, 2012, https://thediplomat.com/2012/06/coping-with-a-g-zero-world/.

貳、美中競合與權力轉移的挑戰

　　中美關係在歐巴馬執政後期摩擦顯著增多，川普就職後進一步加劇。美國與中國在國際金融、雙邊經貿、區域安全事務等擁有許多共同關切，在這些議題上存在衝突又合作之關係，但僅在雙方均有相同利益的問題上，兩國才有某種程度合作的可能性；但是，即便雙方存有共同利益，也並不保證兩國將同心合力確保問題之徹底解決，而是以謀求自身利益極大化為主，例如在北韓核武議題上，北韓近年之行為不僅使中國被迫承擔國際社會的重大壓力，也損及中國與南韓、日本和美國之雙邊關係，更鼓勵南韓加強美國之安全合作。但北韓之存續仍對中共有價值，使美中間不可能達成徹底的合作，而僅能部分的合作，降低北韓造成的威脅。

　　同樣的，中美雙方也會避免因衝突導致利益受到過度損害。兩國在政治價值、市場開放、中國國際責任、解放軍軍力擴張等方面存在競爭關係。由《2017年美國國家安全戰略報告》可知，美國認為中國正在利用經濟上的誘惑和懲罰、施加影響的行動以及潛在的軍事威脅來讓其他國家聽從其政治和安全議程；中國在南海建立軍事基地的做法危害自由貿易，威脅其他國家主權，並破壞地區穩定，[19]並認為在西半球，中國試圖通過投資和貸款將該地區拉入自己的陣營。[20]

　　中美最大的競爭還是出現在經貿領域。早在競選期間，川普即曾多次歸咎過往政府，認為全球自由貿易是造成美國製造業衰退及就業機會流失的元凶，尤其在中國大陸加入世界貿易組織（WTO），以及北美自由貿易協定（NAFTA）、美韓 FTA 等貿易協定生效後，貿易對手國不公平競爭下，美國貿易入超大幅擴增，特別是中美之間的貿易逆差不可接受，誓言扭轉局面，建立更公平的貿易。在就任後，2018 年 3 月川普簽署備忘錄，計劃依據「301 調查報告」對從中國進口的商品大規模徵收關稅，涉及商品規模高達 600 億美

19　The White House, "The Strategy in a Regional Context", *National Security Strategy of the United States of America*（2017），pp. 45-46, https://www.whitehouse.gov/wp-content/uploads/2017/12/NSS-Final-12-18-2017-0905.pdf.

20　The White House, "The Strategy in a Regional Context", *National Security Strategy of the United States of America*（2017），p. 51.

元；[21] 中國在之後也宣布將對原產於美國的大豆、汽車、化工品等 14 類 106 項商品加徵 25% 的關稅。[22] 隔日，川普揚言將再加碼 1000 億美金。[23] 由表 1 可知，中國對美貿易順差是中國對外貿易順差的主要核心，比例超過 88%。關稅戰對美國的代價將是物價提升、對中國出口下降；對中國來說則可能是貿易順差大幅降低，甚至可能使中國轉為入超國。

表 1：中國貿易概況

	出口額	進口額	貿易利得
貨品貿易	15.33	12.46	2.92
服務貿易	7.3	5.95	1.35
總額	22.63	18.41	4.22
對美貿易	5.06	1.30	3.75
對美貿易佔整體比例	22.35%	7%	88.86%

單位：千億美元

資料來源：數據來自〈海關總署介紹 2017 年中國全年進出口情況〉，中華人民共和國商務部網站，2018 年 2 月 10 日，http://www.mofcom.gov.cn/article/tongjiziliao/fuwzn/ckts/201802/20180202711259.shtml。

除關稅戰外，美國亦宣布針對中國中興通訊（ZTE）祭出 7 年出口禁令，這主要是因為中興通訊違反了 2017 年 3 月與美國司法部達成的認罪協議，本協議主要是針對中興違反美國出口管制條例（Export Administration Regulations, EAR），通過第三方公司，向受到美國經濟制裁的伊朗出口美國產品，美國政府就發現中興並未按照當時認罪協議要求，因此直接吊銷中興的出口許可，美國供貨商和合作方立即停止與中興合作。[24]

---

21 Bob Davis, "U.S. to Apply Tariffs on Chinese Imports, Restrict Tech Deals", The Wall Street Journal, March 22, 2018, https://www.wsj.com/articles/u-s-to-apply-tariffs-on-50-billion-of-chinese-imports-1521723078.

22 中華人民共和國國務院關稅稅則委員會，〈國務院關稅稅則委員會發佈對原產於美國部分進口商品加征關稅公告〉，中華人民共和國財政部網站，2018 年 4 月 4 日，http://gss.mof.gov.cn/zhengwuxinxi/gongzuodongtai/201804/t20180408_2862847.html。

23 "Trump proposes $100 billion in additional tariffs on Chinese products", CNBC, April 5, 2018, https://www.cnbc.com/2018/04/05/trump-asks-us-trade-representative-to-consider-100-billion-in-additional-tariffs-on-chinese-products.html。

24 Steve Stecklow, Karen Freifeld, Sijia Jiang, "U.S. ban on sales to China's ZTE opens fresh front as tensions escalate", REUTERS, April 16, 2018, https://www.reuters.com/article/us-china-zte/u-s-bans-american-companies-from-selling-to-chinese-phone-maker-zte-idUSKBN1HN1P1.

一、中國新型大國關係與中國夢

2012 年 11 月，中共十八大報告指出：「我們將改善和發展同已開發國家關係，拓寬合作領域，妥善處理分歧，推動建立長期穩定健康發展的新型大國關係。」[25] 中共一再強調要推動建立「中美新型大國關係」，強調中美可避開冷戰期間美國與蘇聯兩超強對抗的舊型關係，不要落入上述困境。新型大國關係可以三句話來說明內涵 :1. 不衝突、不對抗；2. 相互尊重；3. 合作共贏。[26]

本質上，新型大國關係是在強調歷史上新興大國的崛起往往伴隨著衝突和戰爭，但現在的世界已今非昔比，對抗將是雙輸。「相互尊重」是構建中美新型大國關係的基本原則，這包括相互尊重彼此的核心利益。[27]

在 2012 年 11 月底，習近平也藉著帶領中央政治局常委前往國家博物館參觀時表示：「每個人都有理想和追求，都有自己的夢想。現在，大家都在討論中國夢，我以為，實現中華民族偉大復興，就是中華民族近代以來最偉大的夢想。」第一次提出「中國夢」。[28]「中國夢」所追求的目標是「國家富強、民族振興、人民幸福。」[29] 大陸學者則歸納中國夢是強國夢、政治大國夢、強軍夢。[30]

2013 年 3 月，習近平在第十二屆全國人民代表大會第一次會議闡釋了其實現「中國夢」的「三個必須」路徑：必須走中國道路、必須弘揚社會主義精神、必須凝聚中國力量。[31] 這顯示出中國的自信增強，認為中華民族從來沒有像今天這樣揚眉吐氣，世界任何大問題，如果沒有中國人參與，都很難解決。[32]

25 〈胡錦濤在中國共產黨第十八次全國代表大會上的報告〉，人民網，2012 年 11 月 18 日，http://cpc. people.com.cn/n/2012/1118/c64094-19612151-1.html。

26 孫辰茜，〈外交部：不衝突不對抗、相互尊重、合作共贏是中美兩大國正確相處之道〉，新華網，2017 年 3 月 22 日，http://www.xinhuanet.com/world/2017-03/22/c_1120676126.htm。

27 〈如何構建中美新型大國關係─王毅外長在布魯金斯學會的演講〉，中華人民共和國外交部網站，2013 年 9 月 20 日，http://www.fmprc.gov.cn/web/zyxw/t1078765.shtml。

28 洪向華主編，《民族復興中國夢》（北京：紅旗，2013 年 3 月），頁 1。

29 「習近平在全國人大閉幕會上講話談中國夢（全文）」，人民網，2013 年 3 月 17 日，http:// bj.people.com.cn/n/2013/0317/c349760-18308059.html。

30 杜玲玉，〈習近平「中國夢」之探討〉，《展望與探索》，第 13 卷第 3 期，2015 年 3 月，頁 40。

31 「習近平在第十二屆全國人民代表大會第一次會議上的講話」，新華網，2013 年 3 月 17 日，http:// news.xinhuanet.com/2013lh/2013-03/17/c_115055434.htm。

32 吳建民，《我的中國夢：吳建民口述實錄》（北京：北京大學，2013 年 5 月），頁 69-70。

此種自信的態度擺脫了過去中國對外戰略定位的被動，開始強調在國際體系中的重新崛起，和獲得大國應該享有的國際尊重和地位。

二、美國的孤立主義思想與川普對外政策

在二次世界大戰後，「孤立主義」雖已不再是美國外交政策的選項，但其意識形態仍存在於政策思考中，此即傑佛遜主義（Jeffersonianism），以美國第三任總統傑佛遜（Thomas Jefferson）命名，將美國民主體制和社會利益視為最重要的國家利益，認為政府本身是必要之惡，外交政策亦然。美國可做為其他國家的模範，但不是強加民主於其他國家，更不需為全球民主的推動來服務。[33]傑佛遜主義者深信過度捲入海外事務會損及國內民主，尤其應當減少與外國衝突的風險，推至極端則形成孤立主義。[34]傑佛遜主義在歷史上最重要的影響之一即是反對美國加入國際聯盟。在冷戰初期傑佛遜主義者堅持孤立主義傳統使得傑佛遜主義快速退出外交主流思想，但在廿世紀後半，隨著共和黨雷根政府「新自由主義」（Neo-liberalism）的新右派思想，推動新經濟，反對聯邦政府集權化與專業化，使得其思潮又開始復甦。[35]

從川普就任至今採取的外交行動與作風，諸如空襲敘利亞、退出 TPP 與巴黎氣候協定似乎也展現傑克遜主義（Jacksonianism）的影響。傑克遜主義出自美國第七任總統傑克遜（Andrew Jackson），思想基礎則是美國民粹主義傳統，主張美國政府的一切內政外交政策，無論甚麼手段，都應以維護美國人的實際安全和經濟利益為基礎，竭盡全力增加美國人民的福祉，[36]雷根（Ronald W. Reagan）也是美國傑克森主義的代表人物。因此，在對外政策上，若出現對美國的威脅或挑戰，傑克遜主義者會呈現極度鷹派的立場，將不計代價與其鬥爭，並取得勝利；反之，則會不太感興趣，認為美國無須捲入海外事務，美國是否應介入科索沃之爭論即為一例。

---

33 關中，《意識形態和美國外交政策》（台北：台灣商務印書館，2005），頁 107。
34 沈旭輝，〈200 年前的偶像：特朗普外交與「傑克遜主義」〉，信報，2016 年 12 月 19 日，http://www1.hkej.com/dailynews/article/id/1457590/200。
35 孔誥烽，〈小政府右翼之死〉，端傳媒，2016 年 9 月 16 日，https://theinitium.com/article/20160916-opinion-hunghofung-neoliberalism/。
36 關中，《意識形態和美國外交政策》，頁 112-113。

雖然傑克遜主義反對傑佛遜主義的和平主義，但兩者在後冷戰時期逐漸出現共同點，都認同和平的出現使美國得到機會，擺脫國際承諾，專心處理國內事務。兩者的差別在傑佛遜主義者談論聚焦「和平紅利」（peace dividend）的運用，主張裁軍、縮減情報機構；傑克遜主義則主張維持軍備，但縮減美國在冷戰中所做的政治與經濟讓步。[37]

川普本身的重商主義則是漢米頓主義（Hamiltonianism）的直接表現，將貿易與國內經濟視為國家核心利益，[38] 但川普與傳統漢米頓主義主義者的差異在於對自由貿易的態度，但這可視為對於何種貿易形式對美國有利的認知不同或對於自由貿易的修正。

在川普政府的《2017年美國國家戰略報告》中可以很明顯的看到上述觀點。強調：「我們將保衛我們的國家，保護我們的社區，把美國人民的安全放在首位。」[39]「對手瞄準美國力量的來源，包括我們的民主制度和我們的經濟。」主張：「與世界接觸並不意味著美國應該放棄作為主權國家的權利和義務，更不意味著美國應該在安全問題上做出妥協或讓步。」[40]

三、美中的意識形態衝突

造成中美衝突加劇的原因除物質因素外，「意識形態」也扮演重要角色，特別是近期推動之美中貿易戰與其他領域之競爭之上。美國一直試圖「改造中國」、「打開中國的大門」，其背後除了美中合作可以有效削弱蘇聯勢力的原因外，也包含理想主義的因子，認為在中國的政治民主改革與開放是不可避免的，因此同中國的聯繫一定要持續下去。[41] 在商業領域，長期以來美國國內主流意見認為應該給予中國永久最惠國地位（Most Favored Nation, MFN），支持中國加入 WTO 以融入世界經濟體系，理由除打開中國市場會為美國的企業創

---

37 同上註，頁 124。
38 同上註，頁 100。
39 The White House, "PILLAR I: Protect the American People, the Homeland, and the American Way of Life", *National Security Strategy of the United States of America* (2017), pp. 7-8.
40 Ibid.
41 The White House, "East Asia and the Pacific", *National Security Strategy of the United States of America* (1991), p.9, http://nssarchive.us/NSSR/1991.pdf.

造新商機外，鼓勵中國走向改革也是很大的原因。[42] 反對取消最惠國待遇的主張之一即是擔心取消最惠國待遇會動搖中國政府及共產黨中支持改革的勢力；加強美中經濟交流可強化改革派人士在北京的地位，才能在經改之後出現政治改革。[43]

然而，近年來的發展打破此種以經濟發展改變中國，推動中國民主化的理想。中國不但沒有朝向更民主的方向發展，反而更堅信其中國特色政治體制的有效性，對美國民主制度表現出嗤之以鼻的態度，也對西方試圖「和平演變」中國政治體制加以防範，從各個領域排除西方思想。習近平上台後不久，即力求整頓意識形態，在「8·19」講話中強調「意識形態工作是黨的一項極端重要的工作」[44]；2014 年 9 月 5 日，在慶祝全國人民代表大會成立 60 週年大會的演說上，中共中央總書記習近平否定中國大陸照搬外國政治模式的可能，強調「世界上不存在完全相同的政治制度，也不存在適用於一切國家的政治制度模式」，要固守中國共產黨「一黨專制」。[45] 紐約時報也報導中國教育部長袁貴仁警告，外國思想正在對中國大學校園造成威脅，呼籲禁止使用宣揚西方價值觀的教科書而且不允許課堂上出現批判中國共產黨統治的言論。[46]

更有甚者，美國國內也開始出現認定中國正追求獨霸世界地位的聲浪。例如哈德遜研究所中國戰略研究中心主任白邦瑞（Michael Pillsbury）《百年馬拉松》（*The Hundred-Year Marathon: China's Secret Strategy to Replace America as the Global Superpower*）一書主張中國制訂了長期計畫，追求以中共主導的共產主義體系以替代美國主導的世界秩序，中國絕非美國的合作者，而是挑戰者。

---

42 Robert E. Lighthizer, "Testimony Before the U.S.China Economic and Security Review Commission:Evaluating China's Role in the World Trade Organization Over the Past Decade", U.S.-China Economic and Security Review Commission, p. 3 , June 9,2010, https://www.uscc.gov/sites/default/files/6.9.10Lighthizer.pdf.

43 王一程，美國政要和謀士們關於對華政策的討論，《海峽評論》，第 24 期，1992 年 12 月號，https://www.haixia-info.com/articles/615.html。

44 〈習近平：意識形態工作是黨的一項極端重要的工作〉，新華網，2013 年 8 月 20 日，http://www.xinhuanet.com/politics/2013-08/20/c_117021464.htm。

45 〈習近平在慶祝全國人民代表大會成立 60 周年大會上的講話〉，中國人大網，2014 年 9 月 5 日，http://www.npc.gov.cn/npc/xinwen/2014-09/06/content_1877767.htm。

46 Dan Levin, "China Tells Schools to Suppress Western Ideas, With One Big Exception", *New York Times*, February 9, 2015, https://www.nytimes.com/2015/02/10/world/asia/china-tells-schools-to-suppress-western-ideas-with-one-big-exception.html?_ga=2.119427725.523051701.1524820458-1785415212.1488303706.

因此，其主張美國應正視中國不安於和平崛起的事實，而應正視與應對中國的挑戰。[47]

因此，認同向中國「讓利」以推動中國改革的聲音逐漸淡化，強調美中差異、過往對中過於讓步的聲音開始缺乏制衡力量。在《2017 年美國國家戰略報告》中表明美國將不再對違規、作弊或經濟侵略視而不見，不再縱容不公平貿易往來，現行國際體制是贊成專制的人和贊成自由社會的人之間的政治競爭。[48] 中國媒體與民意在 2016 年美國總統大選中展現較支持川普上台的原因之一是認為在川普的統治下美國會停止對外輸出民主自由，認為從此以後不需要面對美國在人權、民主等議題上施加的壓力。但相對的，川普也要收回過往為推動中國民主所付出的讓利，這則是中國各界未預料到的狀況。

參、未來的世界 G-0 或 G-2?

未來國際結構會如何演變，中美是否會陷入上世紀 60-70 年代初的冷戰狀態？目前看起來仍不太可能。冷戰形成的前提是雙方的相對隔離，各自建立相對獨立的政治與經濟勢力圈。然而，在現今全球化的時代，中美與全球各國間「你中有我、我中有你」的複雜利益羈絆，使得美、中都很難進行單邊作為，而須考慮外溢效果，也會受到本國各種利益集團的反對。因此，未來國際結構之變化主要將聚焦以下兩大因子：

一、誰願意承擔領導世界的責任

在美國退出 TPP 後，中國將承接世界領導者的聲音甚囂塵上。然而，承擔責任不僅只是成為一個霸權，更重要的是必須付出相應代價。在現今全球化時代，領導國必須既願意也有能力把各國帶到談判桌上以應對世界最緊急關切的問題，並能以身作則的帶頭付出，甚至必須超越自己的國家利益。由目前來看，中國在國際議題上的付出仍有限，特別在實際作為上仍遙遙落後於美國與歐盟，例如在氣候變化、伊波拉病毒和北韓核武問題上，中國的投入仍著重於

47 白邦瑞（Michael Pillsbury），《百年馬拉松》（*The Hundred-Year Marathon: China's Secret Strategy to Replace America as the Global Superpower*）（台北：麥田，2015 年 9 月），頁 320。

48 The White House, "Promote Free, Fair, and Reciprocal Economic Relationships", *National Security Strategy of the United States of America* (2017), p. 19.

自身利益的維護，較少以全球利益作考慮做出犧牲與付出。[49]

　　以資訊領域的互聯網治理為例，網際網路在全球每個國家的通訊與商務中都是不可或缺的角色。然而，俄羅斯和中國這樣的獨裁國家倡導建立保障不干涉國家對境內互聯網實施強主權控制的國際協議。中國基於國安，藉由建築網路長城、監控民間輿論的方式進一步監控資訊，此種作法固然對中國對內控制有利，卻有加快全球網際網路邁入所謂「分裂網」（Splinternet）的境地，[50] 亦即將原本單一的網路世界分割得支離破碎，這對於全球化來說將是一大退步與傷害。

　　由國際體系來看，過往的七國集團（G-7）被廿國集團（G-20）所取代，雖然成員更多但一旦討論深入政策細節，能夠達成的一致意見的可能性卻大大減少，且 G-20 各國對政府對在經濟中所應扮演的角色、法治、所有權、透明度等問題仍有不同看法；WTO 內部也因為各國差距過大使得全球貿易自由化推動難有進展。再者，各國政府的權力也因全球化而受到了削弱，權力正在從國家手中逐漸向企業、非政府組織、個人、甚至恐怖主義組織擴散，使得國際問題的解決更形複雜。國際社會需要一個領導者出面領導，但各方勢力往往又各懷鬼胎或過度理想化，使得現有國際體系運作日益艱難。美國的退縮是一個危機也是轉機，一方面可重新界定各國的權利與義務，一方面也可以探索新的互動關係。

二、各國的選擇

　　除了中美兩國外也應考慮其他國家對於國際體系未來發展的看法與行動。長期以來，中國外交運用「和平共處五原則」廣受發展中國家與第三世界的歡迎，近年來，中國在國際關係方面更提出了一系列新概念和新提法，開始逐漸形成一套新的外交理論體系，包括互利共贏、以和平發展為主題的新時代觀、新安全觀、新秩序觀、新文化觀等等，[51] 的確吸引了許多重視國內經濟發展的

---

49 〈專家：把中國吹捧為全球化旗手是個錯誤〉，美國之音，2017 年 1 月 25 日，https://www.voacantonese.com/a/leader-globalization-20170123/3691336.html。

50 Joseph S. Nye, "Internet or Splinternet?", *Project Syndicate*, August 10, 2016, https://www.project-syndicate.org/commentary/internet-governance-new-approach-by-joseph-s--nye-2016-08?barrier=accessreg.

51 裴遠穎，〈和平共處五項原則是中國外交理論核心和實踐指南〉，中國共產黨新聞網，2014 年 6 月 10 日，http://cpc.people.com.cn/BIG5/n/2014/0610/c68742-25127185.html。

開發中國家的支持。這些國家將會是支持中國建立新 G-2 關係的主要支持者。

但中國政治體制也使其難以獲得多數民主國家的長久支持。當代人類社會發展不可能長期漠視民主、民權意識的出現與普及化，特別在已開發國家全數皆為民主國家的情況下，中共的影響力或許可降低其他國家對中國的批判並在特定議題上爭取各國的合作，但要如何克服政治體制的差異形成穩定的領導體系，領導諸多民主國家追求共通價值仍會是一個巨大的挑戰。由已開發國家的內部運作可以了解，安全夥伴關係是建立在廣泛信任的基礎上，而此一基礎則立基於共同價值觀之上。例如《里斯本條約》第 2 條即明確指出寫出歐盟的價值建立在重視人性尊嚴、自由、民主、平等、法治國與維護基本人權、以及少數族群的權利。[52] 民主國家與中共意識形態的差異將是中國意圖領導世界的主要障礙。

再者，也有許多國家滿足於現行國際體系，願意維持現狀，並對中國崛起充滿戒心。在東亞，冷戰雖然結束，但冷戰國際格局遺留至今。美國藉由諸多雙邊關係建構的區域國際體系至今仍有效運作；東協也試圖在美中間保持平衡。對這些國家來說，無論是新的 G-2 或 G-0 都不是最佳解。大多數國家在尋求一個政治和經濟合作的更佳平衡，既能獲得經濟上的利益又能確保政治自主性，因此許多國家採取左右逢源的立場，追求自身最大利益。

三、美國的抉擇

川普領導的美國跟歐巴馬總統的時代有很大的不同，明顯希望從世界事務中進行一定程度的戰略收縮，聚焦國內經濟發展。然而，這並不表示美國會一百八十度的轉向，放棄霸權、放棄國際建制的領導權，反而可視為是「導正」國際秩序，使其更有利美國的做法。以中國的說法，即是「苦練內功，夯實基礎。」

以國際體系來看，美國為首的國際體系架構雖然外部面對諸多新興挑戰但內部仍未發生根本的轉變，特別是以美國為首的同盟體系，包括北約以及美

---

52 Article 2, the Lisbon Treaty, http://www.lisbon-treaty.org/wcm/the-lisbon-treaty/treaty-on-european-union-and-comments/title-1-common-provisions/2-article-2.html.

日、美韓等眾多雙邊同盟關係仍舊維持。冷戰結束迄今沒有國家主動脫離此一體系，顯示其仍保持穩定。再者，由於中國崛起，美國的影響力反而強化了。川普戰略收縮的作法反而在短期迫使各國加深對美國的依賴，也迫使各國正視美國的要求。而在聯合國與其他國際組織中，美國仍期望在政治和安全機構發揮主導作用努力爭取與美國的利益和價值觀相一致的結果，認為聯合國可以幫助解決世界上許多複雜的問題，但必須改革並重新確立其基本原則，強調成員之間的共同責任；美國也期望繼續在 IMF、世界銀行和 WTO 等機構中發揮領導力，美國要求各個國際機構進行改革，「特別是若要求美國為一個機構提供不成比例的支援，則將期望對該機構的方向和努力產生相應程度的影響。」「美國不會把管理權拱手讓與那些要求監管美國公民或與美國憲法框架背道而馳的國家。」[53]

肆、結語

本文的目的是由文獻與行為分析中美對未來國際體系的主觀期許與行動，分析未來國際體系的轉變方向。美國的收縮不應輕易的視為體系瓦解的象徵，而應視作一種主動針對現行體系進行修正的過程。自二戰後美國即自認是世界的領導者，迄今未變。美國優勢逐漸在減少，但美國在過往建立的自由民主價值觀與自身政治、經濟與軍事實力仍是其維持領導地位的核心武器。中國的崛起已成事實，目前，美國已正視中國的挑戰，中美間新一輪的競合將會是決定未來世界體系的關鍵因素之一，除了中美的行動外，各國的選擇將決定未來世界體系的走向，這也會是最大的變數。

---

53 The White House, "Achieve Better Outcomes in Multilateral Forums", *National Security Strategy of the United States of America* (2017), p. 40.

戰略研究的理論與實踐

空權再思考：政治、科技、戰略

蘇紫雲 *

壹、前言

美國為首的北約，2018 年 4 月再次以空中兵力打擊敘利亞的化武戰力，以報復阿賽德（Bashar al-Assad）政府採用化學戰劑攻擊反政府軍，並企圖解除其化學戰力。值得注意的是，參與攻擊行動的英國，前首相梅伊（Theresa May）指稱英軍的作為是藉由精確打擊「避免化武的使用成為常態」（use of chemical weapons to become normalised），而「並非介入內戰、也非要顛覆政權」（This is not about intervening in a civil war. It is not about regime change）。[1] 梅伊的說法堪稱精準的指出現代空中力量與政治間的關鍵連結，也就是在遂行政治目的同時，又要避免或降低政治的負面效應，此種矛盾的需求。

同時，近年來空中力量的展現也更頻繁的被作為國家主權或意志的象徵，在主權爭議區域用來展現國家意志，這在以往多為海軍的專利，但由於科技進步使得具備長程飛行與滯空能力的軍機具有類似海軍的「軍力展示」（show the flag）的作業能力，且具備更高的可操作機動性。類似的行動包括中國軍機的繞行台灣、美國軍機在南海穿越中國人工島礁等，都展現出得空中兵力得以扮演主權工具的戰略角色。前述的作為在冷戰時期雖也有類似行動，但多為近接偵察以試探對手國的防空能力，與目前所能達成的目的及規模都大不同，而此種「戰術行動產生戰略價值」的趨勢，隨著科技進步得以提供決策者更多選擇彈性，因此在未來將更為明顯。特別是精準打擊的細緻化，包括導引精度、以及「微型無人載具」、「群組作戰」技術的成熟，編入作戰序列後將可能帶來另一次的空權革命，並進而連動政治局勢的發展。

可以預見，未來空權的行使將更為多樣化，但另一挑戰卻也隨之而來，由

* 國防安全研究院副研究員兼所長、淡江大學整合戰略與科技中心兼任執行長

1 Jessica Elgot, "'No alternative': Theresa May sends British jets to join airstrikes on Syria," The Guardian, April 14, 2018, https://www.theguardian.com/world/2018/apr/14/theresa-may-britain-air-strikes-syria-chemical-weapons-raf.

於維持空權的成本居高不下，且現代國家各種部門與職能競逐有限經費的情況下，這又造成了政治難題，更遑論各軍種間也存在著資源的競逐關係。如何在空權的發展與政治成本效益之間，藉由科技找出可能的解決途徑，將是維繫國家利益的重要關鍵。

貳、武裝天空：空權的定位爭議

法國飛行部隊在 1915 年首次將兵器裝置並固定於飛機上，[2] 正式開啟了武裝天空的時代，並為後續發展出的空權（air power）奠定最重要基礎，也就是武裝飛機用來驅逐敵人的武裝飛機，以確保己方使用天空的自由。其後歷經二次大戰，主要國家的飛行部隊紛紛由海軍、陸軍附屬兵科獨立出來成為軍種，開啟了現代武裝部隊的全新面貌，並成為戰場的關鍵主角。然而，在 21 世紀的 10 年代卻出現「空權不需獨立空軍」的新論述。

在空中力量作為獨立軍種的地位已然穩固、甚至稱霸戰場的今日，卻仍有不同觀點，認為科技條件與戰爭特性的改變，空軍無法獨力遂行戰爭，因此主張空軍應取消獨立地位，回歸附屬於陸軍、海軍兵科的組織模式，如此才能發揮最大作戰效益。此一看法的代表性人物就是學者法雷（Robert Farley），他認為「美國空軍在 1947 年獲得獨立軍種地位是一大錯誤，空軍不等於空權，也無法替代或視為其他形式的軍事權力。[3]

其主要觀點是認為空軍的戰力無法獨力發揮，需要其他軍種的支援與整合，且維持空軍龐大組織的成本過高，因此不再必要維持獨立的存在。進一步說，所謂的「空權」較諸以往的確更為重要，但空中力量與其他形式的軍事力量整合也同樣重要。例如今日的 B-1B 轟炸機可以提供陸軍或陸戰隊小股士兵的密接支援，而此種程度的（火力）整合在 1947 年是難以做到的。[4] 其所蘊含的意義在於精準彈藥的出現，使得大型軍機得以與地面小兵力整合為一，發揮

---

2 Justin D. Murphy, Matthew A. McNiece, *Military Aircraft, 1919-1945: An Illustrated History of Their Impact* (ABC-CLIO: December, 2008), pp.52-53.

3 Diane Tedeschi, "Is It Time to Abolish the U.S. Air Force? A political scientist says yes," *Air & Space Magazine*. October, 2015, https://www.airspacemag.com/flight-today/air-force-robert-farley-interview-180956612/#yvbzV6KK1mB4TtQp.99.

4 Ibid.

整體戰力。

類似觀點早已出現，另一學者帕博（Robert A. Pape）認為，「若缺乏精確情報，精準武器與精密打擊能力也就無用武之地。[5]

他的主要觀點也是認為空軍無法獨立作業，因此回歸陸軍、海軍航空隊的組織形式，將更有效率也更可節省預算。此類觀點或許甚為極端或前衛，也見仁見智，但對於空中力量的使用，以及空權組成要素的改變，在未來將具有更大的挑戰，也給政軍決策菁英帶來新思維的可能。

參、空權成為新政治工具

傳統上，由於航空器本身的滯空能力、天候耐航性受到約制，因此不若海軍、陸軍可以行使佔領、威嚇等需要兵力長時存在的政治任務。然而隨著科技發展與國際政治環境的改變，軍事遂行政治任務的手段也更加多元，空軍為主的空權也獲得更多的發揮空間。其遂行方式主要為：

一、主權變相延伸

防空識別區（Air Defense Identification Zone, ADIZ）在國際法上並無任何主權的實質效力，設置的原始功能主要是作為防空警戒，做為領空之外的緩衝區，未被視為主權訴求的工具。目前全球劃設防空識別區的國家計有英、美、加拿大、挪威、日、韓、台灣、印度、巴基斯坦，以及中國。以往防空識別區被視為軍事防禦所需，作為防空預警體系的一環，但中國於 2013 年宣布「東海防空識別區」，由於其劃定的區塊與中國東海的主權爭議區高度重疊，對北京而言防空識別區具有主權延伸的潛在意義，因此被認定帶有政治目的。[6]制空權的執行與確保也因此具備高度的主權意涵，此與以往空權的主要目標是排除敵方對空域的使用，並確保其他軍種的存活與任務執行大異其趣。

---

5 Robert A. Pape, "The true worth of Air Power." *Foreign Affairs*, March/ April 2004, https://www.foreignaffairs.com/articles/2004-03-01/true-worth-air-power.

6 Teshu Singh, "China and Air Defence Identification Zone (ADIZ): Political Objectives and International Responses," China and the Air Defence Identification Identification Zone (ADIZ), IPCS, IPCS SPECIAL FOCUS, pp. 14-15.

二、傳遞國家意志

由於航空科技的進步，使得航空器的滯空能力與耐航性能大為提昇，軍機得以對主權爭議所在的陸地、水域遂行接近或繞行的戰術運動，以及迫近對手國的防空識別區、甚至領空。這使得傳統的空權運用，除了純軍事用途外，更產生新的政治意義，用以傳遞政治訊號。

尤其是中國在東海防空識別區設置完畢，加以水門機場啟用後，軍機出海的頻率大為提高，航行軌跡與東海防空識別區高度重疊，用以展現國家意志的意圖甚為明確。以軍事資料相對透明的日本作為主要觀察，其攔防外國軍機的次數在 2016 年達到 1,168 次，其中以中國的成長率最為驚人，2013 年為 415 次，2016 年達 851 次，[7] 成長率達 205%。

表1：日本緊急攔截次數統計（依來源國家／地區統計）

国.地域 / 年度	ロシア		中国		台湾		北朝鮮		その他		合計	
	年度	1四半期	年度	1四半期	年度	1四半期	年度	1四半期	年度	1四半期	年度	1四半期
25	359	31	415	69	1	0	9	9	26	6	810	115
26	473	235	464	104	1	0	0	0	5	1	943	340
27	288	57	571	114	2	0	0	0	12	2	873	173
28	301	78	851	199	8	3	0	0	8	1	1168	281
29	—	125	—	101	—	0	—	0	—	3	—	229

資料來源：統合幕僚監部，《平成 29 年度 1 四半期の緊急発進実施状況について》（東京：國防省，平成 29 年），頁 3。

同時，再由飛行路徑判讀，也可看出中國軍機主要集中於東海，並漸次向第一島鏈外緣進行突穿，而在南邊位置也藉軍機在台灣周邊以「遠海航訓」的方式，分別展開不同機種的組合訓練，一方面建立航行路徑的觀察與規劃，另一方面在台灣東部的航行等同在第一島鏈外緣展現空中兵力的投射能力。這都清楚的突顯北京穩固東海主權、爭奪第一島鏈制海權的戰略意志。

---

7 統合幕僚監部，《平成29年度1四半期の緊急発進実施状況について》（東京：國防省，平成29年），頁3。

三、地緣政治的戰略槓桿

進一步看，以往航空兵力用於防空識別區的主要功能是防空警戒，同時藉由目視接觸（visual identification）方式對不明機做出鑑別、以及其攜載、掛彈等情報，以作為威脅評估。而防空識別區所扮演的角色在目前又更為增加，除防空警戒外，又具備新的功能，扮演地緣政治上的戰略槓桿功能。

例如，除既有的東海防空識別區外，若中國進一步尋求建立南海防空識別區將是純粹的地緣政治需求，更甚於單純的軍事攻防操作性目的，而建立的時機則端賴中國領導者認為的最佳成本效益點。[8]

中國在南海發展強化防空識別區的制空能力並最終設置南海防空識別區，徹底改變南海的政治現狀。依照美軍前太平洋司令哈里斯（Harry Harris）的看法，「中國在南海的一系列軍事設施，將創造一種不需戰爭而使控制南海成為既成事實的機制。[9]

四、精準的政治控制

現代空權可以提供政治決策者微觀的政治管理。[10] 這主要來自於精準打擊武器與載台的深度整合，在技術上可以避免以往空中轟炸造成的大規模殺傷，同時，使得選擇特定目標甚至特定的個人作為攻擊對像成為可能。類似的精準彈藥攻擊，現階段是藉由衛星、雷達、或雷射等定位與導引方式，將彈著點控制在公尺級的範圍，因此得以精準的涵蓋特定目標。進一步更可以調整彈頭種類，甚至裝藥的數量，將破壞力侷限在設定的方向與穿透深度，以限制殺傷效果。

其目的在於只打擊選定的特定目標，避免或減少不必要的殺傷與破壞，減

---

8 Michael Pilger, "ADIZ Update: Enforcement in the East China Sea, Prospects for the South China Sea, and Implications for the United States. US- China Economic & Security Review Commission. Staff Research Report," *US Congress*, March, 2016, p. 7.

9 Sam LaGrone, "U.S. Weighing More Freedom of Navigation Operations in South China Sea Near Reclaimed Islands," *USNI News*, September 17, 2015, http://news.usni.org/2015/09/17/u-s-weighing-more-freedom-of-navigation-operations-in-south-china-sea-nearreclaimed-islands.

10 Bradley J Smith, "On Politics and Airpower," USAWC Strategy Research Project, Carlisle Barracks, Army War College, April 9, 2002, p. 2.

低可能的負面政治效應，此種能力對於未來的都市化戰場極具價值。而對於特定的競爭國家或組織而言，打擊其政治中樞或個人也成為控制武裝衝突升級的重要選項。目前無法預測單靠空權本身，得以造成政權改變或保證其改變正確的方向，[11] 但此一特性在未來將愈趨明顯，並給予政軍決策者更多彈性，然而相對的也得負擔更大責任，一旦目標情報錯誤、技術出現故障，都可能導致嚴重的政治衝擊。

肆、重要空權科技的演變

一、無人飛行載具再創新

　　現代戰場耳熟能詳的無人載具，發展史其實已超過百年。世界上第一架無人飛行器，在 1917 年 3 月問於英國問世，其原型設計的概念與目的在於減少飛行員傷亡，提高防空作戰的成本效益，無人機技術的重要演進，可簡述如后。擴大運用期是在二戰，其中最具代表性的裝備為德國 V1 巡弋飛彈，藉由機械型計算機進行自動導航，可攻擊 250 公里外的面目標。另外由亨謝爾（Henschel Flugzeugwerke AG）開發的 HS-239 遙控滑翔炸彈，則是由搭載母機的操作手以無線電遙控滑翔炸彈用以攻擊約 8 公里外的地面或水面目標，進行點目標的攻擊，這兩者可說是戰史上首次投入戰場實用的遠攻武器。第二階段，後續隨著技術演進，在越戰時期出現的「火蜂」（fire bee, BQM-34A）無人飛行載具，其技術特色是美軍整合光學設備、電子裝備並小型化，將其應用於戰術偵察任務，正式開啟無人機的大規模運用時代。第三階段的重要里程碑則是以色列在 1973 年導入無人機的創新應用，藉誘騙、偵測、干擾埃及防空系統並取得戰略性成功。第四階段則是 1991、2003 兩次波灣戰爭，無人機的發展進入另里程，具有更長的滯空時間、以及即時的影像傳輸能力，甚至可掛載反戰車飛彈等小型武器，可以使後方的操作人員即時辨視目標並發動攻擊。至此，無人機的任務運用方式可說已然齊備。[12]

---

11 Robert A. Pape, "The true worth of Air Power," *Foreign Affairs*, March/ April 2004, https://www.foreignaffairs.com/articles/2004-03-01/true-worth-air-power.

12 蘇紫雲，〈機器人戰爭來臨、台灣勿缺席〉，《新社會政策》，35 期，2014 年 8 月 15 日，頁 49-50。

而前瞻 21 世紀的 20 年代，無人機的類別與運用將更為普及，兩種新型態的運用包括「微型無人機」（micro unmanned vehicle）的問世，以及集群（swarm）的技術運用，結合物聯網、人工智慧的技術，此類微型無人機將可扮演類似蜂群的攻擊模式，對敵人發動組織性、協調性的集群式攻擊。無人飛行載具的持續創新，將使空權的維持與遂行有更多的工具，載人戰鬥機獨霸空權主角的年代，將面臨重大挑戰。

二、鼠象效應：超精準打擊

現代的精準打擊可達到公尺級的精度，倚賴的是衛星的電波解算定位，或雷射光指引，未來則可望達到公分級的精密度，倚賴的將是整合網路通聯技術。打擊對象將可由建築物、載具等大型目標，乃至個人等小型目標，這些多元與多樣性的目標，勢必需要更密集的情報蒐集與整合。因此情報的極度細節分析，都有賴於更高解析度的影像處理與更精確的電子處理裝備。[13]

利用前述的微型無人載具等作戰裝備，在未來也將扮演重要的戰場角色，特別是具備集群攻擊能力，可自主進行協調、分工，對目標進行偵測、發動攻擊。相對的，由於攻擊方具備體積極小、數量眾多，被攻擊目標將喪失有效防禦的機會，存活率極低。目前此一技術軍用部分，原型已由美軍測試，民間則用於無人多旋翼機的群組控制，主要用於夜間燈光的空中排列展演，或更進階的群組空中球類、花式飛行試驗。此一技術更加成熟後，將賦予政軍決策者更多的選擇自由，可執行超精密的打擊任務，其對目標選擇的精確度將可由目前的公尺、縮減至公分等級，且個別微型載具殺傷範圍小，集群攻擊則殺傷威力大，因此可選擇的組合更加彈性。

此種公分級的攻擊，或許可稱為超精準打擊，且由於具備的任務彈性極大，可打擊的對象包括小型目標、或大型目標的特定（弱點）部位，攻擊對象的選擇包括各種武器、固定陣地等，因此在未來戰場上將可創造以小搏大的鼠象效應。

---

13 Brian A. Arnold, Robert P. Vitrikas, *"Effect of Modern Technology on Air Power and Intelligence Support,"* Washington D. C.: National War College, April, 1992, p. 26.

三、匿蹤科技的競合

在軍事史中，匿蹤在空中戰場早已應用，雖不是新技術但仍在持續發展中，70 年代開始發展航空器的低電磁可偵測術可說是初始階段，[14] 至今仍極具潛力可說是處於軍用科技的最前線，不僅美、俄、中競相發展具備匿蹤性能的五代戰機，甚至一些航空後進國家例如印度也宣布投入發展。廣義來說，最早的航空匿蹤技術是「迷彩」塗裝，典型的作法為機背採用地面色系塗裝，機腹則採用藍灰白等天空色系，用於迷惑地面或空中敵機的視覺，從而降低被攻擊或命中的機率，屬於可見光的運用。而在 1970 年代開始出現的匿蹤科技，主要屬於不可見光的隱蔽，包括紅外線的熱訊號、雷達反射訊號的跡訊，以減少被感測器發現的機率、或縮短被偵測的距離。這都可降低飛行器被發現、擊落的機會，也間接提高存活率，並可有較大的成功機會執行任務。

相對的，處於反制的一方，也積極發展新的偵測科技，包括被動雷達（passive radar），[15] 也是藉由飛行器等所發出的電磁訊號，或其飛行動作穿越空中既有的電磁波時產生的訊號擾動進行截收，進而解算出目標所在方位、距離等資訊，以進行偵獲並攔截。可以這麼說，航空器的匿蹤、以及偵測技術的競賽，將是未來空權的重要競爭要素，也決定敵我雙方的作戰任務的成功率。

四、網路結合空戰

網路通訊隨著微電子技術的發展而突飛猛進，軍用的資訊流也由代表性的 Link-16 的網速，未來將轉為高速的軍規 5G 通訊技術。軍用資訊網路最重要的演變方向將是持續無線化的擴大運用，涵蓋的裝備與物件將更為廣泛，例如個別人員、甚至微型無人載具等以往被視為戰鬥層級的領域。而在頻寬以及傳輸速度上都大幅改進，以能將戰場訊息完整而迅速的分享給指定的友軍。此種網路行為模式，與民間網路通訊的「群組」、「社群」可說相當類似，但是在內容的精細度與正確度卻大異其趣。主因是其需具備完整的決策輔助功能，甚至直接饋入火力控制系統，以指揮武器對目標發動攻擊。

---

14 Ibid., p. 12.

15 "Ila2018: Hensoldt hights passive radar," *Air & Cosmos International*, April 26, 2018, http://www.aircosmosinternational.com/ila-2018-hensoldt-highlights-passive-radar-110229.

　　同樣重要的是通訊過程的保密、加密，基本目的在於確保己方資訊只傳達給正確的接收者運用，避免遭到未授權者截收，甚至為敵方所知悉，導致己方的行動計畫遭到破壞。進一步來看，則是防範己方資訊為敵方插入編碼產生假訊息，其所造成的影響更為廣泛，不僅特定行動計畫將遭到破壞，甚至會產生外溢效果，誤導其他的友軍、或作戰行動相互牴觸，破壞性更為深廣。二次大戰中的英倫空戰，英軍地面戰管人員模擬德軍戰管的語音指揮，造成德軍攻擊機隊的混亂就是經典案例。此在現代的網路資通中的保密加密技術雖然不可同日而語，但邏輯卻是完全相同。

　　而運用在未來空中作戰的代表性戰具，當數第五代戰機的代表性機種 包括 F22、F35 皆大幅利用網路的技術優點，在偵測與接敵的流程都可利用資訊分享方式與戰場上的友軍傳遞目標動態、友軍部署、戰場地形等即時資訊，並可由最具優勢條件的友軍單元進行接戰。此種網路通訊科技使得戰場透明度得以最佳化、創造最大的火力效果、同時使匿蹤戰機可以在不暴露主動電磁訊號的情況下發動攻擊，得到最大的戰術與戰略效益。

伍、結語：整合式空權將成重要選項

　　空權未來仍將是實體戰場中的關鍵要素，但由於科技的進步，掌握空權的方式與工具將更為多元，不再只限於傳統的載人戰機。因此，綜合考量科技、成本、效益，進而提出有效的對應戰略將是政治菁英與軍事計畫者的重要課題。主要的關鍵包括：

一、成本是傳統空權最大挑戰

　　傳統空權的建構依賴載人戰機為主力，最關鍵的挑戰就在成本。一般而言，新式戰機開發的投資成本極為高昂，研發計畫由概念發展階段至投入量產約需耗時 10-15 年，[16] 且後續的維護成本也極高。例如，歐洲戰機研發生產的總成本高達 540 億歐元，[17] 而具匿蹤功能的 F-22、F-35 等戰機開發成本更高，

---

16 Brian A. Arnold, Robert P. Vitrikas, "*Effect of Modern Technology on Air Power and Intelligence Support*," Washington D. C.: National War College, April, 1992, p. 35.

17 Markus N. Heinrich, "The Eurofighter Typhoon programme: economic and industrial implications of collaborative defence manufacturing," *Defence Studies*, 2015, pp. 341-342.

以美國國力之豐裕，仍將 F-35 的開發交由多國合作方式進行，以降低負擔。另一方面則是匿蹤戰機的維護成本較傳統戰機更高，例如其雷達波吸收塗層在執行任務時便會因為空氣摩擦、雨霧、紫外線等因素造成耗損、減低匿蹤性能，因此需定時重新噴塗，即便國防資源龐大的美軍也為此傷透腦筋，[18] 甚至設計出維修專用的「塗裝機」（coating），此一塗裝設備號稱在研發與設計的初期階段，便投入大量時間，集結維修人員、環境、技術專家協同並蒐集數據以了解並辨識 F-35 保修團隊的需求。目標是既有商業技術的最大化運用、並藉自動化、先進技術的投入以在全壽命週期得以降低保修成本。[19] 由此可看出空權的維護成本將逐步攀高，若採「制空」為主的方式，也就是以戰機為空權維護的主力將造成國防資源的嚴重負擔，因此成本較低的「防空」手段，將成為守勢國家重要的搭配選項之一。

二、防空戰力的優勢

進一步來說，空權的組成基本架構主要包括制空、防空兩大能力，而對主權國家而言，制空與防空戰力的組合比例，主要依據國家戰略目標、經濟能力、地緣環境等條件的綜合考量。而未來在先進戰機造價逐步高漲的趨勢下，以守勢為主，不追求向外兵力投射的國家，明顯的將以防空能力的建構較具成本優勢，特別是現代防空飛彈的性能大幅增加，命中率、攔截高度與射程都具有極佳表現，因此在防空兵力的組成具有極多選擇，且可提高任務的運用彈性。

以本土防空為例，以防空兵力、制空兵力的基本構成單元比較，防空部隊主要的火力為地對空飛彈，屬於不可重複使用的消耗品。但若考慮制空兵力的戰機，其主要火力亦同屬消耗品的飛彈，兩者載具的成本即可快速的作比較，並明確的選擇出優先投資順序。

換句話說，以國防投資的有限成本，追求高「交換比」的核心邏輯不變的原則下，飛行部隊、防空部隊的比例或許可以有更佳的組合以創造更大戰略價

---

18 〈美空軍也吃不消，F-35 戰機恐砍單 1/3〉。蘋果日報，2018 年 3 月 29 日，https://tw.appledaily.com/new/realtime/20180329/1324971/。

19 Marisa Alia-Novobilsk, "Next generation coatings booth poised to save Air Force millions in energy," *Air Force Research Laboratory*, March 27, 2017, http://www.wpafb.af.mil/News/Article-Display/Article/1130863/next-generation-coatings-booth-poised-to-save-air-force-millions-in-energy/.

值，達成以小博大的防衛戰略優勢。

三、地空整合的趨勢

以觀念不斷創新的美軍為例，提高未來作戰效率的真正方式是整合而非單一的轉型。例如精密打擊的革命來自於武器結合微電子技術，而下一步是將分散於不同兵種的複雜系統整合起來，空中力量與地面力量結合，特別是兩者皆擁有精準戰力的武器。[20]

同時，未來的軍兵種之角色，將走向複合作戰的樣態。傳統陸海空的軍種分工，隨著科技應用的穿透性、以及多元化，作戰功能勢必重新調整，就如同市場變化，商品、服務也須隨著調整一。其中最可能的轉變就是複合戰力。例如前文所述的網路、飛彈科技進化，使得制空權得以不必單獨仰賴空軍，而機器人集群作戰的科技，則可在特定條件下，不必仰賴地面部隊。此種科技應用的特性，使得陸海空戰場的空間，在特定條件下都可能由單一軍兵種獨力擔負，走向複合作戰的三軍，不再是相互倚賴的聯合作戰，而是真正的重層戰力，並可發揮最大的加乘效益。

因此，無論未來的空軍定位是否走向取消獲保持獨立軍種，其在空權所扮演的角色將被部分取代或稀釋。隨著科技的發展，在軍事上賦予地面部隊、水面部隊擁有更廣的防空射程、更精準的偵測、攻擊能力，更能朝向深遠後方的機場進行火力投射，以消滅或壓制對手的空軍。海陸軍種橫跨空權的作戰潛力就賦予政治決策的新意義，也就是在前端的預算投資，可能優先分配予成本效益相對較高的防空項目，而在後端的任務執行部分，空軍的深遠打擊角色也將縮減，由無人機或遠程攻陸飛彈進行，以減少風險。此種趨勢都很清楚的標明空權的發展方向，特別對資源有限的中小型國家而言，空權的維繫將擁有更大且更有效的政策選項，此一特性對企圖尋找不對稱戰力以抵消對手威脅的中小型國家而言特別值得評估，也是需採行的途徑。

---

20 Robert A. Pape. "The true worth of Air Power," *Foreign Affairs*, March/ April 2004, https://www.foreignaffairs.com/articles/2004-03-01/true-worth-air-power.

兩大強權夾縫中的生存之道：

從「第三鄰國」政策看後冷戰時期蒙古的外交戰略

巫彰玫 * 白兆偉 **

壹、前言

　　由於蒙古為中亞內陸國，冷戰時期曾為隸屬於蘇聯的一個附庸國。冷戰結束後，脫離蘇聯掌控的蒙古也開始進行變革。首先，1990 年 5 月蒙古政府頒布《蒙古人民共和國政黨法》，承認反對黨的合法性，實行多黨民主制。[1]接著，1992 年通過《新憲法修正案》，改國名為「蒙古國」，實施總統議會制，成功轉型。1994 年 6 月蒙古議會通過《蒙古國對外政策構想》決議，奠定冷戰後蒙古對外政策的基調。這是蒙古建國以來第一份獨立不結盟的外交政策。蒙古將採行「不結盟、等距離、全方位」的多支點外交政策。在此原則下，除著重與中俄兩國的關係外，並會積極發展與美、日、印等「第三鄰國」的關係，以期在與「第三鄰國」的外交互動中，逐步形塑「第三鄰國」外交政策。2010 年，蒙古以法律的形式將這一政策固定下來。2011 年，蒙古頒佈新《對外政策構想》，明確將「第三鄰國」政策列入。在此文件中蒙古強調「全方位」、「多支點」、「等距離」的外交戰略。

　　草原上的民主國家蒙古已開始籌備成為東亞唯一的「永久中立國」。由於蒙古夾在俄羅斯和中國兩個大國之間，該國外交專家稱這是其「唯一的生存之道」，蒙古總統額勒貝格道爾吉（TS. Elbegdorj）等人率先在外交層面上協調，並推進國內相關法制的完善工作。據媒體報導[2]，該國總統 2015 年 9 月底在美國紐約舉行的聯合國大會上演講時強調，蒙古的永久中立化將「為促進全球的

---

* 國立中興大學國際政治研究所博士候選人
** 國立中山大學中國與亞太區域研究所博士

1 馬立國、黃鳳志，〈 "冷戰" 後蒙古國亞太外交戰略探析〉，《內蒙古大學學報》，第 44 卷第 6 期，2012 年 11 月，頁 34。

2 日本一般社團法人共同通訊社（以下簡稱共同社）成立於 1945 年，由日本全國的報社及日本放送協會（NHK）等媒體共同設立，獨立於政府，旨在準確公正地報道國內外的新聞，在滿足國民知情權的同時，為增進國際社會的相互理解做貢獻。

和平、安全和繁榮做出貢獻」，並呼籲支持該政策。據外長普日布蘇倫（Lundeg Purevsuren）稱，該國政府為締結有關永久中立的協定正在與中俄兩國進行磋商，並表示「兩國都對蒙古國的立場表示理解」。在上世紀 1960 年代至 1970 年代中蘇對立時，蒙古接受了前蘇聯軍隊的大規模駐軍。人口約 300 萬的蒙古無法和軍事力量強大的鄰國相較量。雖然從 20 世紀上半期起就曾有中立化的構想，但中俄關係現在正是「宣佈中立的絕佳時機」。[3]

　　蒙古發展出所謂的「第三鄰國」（the third neighbor policy）外交，企圖透過引進多邊的力量以強化自身安全。身為中小型國家，在地緣政治及地緣經濟的條件中，蒙古如何考量符合國家利益的前提下進行其國家安全戰略的規畫及設定，以在兩強夾縫中生存是值得探究的。期望由歐美國家獲得軍事及科技的協助，冷戰時期以來對蘇聯意識型態的認同，以及因歷史因素對中國的疑慮但又必須仰賴中國的經濟發展，使得蒙古所謂的全方位、多支點，以及等距離外交戰略必須保持平衡。在傳統的抗衡與扈從之外，面對強權的威脅，小國事實上還可以採取避險（Hedging）的策略[4]。

貳、避險策略

　　瓦特（Stephen Walt）認為，一國在追求其生存之目標時係在抗衡與扈從兩者之間做抉擇，但認為所謂的抗衡，國家所制衡的並非「權力」，而是「威

3　楊家鑫，〈在中俄間求生存 蒙古尋求成為永久中立國〉，中國時報，2015 年 10 月 16 日，http://www.chinatimes.com/realtimenews/20151016002257-260408。

4　牛津大學國際關係講師吳翠玲（Evelyn Goh）撰文指出：（一）「避險（hedging）」一辭現已成為美國戰略論述的熱門字眼，特別是在論述與中共相關的戰略問題時更是如此。其含義是將「交往與融合機制」以及「務實的制衡」結合，而制衡之道是與亞洲各國建立安全合作關係，並推動自身軍事現代化。簡言之，避險就是將圍堵與交往結合，以針對他國現在和未來的意圖形成保險。但若將避險一辭用於美國的對中政策，則在概念上會有問題，可能會造成政策混淆與誤解。（二）因此避險概念必須加以適當界定，以使其含義一貫，並與制衡、圍堵、順勢而為和推諉等比較直接的戰略選擇有所區別。比如有人也許會說避險戰略包含制衡或圍堵在內，但此戰略必須與制衡和圍堵區隔，而區隔的作法可以是納入交往或安撫等要素，也可以將圍堵戰略用作達成目標的手段，而非目標本身。（三）對避險比較妥當的界定是，「一套旨在避免某些狀況出現的戰略，而在這些狀況下，各國無法就比較直截了當的制衡、順勢而為或中立等替代性作法作出決定。避險是要使各國嚴守中庸，以免必須為選擇某一直截了當的政策立場而放棄另一個立場」。就瞭解或制定美國對中政策而言，避險並非特別有用的字眼，美國不如重視並擴大與中共的協議及合作範圍，但也坦白承認，彼此存有歧見和猜忌，以作為展開坦白對話，並談判出亞洲安全新秩序的第一步。吳翠玲，〈析論亞太安全戰略中的避險概念〉，美國國際戰略研究中心，2006 年 9 月 18 日，http://www.csis.org/media/csis/pubs/pac0643.pdf。

脅」，這就是所謂的威脅平衡理論（balance of threat theory）之要旨。簡言之，「權力總量」、「地理位置鄰近」、「攻擊能力」，以及「侵略意圖」等四個客觀變數的值越高，代表威脅之程度越高，一國採取抗衡策略以因應外部威脅的可能性也越高。[5] 其中，以侵略意圖對威脅程度之影響最為關鍵。舉例來說，擴張性的對外政策象徵著一國所具有之侵略性，因而會被其他國家視為具有極高之侵略意圖。

大多數現實主義的均勢理論家皆主張，在無政府狀態下，抗衡政策是國家為了維繫生存所必然（至少是非常可能）會採取的策略選擇。國家將透過內部動員或結盟（或者兩者兼施）的途徑增強軍事力量，以預防或挫敗外部國家或聯盟的領土佔領、政治和軍事控制等企圖。一般來說，抗衡所必然帶來的成本即為國內生產要素與人力資源的動員。一國勢必必須配置相當程度之資源於安全防衛之上，甚至需要為了達成戰略目標而承受國內生產要素配置的失衡或者經貿利益之犧牲。若從安全困境理論（security dilemma theory）的角度來看，抗衡還會帶來惡化安全困境之弊端。

然而，純粹的均勢或屈從策略很難成為國家所欲施行的政策，因此，有學者提出另一個主要的策略—「避險」，並有不同的定義。[6] 基本來說，避險是一種透過多種政策工具使國家能處理其夥伴國家未來行為的不確定性，以因應夥伴國家可能出現的潛在安全威脅。[7] 是故，避險是一種在高度不確定性和高風險下尋求保險的行為，其中理性行為者尋求採用多種抵制行為，以便不管最終的結果如何，矛盾的行為將有助於消除彼此的影響，從而避免完全接觸的風險，

---

5 Stephen M. Walt, *The Origins of Alliances* (Ithaca: Cornell University Press, 1987), pp. 24-26.

6 例如，避險被定義為：(1) 在高度不確定和高風險的情形下，國家採取多種政策選項以製造相互作用的行為；(2) 為避免陷於無法決定直接採取平衡、屈從或中立等策略的情形所採取的行為；(3) 在預期未來可能有安全威脅的情形下，保持多種戰略選項；(4) 一種具有雙重意義的國家策略，一方面強調交往與整合機制，另一方面則強調透過與他國的安全合作和軍事現代化計畫，取得現實主義式的平衡。引自 Le Hong Hiep，〈建交後越南對中國的避險戰〉，《全球政治評論》，第 49 期，2015 年，頁 149-180。參見 Kuik Cheng-Chwee, "The Essence of Hedging: Malaysia and Singapore's Response to a Rising China," *Contemporary Southeast Asia*, Vol.30, No.2, 2008, p. 163; Evelyn Goh, "Understanding 'Hedging' in Asia-Pacific Security," *PacNet*, No.43, August 31, 2006; Denny Roy, "Southeast Asia and China: Balancing or Bandwagoning?," *Contemporary Southeast Asia*, Vol.27, No.2, 2005, p. 306; Evan Medeiros, "Strategic Hedging and the Future of Asia-Pacific Stability," *The Washington Quarterly*, Vol.29, No.1, 2005, p. 45.

7 Kuik Cheng-Chwee, "The Essence of Hedging: Malaysia and Singapore's Response to a Rising China," p.163.

並保護行為者的長遠利益。[8]

　　避險是一個混合性的策略，兼容抗衡與扈從的部分特徵，是與潛在威脅維持暨對抗又合作之關係。事實上，避險策略最大的好處即是國家可以試圖在安全與利益兩個層面上達到雙贏的局面。在安全防衛方面，避險策略有抗衡之部分特徵，對突發之危急情況不至於不具反應能力，但對抗性的程度又不如純然的抗衡，因而可以配置較少資源於軍事之上，同時又較能避免抗衡策略使安全困境成為「自我實現之預言」。在利益的取得方面，避險策略有扈從之部分特徵，因而能夠與崛起中的強國保持相對和緩的關係，確保經濟利益之實現，又可以避免純然的扈從政策所可能導致的政策自主性喪失。除了以上的討論之外，避險還包括兩項潛在的成本，一個是必須承擔壯大具有不良企圖的潛在外部威脅之風險，另一個是在避險策略中抗衡部分的程度若有所不足，很可能被盟友視為一種背叛，從而危及既存的同盟體系。不過，倘若崛起中之強權之政策以合作為主，並不涉及強制力之使用，甚至在領土主權爭端等與國家根本利益高度相關之議題上採取妥協性的姿態，便可以大幅降低第一項潛在的成本。[9]

　　以下將以這兩種政策選項分別檢視蒙古對中國、俄羅斯及第三鄰國（以美國為主的西方國家）的外交政策。

---

8 Kuik Cheng-Chwee, "Light or Heavy Hedging: Positioning between China and the United States," Joint U.S.-Korea Academic Studies, Korea Economic Institute of America, February 26, 2016, http://www.keia.org/sites/default/files/publications/introduction_-_light_or_heavy_hedging.pdf.

9 媒體小農，〈東亞國家如何在南海議題避險〉，洞見國際事務評論網，2012 年 7 月 31 日，http://www.insight-post.tw/editor-pick/asia-pacific-strategy/20120731/36。

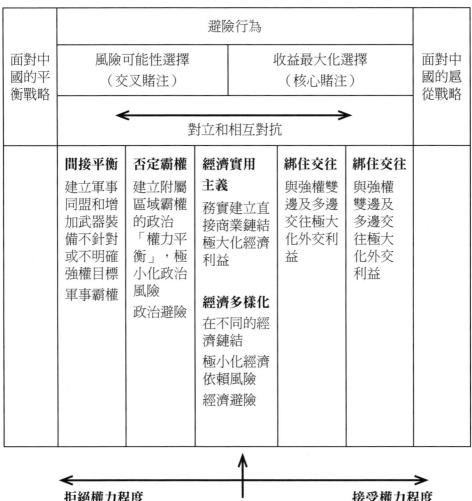

面對中國的平衡戰略	避險行為					面對中國的扈從戰略
	風險可能性選擇（交叉賭注）		收益最大化選擇（核心賭注）			
	對立和相互對抗					
	間接平衡 建立軍事同盟和增加武器裝備不針對或不明確強權目標軍事霸權	**否定霸權** 建立附屬區域霸權的政治「權力平衡」，極小化政治風險 政治避險	**經濟實用主義** 務實建立直接商業鏈結極大化經濟利益 **經濟多樣化** 在不同的經濟鏈結 極小化經濟依賴風險 經濟避險	**綁住交往** 與強權雙邊及多邊交往極大化外交利益	**綁住交往** 與強權雙邊及多邊交往極大化外交利益	

拒絕權力程度　　　　　　　　　　　　接受權力程度

中立點

圖 1：權力—回應 範圍

資料來源：Kuik Cheng-Chwee, "Light or Heavy Hedging: Positioning between China and the United States," Joint U.S.-Korea Academic Studies, Korea Economic Institute of America, February 26, 2016, http://www.keia.org/sites/default/files/publications/introduction_-_light_or_heavy_hedging.pdf.

參、蒙古與中、俄及第三鄰國的關係

以往，蒙古一直扮演著中國與俄羅斯爭奪亞洲大陸霸權的關鍵角色。除現實地緣政治的考量外，蒙古與中國、俄羅斯間也存在著複雜又矛盾的歷史恩怨。而美國在非其傳統勢力範圍的中亞建立正式的盟邦體系，卻極度需要可以協助維持區域穩定的可靠安全伙伴存在，因此積極發展與蒙古的關係。[10] 自蒙美軍隊在 2004 首度舉行「可汗探索」（Khaan Quest）軍演以來，中國也越來越重視發展與北方鄰國的關係。2015 年 8 月，中國與蒙古的國防部長就在烏倫巴托舉行了會談，目的是提升雙方在防務領域上的合作。[11]

另外，2013 年 3 月，日本首相安倍晉三（あべしんぞう）前往蒙古，雖表示此行重點是經濟，但日本媒體報導蒙古是戰略鄰國，訪問蒙古旨在遏制中國在這一地區的影響力的增長。其實，大陸學者表示安倍的訪問凸顯了蒙古外交的平衡。2014 年，蒙古總統額勒貝格道爾吉（Цахиагийн Элбэгдорж）回訪日本，雙方就自由貿易達成原則性協議。蒙古也表示說如果日本與中國是蒙古的稀客，美國則是蒙古的常客。2014 年 4 月，美國國防部長哈格爾（Chuck Hagel）前往烏蘭巴托，重點即在增進兩國軍事關係。

在上世紀 60 年代至 70 年代中蘇對立時，蒙古接受了前蘇聯軍隊的大規模駐軍。蒙古學者指出，「作為不想被捲入大規模紛爭的蒙古，那是一段再也不想重演的歷史」。人口約 300 萬的蒙古無法和軍事力量強大的鄰國相較量。雖然從 20 世紀上半期起就曾有中立化的構想，但學者分析稱，中俄關係良好的現在正是蒙古「宣佈中立的絕佳時機」。[12]

---

10 Peter Bittner, "China, Russia, Mongolia Sign Long-Awaited Economic Partnership Agreement," The Diplomat, June 28, 2016, http://thediplomat.com/2016/06/china-russia-mongolia-sign-long-awaited-economicagreement/.

11 Peter Bittner, "China, Russia, Mongolia Sign Long-Awaited Economic Partnership Agreement,"；中國人民大學國際關係教授時殷弘說，高毅，〈分析：習近平為何對蒙古「點對點訪問」？〉，BBC 中文網，2014 年 8 月 21 日，http://www.bbc.com/zhongwen/trad/world/2014/08/140821_mongolia_china_xi_visit。

12 據共同社報導地政學研究所所長喬因戈爾（Jogh Gore）表示，見楊家鑫，〈在中俄間求生存 蒙古尋求成為永久中立國〉，中國時報，2015 年 10 月 16 日，http://www.chinatimes.com/realtimenews/20151016002257-260408。

一、蒙古與中國

1980 年代末，蒙古開始恢復與中國的關係，自 1990 年代開始政治經濟轉型，在外交方面，蒙古開始改變策略，原定「一邊倒」改變為「多支點」，在保證和中國與俄羅斯兩個國家交往的同時，採取「第三鄰國」的策略，蒙中關係開始了新的發展。[13] 1994 年，蒙古與中國恢復睦鄰友好關係，雙方重新簽署《中蒙友好合作關係條約》，1998 年，蒙古國總統巴嘎班迪（Нацагийн Багабанди）訪問中國時表示，蒙古政府高度重視對華關係，蒙古致力於發展兩國關係，[14] 同時表示要建立面向 21 世紀相互信賴的友好關係。2000 年 10 月中國共產黨書記處書記訪問蒙古。雙方表示政黨間關係是國家關係的重要組成部分，儘管蒙古和中國有著不同的社會制度，兩黨合作的想法和思維也略有不同，但並不阻礙雙方友好關係。[15]

自 1999 年以來，中國成為蒙古最大的貿易伙伴和最大的投資國。2003 年雙方領導人「關於相互信任睦鄰伙伴關係的建立和發表聯合聲明」出發，蒙古和中國正式宣布建立蒙中兩國睦鄰友好伙伴關係。[16] 蒙古與中國自實現關係正常化以來，兩國關係平穩發展，高層互訪頻繁，政治與經濟互信繼續加深。蒙古國家大呼拉爾主席貢其格道爾吉（Раднаасүмбэрэлийн Гончигдорж）、總理恩赫巴亞爾（Намбарын Энхбаяр）等政府高層領導人相繼訪問中國，以維繫雙方友好關係（雙方互訪如表 1）。2005 年 11 月，蒙古總統恩赫巴亞爾回訪時表示，「中國是蒙古偉大和友好的鄰邦。兩國建立了睦鄰互信夥伴關係，推動蒙中關係在新世紀進入新的更快的發展階段」。[17]

在 2006 年，蒙古革命黨總書記巴亞爾（Санжийн Баяр）為團長的蒙古人民革命黨代表團到中國浙江省訪問，蒙古民主社會主義青年聯盟主席贊達沙塔爾（омбожавын Занданшатар）隨團到訪中國。巴雅爾總書記表示蒙中之間的

---

13 思美，〈蒙古國的「多支點」和「第三鄰國」外交政策下的蒙中關係研究〉（碩士論文：華東師範大學，2016 年），頁 25。

14 中華人民共和國外交部政策研究室，《中國外交》（北京：世界知識，1999 年）。

15 〈蒙古國外交部長包勒德德與中國國務委員戴秉國會見〉，http://www.mfa.gov.mn/?page_id=32253。

16 娜琳，〈中蒙關係 45 年 1949-1994〉（內蒙古：呼和浩特，1998 年），頁 47。

17 Robert Shutter, *Chinese Foreign Relations: Power and Policy Since the Cold War.* (Landham: Rowaman & Littlefields, 2012), p. 555；〈胡錦濤與蒙古總統恩赫巴亞爾舉行會談〉，中華人民共和國外交部網站，2005 年 11 月 28 日，http://www.fmprc.gov.cn/chngxh/tyb/zyxw/t223727.htm。

發展對於兩國人民有很大益處，人民革命黨會繼續做出努力，為蒙中的關係進步貢獻。[18] 2009 年，蒙中兩國軍隊舉行維和聯合軍事訓練。

2010 年 6 月，國務院總理溫家寶出訪蒙古期望兩國從各自國家長遠發展和地區持久繁榮穩定出發，加強交往與合作，推動蒙中睦鄰互信夥伴關係不斷向前發展。[19] 2011 年 6 月 17 日，蒙古總理巴特包勒德（Сүхбаатарын Батболд）對中國進行國是訪問。在華訪問的蒙古總理巴特包勒德在清華大學發表演講時表示 60 多年來，蒙中關係經受住了國際風雲變幻的考驗，不斷得到鞏固和發展。中國國務院總理溫家寶在北京人民大會堂舉行了「關於建立戰略伙伴關係的聯合宣言」的會談，決定將「蒙中睦鄰互信伙伴關係」提升為「蒙中戰略伙伴關係」，加強和發展兩國戰略伙伴關係。[20]

2014 年是蒙中建立外交關係第 65 年，中國國家主席習近平對蒙古進行國是訪問。到訪當天，中蒙簽署聯合宣言，宣佈將兩國關係提升為全面戰略伙伴關係。蒙古報紙認為，訪問的重點仍是經濟。中國連續多年是蒙古國最大貿易伙伴和投資來源國，儘管中國需要蒙古的煤等資源，但相對來說，「蒙古更需要中國」。[21]

表 1：蒙中關係領導人互訪簡史

編號	時間	事件
1	1994 年	1. 雙方重新簽署《中蒙友好合作關係條約》，為兩國關係健康、穩定發展奠定了政治、法律基礎。 2. 李鵬訪問蒙古 會見蒙古國大呼拉爾主席。
2	1997 年 7 月 15 日	江澤民對蒙古國進行國是訪問 增進中蒙互信。

---

18 思美，〈蒙古國的「多支點」和「第三鄰國」外交政策下的蒙中關係研究〉，頁 33。
19 明金維、阿斯剛，〈溫家寶與蒙古總理會談〉，人民日報海外版，2010 年 6 月 2 日。
20 思美，〈蒙古國的「多支點」和「第三鄰國」外交政策下的蒙中關係研究〉，頁 26；侯麗軍、趙超越，〈蒙古國總理對話清華學子 展望未來蒙中關係〉，人民網，2011 年 6 月 17 日，http://www.people.com.cn/2011/0617/c25408 － 2534377640.html。
21 蒙古英文報紙《UB 郵報》（UB Post）高級編輯哈什 - 額爾德尼（Khash-Erdene）認為。高毅，〈分析：習近平為何對蒙古「點對點訪問」？〉，BBC 中文網，2014 年 8 月 21 日，http://www.bbc.com/zhongwen/trad/world/2014/08/140821_mongolia_china_xi_visit。

編號	時間	事件
3	1998 年	蒙古國總統那楚克・巴嘎班迪訪華時明確表示「蒙古政府高度重視對華關係，致力於進一步發展兩國關係」。
4	2000 年 10 月	中國共產黨書記處書記訪問蒙古。雙方表示，政黨間關係是國家關係的重要組成部分，儘管蒙古和中國有著不同的社會制度，兩黨合作的想法和思維也略有不同，但這並不阻礙雙方友好關係。
5	2003 年 6 月	胡錦濤主席訪問蒙古，宣佈在目前友好關係的基礎上把兩國關係提升為睦鄰互信夥伴關係。
6	2004 年 7 月	蒙古總統巴嘎班迪對中國國是訪問，雙方發表中蒙聯合聲明。
7	2005 年 11 月	蒙古總統恩赫巴亞爾訪華時表示「中國是蒙古偉大和友好的鄰邦。兩國建立了睦鄰互信夥伴關係，推動蒙中關係在新世紀進入新的更快的發展階段」。
8	2006 年	1. 蒙古革命黨總書記桑吉・巴亞爾 (Санжийн Баяр) 為團長的蒙古人民革命黨代表團到中國浙江省訪問，蒙古民主社會主義青年聯盟主席貢布扎布・贊達沙塔爾（омбожавын Занданшатар）隨團到訪中國，中國國家副主席曾慶紅接見。瑟・巴雅爾總書記表示蒙中之間的發展對於兩國人民有很大益處，人民革命黨會繼續做出努力，為蒙中的關係進步貢獻。 2. 中蒙建立防務安全磋商機制。
9	2008 年	中國國家副主席習近平訪問蒙古。
10	2009 年	蒙中兩國軍隊舉行維和聯合軍事訓練。
11	2010 年 6 月	中國國務院總理溫家寶訪問蒙古。
12	2011 年 6 月 17 日	在華訪問的蒙古總理巴特包勒德對中國國是訪問在清華大學發表演講：「60 多年來，蒙中關係經受住了國際風雲變幻的考驗，不斷得到鞏固和發展。現在，我們又開啟了一個新的 60 年。隨著兩國戰略夥伴關係的確立，我相信未來兩國關係的發展將會更加充滿活力」。
13	2014 年	中國國家主席習近平，對蒙古進行為期兩天的「點對點」的國是訪問。到訪當天，中蒙簽署聯合宣言，宣佈將兩國關係提升為全面戰略伙伴關係。
14	2016 年	中國國務委員楊潔篪訪問烏蘭巴托。

蒙中全面戰略合作伙伴協定：第一，政治上，雙方會按照《友好合作條約》（1994）和《聯合宣言》（2014）的基本要求，擴大友好合作的發展，中國支持蒙古領土完整，也尊重蒙古的發展道路。兩國將加強黨政之間的合作和交流。同時各層級各部門的交流合作也將逐步展開。外交對話機制也開始啟動。[22]蒙中還會在立法機構、黨政之間、政府之間保持交往合作。雙方議會建立的定期交流機制是兩國政治關係的重要組成部分，雙方將會利用這個機制加深各層級、各單位的交流合作。[23]

第二，蒙中之間的經貿合作是全面戰略合作伙伴關係的重要保障。中國的「一帶一路」和蒙古的「草原之路」是合作的機會，雙方將在這個框架下進行各方面的合作。同業投行、絲路基金等金融組織合作，一起努力為地區的基礎設施建設融資投資。[24]蒙中的經貿合作還會繼續擴大。在改善貿易結構和擴大投資方向繼續一些合作。在發展合作的同時，也要修補合作的機制。蒙中將會根據各自的特點和要求，在電力和可再生能源上進一步合作。[25]

第三，蒙中會在民眾交流的方式上開拓新的方式，這可以增進了解和相互信任。支持雙方的新聞和醫療、體育等領域的合作。在留學領域，中國將會進一步擴大蒙古留學生的獎學金範圍，同時加強學術之間的交流和合作。[26]

綜合所述，在蒙古與中國的關係發展上，除了維持與中國的地緣政治的良好關係外，蒙古也期望藉由中國的發展讓蒙古能夠「搭便車」（free rider）獲得更實值的利益，而又不會落入中國的併吞。

二、蒙古與俄羅斯

為了在新形勢下發展蒙俄新型關係，得到俄羅斯的支援和援助，蒙古總統奧其爾巴特（Пунсалмаагийн Очирбат）於 1993 年兩次造訪莫斯科，兩國簽訂

---

22 思美，〈蒙古國的「多支點」和「第三鄰國」外交政策下的蒙中關係研究〉，頁 28。
23 〈中華人民共和國和蒙古國關於深化發展全面戰略伙伴關係的聯合聲明〉，中國政府網，2015 年 11 月 11 日，http://news.sina.com.cn/c/2015-11-11/doc-ifxkrwks3768574.shtm。
24 思美，〈蒙古國的「多支點」和「第三鄰國」外交政策下的蒙中關係研究〉，頁 29。
25 中華人民共和國外交部，〈中華人民共和國和蒙古國關於深化發展全面戰略伙伴關係的聯合聲明〉，2015 年 11 月 12 日。http://surabaya.china-consulate.org/chn/sbxw/t1314186.htm。
26 同上註。

新的《蒙俄友好關係與合作條約》，確立了面向 21 世紀的友好合作關係，由原來的依附性關係轉向平等夥伴關係。此後兩國高層互訪頻繁，政治、經濟和軍事安全聯繫不斷加強。蒙古獲得來自俄羅斯的大規模經濟援助和政治支持。1996 至 1998 年，兩國在蒙古欠俄債務問題上達成諒解。[27]

　　蒙古在 1921 年至 1989 年的共產主義時代，是蘇聯的衛星國，外交與經濟政策皆由蘇聯控制，也受到蘇聯與中國多變的關係影響。[28] 蒙古在 1990 年春經歷了反共產主義的和平革命。蒙古人民革命黨（Mongolian People's Party, MPP）[29] 察覺到國際情勢因蘇聯和東歐的瓦解以及中國 1989 年天安門事件（槍桿子維繫共產主義）影響，而產生急遽變化，決定摒棄其蘇聯式共產哲學，並實施對外開放。蒙古總理邊巴蘇倫（Дашийн Бямбасүрэн）於 1991 年 2 月訪問莫斯科，與俄羅斯簽訂新的友好睦鄰合作宣言，內容包含經濟合作協議，但很快地取消了蘇聯時代的建設計畫。在 1990 年代初期，蒙古與俄羅斯簽訂了合作聯合宣言以及雙邊貿易協定，接著又在 1993 年簽訂，建立兩國平等關係新基礎的友好合作條約。蒙俄兩國政府繼續共同擁有蒙古境內唯一的主要南北向鐵路以及產量大而獲利高的額爾德尼銅礦（Erdenet Mine）。在蒙古轉型初期，一些特殊的蒙古人，特別是反對蒙古人民革命黨的年輕領袖（常被稱為「民主」領袖），認為蒙古應立刻與冷戰的贏家—美國建立深厚的政治關係。對他們來說，美國是民主蒙古國家安全的「主要支柱」。[30]

　　2000 年 11 月，時任俄羅斯總統的普丁（Влади́мир Влади́мирович Пу́тин）訪問蒙古，雙方簽署《烏蘭巴托宣言》（ulaanbaatar declaration），為兩國關係的恢復奠定了法律基礎都是界定蒙古與強鄰的新平衡關係的基礎文件。[31] 普丁努力活化俄羅斯與蒙古的關係，他在訪問期間所簽署的《烏蘭巴托

27　宋效峰，〈冷戰後蒙古國的多支點外交及其影響〉，《世界經濟與政治論壇》，第 2 卷第 2 期，2011 年 3 月，頁 124。

28　Dr. Alicia、J. Campi 著，許馨譯，〈蒙古對於俄羅斯與中國在蒙古未來發展上所扮演角色的戰略觀〉，《蒙藏季刊》，第 21 卷第 2 期，100 年 12 月 5 日，頁 96。

29　Since 2011, the party has used the name MPP or Mongolian People's Party.

30　Dr. Alicia, J. Campi 著，〈蒙古對於俄羅斯與中國在蒙古未來發展上所扮演角色的戰略觀〉，頁 97-100。

31　N.Altantsetseg, "Russian-Mongolian and Sino-Mongolian Relations since nineties," in The Geopolitical Relations between Contemporary Mongolia and Neighboring Asian Countries: Democracy, Economy and Security (Taipei: Mongolian and Tibetan Affairs, 2003), pp. 356-369.

宣言》勾勒出 21 世紀蒙俄合作計畫。後來普丁又訪問烏蘭巴托幾次，提供經濟與軍事援助，也試圖影響蒙古選舉的結果，希望前共產政黨蒙古人民革命黨能夠勝出。2006 年，蒙古總統恩赫巴亞爾（Намбарын Энхбаяр）訪俄，雙方簽署《莫斯科宣言》（Moscow Declaration），蒙俄關係進入全面恢復階段。俄羅斯對蒙古積極的態度，在 2009 年更是付諸行動，運用升級蒙俄聯合鐵路系統的 1.88 億美元，建議要興建新鐵路支線，從蒙古礦區城市直接通往俄羅斯。[32] 蓄意阻擋美國透過千禧年（millennialism）的公司投資挑戰。

2009 年稍早，普丁與蒙古政府簽署協定，蒙古邀請俄羅斯核能機構 Rosatom 獨力開發蒙古的鈾礦，換取俄羅斯提供蒙古 3 億美元貸款，協助蒙古的農業部門。蒙古比較能從這個局面獲益，因為蒙古已經拿到貸款了，鈾礦開採合作卻是到現在都還沒展開，這也反映了蒙古有些人希望能吸引其他開發鈾礦的外國伙伴。[33] 同年，蒙古總統巴亞爾（Намбарын Энхбаяр）兩次訪問俄羅斯，俄國政府高層和國家元首也在一年內連續訪問蒙古，讓蒙俄關係躍上新台階。2010 年 12 月，蒙古總理巴特包勒德（Сүхбаатарын Батболд）訪問俄羅斯，就蒙俄兩國關係發展現狀和未來前景與俄總理普丁及議會兩院主席進行了會談，雙方簽署了一系列合作和《蒙俄關於 2011 年至 2015 年經貿發展綱要》（Набросок Монголии развития экономики и торговли с 2011 по 2015 год）的聯合公報。俄國決定免除蒙古的部分欠款並承諾向兩國合資企業注資。兩國傳統經貿關係的日益恢復和緊密聯繫是兩國關係未來發展堅實的物質基礎。[34]

---

32 Dr. Alicia, J. Campi 著，〈蒙古對於俄羅斯與中國在蒙古未來發展上所扮演角色的戰略觀〉，頁 101。

33 〈千禧年挑戰公司〉，檢索於 2017 年 3 月 31 日，http://www.mcc.gov/pages/countries/program/mongolia-compact。

34 馬立國、黃鳳志，〈"冷戰"後蒙古國亞太外交戰略探析〉，頁 34。

表 2：蒙俄關係領導人互訪簡史

編號	時間	事件
1	1988 年	蘇聯戈巴契夫決定自蒙古撤軍。
2	1992 年	蘇聯約 7 萬常駐軍隊全部撤出蒙古境。
3	1993 年	蒙古總統彭・奧其爾巴特於 1993 年兩次造訪莫斯科，兩國簽訂新的《蒙俄友好關係與合作條約》，確立了面向 21 世紀的友好合作關係，由原來的依附性關係轉向平等夥伴關係。
4	1994 年 4 月 7 日	蒙古總理札色萊雙方就文化、科技、郵電通訊、高等教育、醫療衛生、水文氣象、環境監測、兩國公民簽證制度等合作簽一系列協訂。
5	1994 年 6 月 30 日	蒙古大呼拉爾通過《蒙古國外交政策構想》，用法律形式決定新的外交戰略方向和政策。提出實行不結盟、等距離、全方位的多支點外交方針。
6	2000 年	俄總統的普京訪問蒙古國，雙方簽署《烏蘭巴托宣言》，為兩國關係的恢復奠定了法律基礎。
7	2002-2003 年	兩國總理互訪，確立蒙俄「睦鄰傳統夥伴關係」。
8	2006 年	蒙總統恩赫巴亞爾訪俄，雙方簽署《莫斯科宣言》，蒙俄關係進入全面恢復階段。
9	2008 年	巴亞爾總理訪俄，開啟兩國的全面合作。 基於彼此的利益需要，2008 年俄蒙正式恢復軍事交流，俄羅斯同意向蒙古提供 6,000 萬美元武器裝備，此後每年舉行一次「達爾汗」(意即神聖)聯合軍演。
10	2009 年	巴亞爾兩次訪問俄羅斯，俄政府首腦和國家元首也在一年內接連訪蒙。普丁與梅德韋傑夫先後訪蒙，蒙俄兩國元首簽署《發展戰略夥伴關係宣言》，確定高層定期交流，使雙邊關係提升至戰略層面。
11	2010 年 12 月	蒙古總理巴特包勒德訪問俄羅斯，就蒙俄兩國關係發展現狀和未來前景與俄總理普京及議會兩院主席進行了會談，雙方簽署了一系列合作檔和《蒙俄關於 2011 年至 2015 年經貿發展綱要》的聯合公報。

編號	時間	事件
12	2011 年	俄羅斯總統梅德韋傑夫的邀請，蒙古國總統額勒貝格道爾吉展開五天的正式友好訪問。這是在俄總統于 2009 年 8 月訪問蒙古之後，兩人進行的再次會晤。在上次俄總統訪蒙期間，俄蒙雙方發表了戰略合作夥伴關係聲明。
13	2014 年 9 月	普丁訪問蒙古，在兩國總統 9 月 3 日蒙古談判後簽署一系列聯合文件，包括給予烏蘭巴托無償軍事技術援助和現代化改造蒙古國鐵路。
14	2014 年 11 月	2014 年 11 月，中國、蒙古國、俄羅斯在呼和浩特市舉行了首次中蒙俄旅遊聯席會議。
15	2016 年 7 月	俄國總理梅德韋傑夫訪問蒙古烏蘭巴托。

資料來源：整理自范麗君，〈蒙古與俄羅斯雙邊關係綜述〉，《內蒙古財經學院學報》，第 6 期，2011 年，頁 48。

因此，蒙古認定，今後在捍衛國家獨立和經濟建設方面仍有必要積極尋求俄羅斯的支持和幫助。[35] 為了在新形勢下發展蒙俄新型關係，得到俄羅斯的支援和援助，兩國高層互訪頻繁，政治、經濟和軍事安全聯繫不斷加強（雙方互訪如表 2）。與此同時，隨著美國對蒙古的戰略提昇，日益感覺到壓力的俄羅斯開始提高警惕維護自己的戰略利益，重返蒙古的步伐不斷加快，而這一進程的加快又符合了蒙古的戰略調整意圖。

蒙古人對俄國人是同情的，甚至可以說是持正面的看法，這點跟大部分的前蘇聯附庸國很不一樣。現在回顧蘇聯時代，蒙古人不管政治傾向為何，都覺得克里姆林宮（Московский Кремль）的繼承人是有價值的潛在伙伴，而非不值得信任的壓迫者。蒙古人持這種態度，是因為蒙古人認為俄羅斯人捍衛蒙古獨立、在社會主義期間提供蒙古大量援助，又防止蒙古像內蒙古那樣被中國併吞。[36]

蒙古樂見與俄羅斯恢復軍事合作，但是要想真正強化兩國關係，關鍵在於俄羅斯必須提供蒙古更多貿易與投資，以與中國抗衡。2011 年 6 月 1 日的蒙俄

---

35 娜琳，〈蒙俄積極修復雙邊關係的背景剖析〉，《當代亞太》，2002 年，第 12 卷，頁 42-46。
36 Dr. Alicia, J. Campi 著，〈蒙古對於俄羅斯與中國在蒙古未來發展上所扮演角色的戰略觀〉，頁 99-100。

跨政府協議，將軍事、經濟領域以及互相保護智慧財產等項目納入雙邊軍事技術合作，以及恢復在蒙古東部建立合資的鈾礦企業的協議。

因為蒙古與中國的經濟關係似乎帶來了發展與財富，卻也威脅到蒙古的國家認同與領土完整。反之，蒙俄關係似乎比較少威脅性，但經濟效益也較低。蒙古知道要更努力調和與鄰國之間的關係，加深與俄羅斯的關係，以抗衡中國日益增加對蒙古貿易與內部政治的控制。蒙古的政策制訂者確實仍對西方國家與亞洲民主國家的直接投資開放，也願意與這些國家發展多邊關係，然而在外交政策的算計當中不是軟弱或樂觀。他們越來越明白要想穩定蒙古的經濟與政治前途，就必須靠自己向「第三鄰國」靠攏。[37]

三、蒙古與第三鄰國外交

1987 年蒙美正式建交。1991 年貝克（James Addison Baker III）在訪問蒙古時首次將蒙古看作是美國的「第三鄰國」；1998 年奧爾布賴特（Madeleine Jana Korbel Albright）再次強調，從民主觀和經濟模式來看兩國是鄰國。在「911 事件」之後，蒙美關係有了實質性的進展。美國繼續加強與蒙古政治和軍事合作，把蒙古作為其在亞太的重要基地。2004 年兩國確立了「全面夥伴關係」。2005 年美國總統布希訪問蒙古，將蒙美關係推到了一個新階段。2009 年和 2010 年，蒙古對其 1994 年制定的《對外政策構想》做了重要修改和補充，把「第三鄰國」外交政策以法律形式加以強化。蒙古總統額勒貝格道爾吉還強調到「美國是其第一首選第三鄰國」。[38]

蒙古發展對美關係首先是尋求更多安全保障，希望加入美國主導的亞太安全體系。美國支持蒙古加入北約「和平夥伴關係計畫」，蒙古則積極支持美國的全球反恐行動，並參與了伊拉克、阿富汗等地的維和行動。蒙美雙邊軍事合作始於 1996 年，兩國著手舉行聯合民防演習。2000 年，兩國建立安全磋商機制，蒙軍首次出境參與美國主導的維和軍演。2003 年，蒙美兩國在蒙古境內舉行「可汗搜索」聯合軍演；自 2006 年起該項演習發展為多國，日、韓、印等

---

37 同上註，頁 101-107。
38 "US President Barack Obama: I Admire Mongolia For Its Success in Strengthening Democracy," President. mn, June 16, 2011, http://eng.president.mn/newsCenter/viewNews.php?newsId=560.

國也逐漸參加進來。[39] 美國國防部長哈格爾（Chuck Hagel）2014 年在烏蘭巴托簽署一項協議，美國和蒙古將加強軍事關係。美國為蒙古提供價值大約 3 百萬美元軍售和訓練。[40] 因此，蒙美之間的「第三鄰國」關係已全面建立，其核心是軍事、政治合作（雙方互訪如表 3）。[41]

蒙古「第三鄰國」中，日本的重要性僅次於美國。1989 年日本外相宇野宗佑（うの そうすけ）訪問蒙古和 1990 年蒙古部長會議主席索德諾姆（Думаагийн Содном）訪日，蒙古與日本的高層互訪。1991 年 8 月日本首相海部俊樹（かいふ としき）應邀訪問蒙古，這是日本最高領導人首次訪問蒙古。1997 年 2 月蒙古總理訪問日本時，雙方確定建立「面向 21 世紀的全面伙伴關係」。1998 年 5 月蒙古總統巴嘎班迪訪問日本和 1999 年 7 月日本首相小淵惠三（おぶち けいぞう）訪問蒙古，兩國再次重申建立全面伙伴關係。2000 年恩赫巴爾亞（Намбарын Энхбаяр）當選蒙古總理和 2006 年恩赫包勒德（Миеэгомбын Энхболд）當選蒙古總理，首次出訪的都是日本。2007 年 2 月蒙古總統恩赫巴亞爾訪問日本，並與日本簽署了兩國關係未來十年發展綱要。2010 年 11 月蒙古總統額勒貝格道爾吉訪日，兩國發表聯合聲明，表示將建立戰略伙伴關係。[42]

表 3：蒙美關係歷史發展

編號	時間	事件
1	1987 年	蒙美正式建交。
2	1991 年	美國國務卿貝克在訪問蒙古國時，首次將蒙古國看作是美國的「第三鄰國」。
3	1996 年	蒙古與美國簽訂雙邊軍事協定，賦予美國軍隊進入蒙古的權利。

---

39 「集中 -2000」維和軍演。宋效峰，〈冷戰後蒙古國的多支點外交及其影響〉，《世界經濟與政治論壇》，第 2 卷第 2 期，2011 年 3 月，頁 37。

40 美國之音，〈美國和蒙古將加強軍事關係〉，美國之音，2014 年 4 月 11 日，https://www.voacantonese.com/a/cantonese-7439834-hagel-mongalia-0410-ry/1890317.html。

41 魏力蘇，〈「第三鄰國」：蒙美關係的地緣戰略視角〉，《赤峰學院學報》，第 33 卷第 6 期，2012 年 6 月，頁 41。

42 申林，〈蒙古「第三鄰國」外交析論〉，《當代世界》，2013 年 4 月，頁 45。

編號	時間	事件
4	2000 年	兩國建立安全磋商機制，蒙軍首次出境參與美國主導的「進中 -2000」維和軍演。
5	2001 年	蒙古開始以觀察員身份參加美、泰等國舉行的「金色眼鏡蛇」聯合軍演。
6	2003 年	蒙美每年都在蒙古境內與行代號「可汗探索」的聯合軍事演習。
7	2004 年	正式參加「可汗探索」該項演習。 兩國確立了「全面夥伴關係」。
8	2005 年	布希訪蒙期間即宣佈，美向蒙提供價值 1,100 萬美元的軍事援助，以提高蒙古的反恐裝備水準。
9	2006 年	蒙美代號「可汗探索」的聯合軍事演習，包括大到日本、印度等多國參加的演習。
10	2011 年 6 月	蒙古總統額勒貝格道爾吉訪問美國時稱：「我們將美國視作是我們的首選『第三鄰國』，我們希望提升兩國關係。」美蒙關係迅速發展也是雙方互有所需的必然結果。
11	2014 年	美國國防部長在烏蘭巴托簽署一項協議，美國和蒙古將加強軍事關係，美國為蒙古提供價值大約 300 萬美元軍售和訓練。
12	2016 年 6 月	美國國務卿克里對蒙古進行了短暫訪問，讚揚蒙古面對俄羅斯和中國的壓力，仍然扮演該地區民主綠洲的角色。

資料來源：整理自申林，〈蒙古「第三鄰國」外交析論〉，《當代世界》，2013 年 4 月，頁 45-47。

　　印度也是蒙古「第三鄰國」的重要國家之一。冷戰結束後，蒙印關係有了新發展。1994 年兩國簽署友好關係與合作條約。但在 2000 年之前，蒙古與印度的來往有限。進入 21 世紀後，兩國關係密切起來。2001 年蒙古總統巴嘎班迪訪問印度，與印度簽署《蒙印聯合宣言》（Joint Statement for India-Mongolia Strategic Partnership）、《國防合作協議》（India-Mongolia defence relations）等多項重要文件。2004 年 1 月蒙古總理恩赫巴亞爾訪問印度，兩國一致同意深化和加強軍事交流合作。2006 年 5 月和 2007 年 7 月，印度國防部長拉久（A.K Antony）和國家安全委員會聯合情報委員會主席普拉丹（S.D. Pradhan）相繼訪

問蒙古。[43]

在與歐盟的關係方面，1989 年蒙古與歐盟的前身歐洲共同體建立外交關係，此後蒙古與歐盟的關係不斷的向前發展。在與北約的關係方面，20 世紀 90 年代蒙古就要求加入北約「和平伙伴關係計劃」（Partnership for Peace, PFP），2011 年 10 月蒙古正式向歐安組織提出加入申請，希望歐安組織成為其第三鄰國。2012 年 3 月蒙古與北約正式簽署了合作伙伴協議，雙方關係有了實質性的突破。2012 年 5 月北約峰會在美國芝加哥召開，蒙古首次以「全球伙伴關係」框架內的和平伙伴關係國與會。2004 年蒙古成為歐安組織的亞洲伙伴國。[44]

積極參與雙邊與多邊國際合作是蒙古「多支點」外交政策的目標之一。不結盟、全方位、等距離平衡也是蒙古與亞太和世界其他地區的國家、各個國際組織發展關係的主要原則。蒙古實現經濟社會轉型以來，外交活動日益活躍。蒙古高層領導頻繁出訪亞太地區和歐洲、拉美、中東等地區的國家，在政治、經濟、文化等領域與相關國家進行廣泛地交流與互動，取得了豐碩的成果。蒙古積極爭取亞洲開發銀行（Asian Development Bank, ADB）、國際貨幣基金組織（International Monetary Fund, IMF）、歐洲復興發展銀行（European Bank for Reconstruction and Development, EBRD）和聯合國糧農組織（Food and Agriculture Organization of the United Nations, FAO）等國際組織的援助，為其經濟社會發展獲取更多的資金支持。在亞太地區，蒙古於 2000 年正式加入太平洋經濟合作理事會（Pacific Economic Cooperation Council, PECC）。2004 年蒙古成為上海合作組織（The Shanghai Cooperation Organization, SCO）觀察員國，蒙古總統、總理和部長多次出席會議並積極參加各項活動。2005 年，蒙古國加入了《東南亞友好合作條約》（Instrument of Extension of the Treaty of Amity and Cooperation in Southeast Asia）。蒙古的多邊外交政策最大的目標是建立東北亞合作對話機制，借此保障國家安全，帶動經濟發展。[45] 同時也積極

---

43 〈蒙古總統表示：蒙古視印度為重要伙伴〉，中國經濟網，http://www.ce.cn/xwzx/giss/gdxw/200707/31/t20070731-12373752.shtml；申林，〈蒙古「第三鄰國」外交析論〉，頁 45-46。
44 申林，〈蒙古「第三鄰國」外交析論〉，頁 46。
45 郝利鋒，〈世界看到蒙古國越來越活躍〉，新華網，2007 年 11 月 28 日，http://news.xinhuanet.com/world/2007 － 11/28/content_7159494.Htm。

參與地區安全問題的解決。另外積極發展與亞太和世界其他地區的國家和國際組織的關係，為其經濟社會發展打造了良好的外部國際環境，同時也讓世界看到了一個充滿活力的、開放的、積極參與國際事務、渴望融入國際社會的良好的國家形象。[46]綜上所述，蒙古期望積極走入國際社會，結合「第三鄰國」、「多支點」（不結盟、等距離、全方位），達到最後目標「永久的中立國」。

肆、以避險策略檢視蒙古的對外關係

蒙古高層領導頻繁出訪亞太地區和歐洲、拉美、中東等地區的國家，在政治、經濟、文化等領域與相關國家進行廣泛地交流與互動，取得了豐碩的成果。冷戰、世界社會主義制度與蘇聯的瓦解後，兩極化的世界架構瓦解，徹底改變了蒙古的外部環境。蒙古的「外交政策綱領」應運而生。此外，蒙古兩大鄰國接續遭遇重大改變，也直接影響蒙古的外部環境。因此蒙古目前正在重整與改革政治、社會與經濟制度，確認國家利益的優先次序，以利實施以現實主義為基礎的外交政策。[47]

1994 年 6 月，蒙古大呼拉爾通過的《蒙古國對外政策構想》明確提出了蒙古亞太外交新戰略。其主要內容包括三個方面：一是與中俄兩國為保持等距離交往，把發展與中俄的睦鄰友好關係作為蒙古對外戰略的首要目標；二是積極發展與美、日等西方發達國家的外交關係，確立所謂的「第三鄰國」，從而獲得政治上的支持和經濟上的援助。三是與亞太及世界其他地區的國家和一些國際組織開展交流與合作，提高本國在亞太乃至世界舞台上的影響力。[48]有學者認為，1990 年代的蒙古外交戰略採用的是「全面牽制策略」（Omni-enmeshment strategy），而 2000 年後則是「全面牽制及影響力平衡的混合」（Amalgam of omni-enmeshment），且隱含中國是蒙古的最大安全威脅。[49]

---

46 馬立國、黃鳳志，〈 "冷戰" 後蒙古國亞太外交戰略探析〉，頁 34。

47 Dr. Alicia、J. Campi 著，〈蒙古對於俄羅斯與中國在蒙古未來發展上所扮演角色的戰略觀〉，頁 105。

48 Vaishali Krishna, "Mongolian Foreign Policy Implications for Russia and China," Mongolian Journal of International Affairs, Vol.19, 2014, pp. 69-71；馬立國、黃鳳志，〈 "冷戰" 後蒙古國亞太外交戰略探析〉，頁 34。

49 Vaishali Krishna, "Mongolian Foreign Policy Implications for Russia and China," pp. 71-72；馬立國、黃鳳志，〈 "冷戰" 後蒙古國亞太外交戰略探析〉，頁 34。

　　成為民主國家二十年之後，蒙古重新檢討區域安全以及「第三鄰國」政策，結果發現第三鄰國政策並非全然成功，需要重新結盟，因此在 2010 年底公布全新的「外交安全綱領」以及「外交政策綱領」，作為國家未來二十年的施政方針。這兩項政策綱領經由蒙古國會通過，對俄羅斯、中國以及其他國家政策而言，是很重要的指標。2010 年發布的「國家安全綱領」列出七項促進蒙古政治與經濟發展的主要國家安全「綱領」，其中較重要的即在制訂國家安全策略，保障國家領土完整；維持各鄰國投資的平衡，避免蒙古成為其他國家的「附屬國」，追求蒙古境內的外國投資與企業更加多元化；「國家安全綱領」的其他重點包括蒙古財政部有權核准蒙古政府所提出的刺激經濟方案。[50]

　　基於謀生存、求發展的戰略考慮，蒙古制定了以「第三鄰國」制衡中俄的戰略政策。從理論到實踐考慮均勢理論和「第三鄰國」戰略理論是蒙古不結盟、等距離、全方位的「多支點」外交政策的理論基礎。[51] 蒙俄、蒙中關係的友好發展是主流，但由於歷史上的原因，蒙古一直對兩大鄰國保持著不同程度的戒備。蒙方對中國快速的發展存在不安全感，怕中國會對其主權形成威脅。同時，蒙古也對俄羅斯心存戒備。過去 70 年的特殊關係促使其對俄保持警惕。[52]

　　「不結盟、等距離、全方位」的多支點外交方針。在這一方針指導下，蒙古除優先發展與中俄兩國的關係外，還積極發展與美、日、印等「第三鄰國」關係，並在與第三鄰國的外交活動中，逐漸形成了「第三鄰國」外交政策。[53] 蒙古最初將「第三鄰國」界定為美國，後來又發展到西方國家，再後來又發展到「援蒙國家」，把日、韓、印等亞洲國家也包括進來。[54] 除了國家行為體外，

---

50 2010 年發布的「國家安全綱領」列出七項促進蒙古政治與經濟發展的主要國家安全「綱領」：1. 廢除目前的總統制，改行議會民主制，以維護自由民主。2. 制訂國家安全策略，保障國家領土完整。3. 限制外商投資上限不得超過該公司所有產品市值的三分之一。4. 維持各鄰國投資的平衡，避免蒙古成為其他國家的「附屬國」，追求蒙古境內的外國投資與企業更加多元化。5. 以蒙古貨幣（圖格里克）進行所有金融交易。6.「國家安全綱領」的其他重點包括蒙古財政部有權核准蒙古政府所提出的刺激經濟方案。7. 以及蒙古將成立一家蒙古開發銀行，提供中長期貸款與其他服務，以促進優先產業 發展、增加附加價值出口量，並提升蒙古的經濟競爭力。Dr. Alicia J. Campi，〈蒙古對於俄羅斯與中國在蒙古未來發展 上所扮演角色的戰略觀〉，頁 104-105。

51 魏力蘇，〈「第三鄰國」：蒙美關係的地緣戰略視角〉，頁 42。

52 Dr. Alicia, J. Campi 著，〈蒙古對於俄羅斯與中國在蒙古未來發展上所扮演角色的戰略觀〉，頁 98。

53 申林，〈蒙古「第三鄰國」外交析論〉，頁 45。

54 〈蒙古推行「第三鄰國」外交〉，環球網，2011 年 9 月 9 日，http://world.huanqiu.com/roll/2011-09/1988731.html。

蒙古的「第三鄰國」還包括一些國際組織，主要包括北約、歐安組織、歐盟等。[55]

在蒙古的「第三鄰國」中，美、日、印是其中比較重要的國家，歐盟、北約和歐安組織（Organization for Security and Co-operation in Europe, OSCE）是比較重要的國際組織。在「第三鄰國」中，美國是最重要的國家。在第三鄰國中，美國被蒙古視為最重要的安全伙伴。在蒙古看來，美國是能夠對中俄兩個鄰國產生重要影響的唯一國家，是保障蒙古安全的主要依托力量，使蒙古不致於淪為中俄附庸的重要保障。[56] 與美國的密切關係使得蒙古對自身的防衛安全倍感自信。2005 年當蒙古國防部長被隨行的美國記者問及，地處中國和俄羅斯兩個大國之間的蒙古是否感到擔心與害怕的時候，國防部長的回答，以蒙美目前的軍事關係，對兩個鄰國一點也不感到擔心。[57] 除了美國外，北約也被蒙古視為重要的安全伙伴，這就是蒙古熱衷發展與北約關係的根本原因。除了對國家獨立與安全的政治考慮外，蒙古還想通過「第三鄰國」外交獲得經濟上和軍事上的援助。[58]

以美、日為例，美國從 1994 年起每年對蒙古的無償援助折合約 1200 萬美元，截至 2004 年底，美國對蒙古的無償援助已達到 1.5 億美元。[59] 2005 年美國宣布向蒙古提供 1100 萬美元的無償軍事援助。2007 年美國總統布希批准向蒙古提供 2.85 億美元的無償援助。2010 年美國再次向蒙古無償援助 950 萬美元。除了無償援助外，美國還給蒙古長期最惠國待遇。另外美國還幫助蒙古改進現有的軍事裝備。日本方面，從 1991 年到 2006 年，平均每年對蒙古援助 1 億美元，其中 3000-3500 萬美元為無償援助，另有約 1000 萬美元為技術合作。2009 年日本宣布向蒙古提供 9.4 億日元無償援助。日本還將提供 288 億日元貸款用於蒙古修建國際新機場。[60]

綜上所述，蒙古由於其地緣政治及國家發展利益的考量，對於中俄的威脅

55 申林，〈蒙古「第三鄰國」外交析論〉，頁 45。
56 娜琳，〈蒙美關係發展的動力是互需〉，《東北論壇》，第 3 期，1999 年，頁 10-15。
57 徐冰川，〈美國欲拉蒙古加入對華包圍圈〉，法制網，http://www.legaldaily.com.cn/zbzk/sjb/116/wz06.htm。
58 申林，〈蒙古「第三鄰國」外交析論〉，頁 46。
59 木東、翁天成、牛雨辰、薛小樂，〈布希為何去蒙古〉，人民網，2005 年 11 月 18 日，http://world.people.com.cn/GB/3867623.html。
60 同上註，頁 46-47。

既無法採取抗衡，亦無法進行扈從的外交政策，故而避險是最符合蒙古整體戰略考量的策略選項。

根據前節的討論，就庫成輝的避險工具分類，可約略整理成表4。

表4：蒙古針對中俄採取的第三鄰國避險措施

針對國家			中國	俄國	第三鄰國
平衡策略（純粹形式）					
避險策略	偶發風險選項	間接平衡（軍事避險）	V	V	
		阻絕主導（政治避險）	V	V	
		經濟多樣化（經濟避險）			V
	回報極大化選項	經濟實用主義（經濟避險）	V	V	V
		約束交往	V	V	V
		限制性扈從			V
扈從策略（純粹形式）					

伍、結論

蒙古媒體表示，習近平訪問蒙古主要是出於經濟利益外也考慮到地緣政治因素。蒙古像三明治一般地被夾在俄羅斯和中國兩個大國之間，蒙古不希望過度依賴兩個大國，因此推行「第三鄰國」政策，這讓蒙古與美國、日本、加拿大、歐盟等提升戰略關係。同時，蒙古對中國的印象毀譽參半，大約60%的人對中國保持戒心，他們認為中國企業攫取蒙古資源，中國人也搶去了他們的工作機會，大約40%的人對中國持中立或正面的看法。與世界其它地方一樣，蒙古媒體表示，蒙古人日常使用的商品有80%的來自中國，中國因素已滲透至百姓生活當中。媒體認為，蒙古最需要的就是中國幫助提升蒙古的基礎設施建設。[61]

---

61 英文報紙《UB郵報》（UB Post）高級編輯哈什-額爾德尼（Khash-Erdene）對BBC中文網表示，

　　雖然蒙古與中國關係表面上發展良好，蒙古人民與政策制訂者卻非常擔心蒙古在經濟上越來越倚賴中國，對蒙古的國家安全與均衡經濟發展並不見得是好事。過往的不信任與敵意仍舊影響蒙古的決策。而且蒙古也擔心中國商人、移民與觀光客的湧入可能會超過蒙古的人口數，這種憂慮也瀰漫到蒙古國內政治。[62] 蒙古官員私下表示，「中國沒有興趣發展蒙古的經濟，而僅在開掘我們的天然資源…中國有可能會完全併吞我們」。[63]

　　其次，儘管俄羅斯再度重視蒙古的戰略緩衝作用，但並不意味著蒙古與俄重新結盟，兩者的戰略夥伴關係對於蒙古而言首先是工具意義上的。2009年，恩赫巴亞爾兩度訪問俄羅斯，普丁與梅德韋傑夫（Дми́трий Анато́льевич Медве́дев）先後訪蒙，拉攏蒙古從俄羅斯的外交邊緣重返中心位置。蒙俄兩國元首簽署《發展戰略夥伴關係宣言》（Strategic Partnership Declaration, 2009），確定高層定期交流，使雙邊關係提升至戰略層面。俄羅斯強勢重返蒙古這一「後院」，既是對日益上升的美國對蒙古影響的反制，也是在某種程度上出於對中國迅速發展的隱憂。[64]

　　小國外交在國際關係中本來就因種種條件而受限。蒙古經後冷戰時期20餘年以來的發展，對於其自身處境及國際情勢，自有其解讀及因應之道。由於蒙古夾在俄羅斯和中國兩個大國之間，蒙古已開始籌劃成為東亞唯一的「永久中立國」，該國外交專家稱這是其「唯一的生存之道」。蒙古總統 2015 年 9 月底在美國紐約舉行的聯合國大會上演講時強調，蒙古的永久中立化將「為促進全球的和平、安全和繁榮做出貢獻」，並呼籲支持該政策。該國政府為締結有關永久中立的協定正在與中俄兩國進行磋商，並表示「兩國都對蒙古的立場表示理解」。儘管國內存在反對意見，但額勒貝格道爾吉在 2015 年 9 月蒙古的國家安全保障評議會上確定了永久中立的方針。若成為永久中立國，蒙古與他國進行軍事演習將受到限制，聯合國維和行動（PKO）中直接行使武力的活動也將受限。學者認為蒙古將來應發揮永久中立的立場，「最理想的是使蒙古

　　詳見高毅，〈分析：習近平為何對蒙古「點對點訪問」？〉，BBC 中文網，2014 年 8 月 21 日，http://www.bbc.com/zhongwen/trad/world/2014/08/140821_mongolia_china_xi_visit。

62　Dr. Alicia, J. Campi 著，〈蒙古對於俄羅斯與中國在蒙古未來發展上所扮演角色的戰略觀〉，頁 103。

63　Robert D.Kaplan, "The Man Who Would be Khan," The Atlantic Monthlym, March 2004, pp. 55-61.

64　每年舉行一次「達爾汗」（意即神聖）聯合軍演；Robert D.Kaplan, "The Man Who Would be Khan,".

成為防止亞洲紛爭等的『對話中心』」。[65]

　　對於蒙古是否能成為永久中立國，仍有待時間的驗證。避險策略的適當運用無異是蒙古當下最佳的選擇。蒙古的「第三鄰國」外交之所以開展的比較順利，關鍵在於得到了這些鄰國的積極響應，而這些鄰國之所以積極響應，是因為他們看好了蒙古的特殊價值。這為蒙古推行「第三鄰國」外交創造了有利條件。[66] 由於中國和俄羅斯是蒙古的兩個強鄰，所以無法擺脫對中俄的芥蒂；美國、日本和印度等「第三鄰國」領土未與蒙古接壤，因而蒙古對他們的戒心就小得多。由於這種心態，所以蒙古的三方平衡政策在有意無意之中就朝向「第三鄰國」偏斜的政策轉移。[67]

　　2011 年 11 月蒙古正式向歐安組織提出加入申請，2012 年 3 月蒙古與北約正式簽署合作伙伴協議。中俄兩國出於維護其與蒙古關係大局考慮，中俄在一定程度上容忍蒙古「第三鄰國」外交政策。但蒙古超越中俄兩國容忍度，該兩國可能會採取某些因應措施。畢竟蒙古包圍在中俄之間，沒有其他對外通道，如果兩國禁止蒙古經由自己的領土和「第三鄰國」進行交往，蒙古就會陷入極大的困境。[68]

　　蒙古也意識到這一點，所以「第三鄰國」外交雖然有所越位，但還是在一定限度內。比如蒙古允許美國軍隊進入其本土，每年都和美國在本土舉行聯合軍演，甚至允許美國在其領土上監聽中俄情報，但並沒有提出或准許美國在蒙古駐軍。再如，蒙古深知加入北約的利害，因此只與北約建立密切的關係，並沒有申請要加入北約。綜上所述，蒙古出於對中俄兩國的戒心而使其「第三鄰國」外交有所越位，但被夾在中俄之間的客觀現實又決定了其越位又不會太遠。這樣蒙古避險策略的外交支點還能維持大致的平衡，在國際局勢未發生大規模變化的情況下，其「第三鄰國」外交可能繼續維持。[69]

---

65 蒙古地政學研究所所長喬因戈爾（音譯）。楊家鑫，〈在中俄間求生存 蒙古尋求成為永久中立國〉，中國時報，2015 年 10 月 16 日，http://www.chinatimes.com/realtimenews/20151016002257-260408。
66 申林，〈蒙古「第三鄰國」外交析論〉，頁 47。
67 同上註，頁 47。
68 同上註，頁 48。
69 同上註。

由網路社會暨資訊流成長初探新世紀戰爭

孫立方 *

壹、前言

人之初，性本善；性相近，習相遠。

——《三字經》，13 century。

研究顯示，遺傳造成的兩性差異很小。是社會化過程，拉大了兩性差異。

——《社會心理學》，1958。

戰爭就是一種以迫使實現我方意志為意圖的暴力行為。

——《戰爭論》（克勞塞維茨 1996）

這是一場軍備競賽。他們會日益精進，我們也必須投資改善。[1]

　　三字經概約自 13 世紀流傳下來，幾百年來，常用於教育幼童，其內容除了相當程度地反映儒家思想；更重要的，是點出了後天社會、文化的影響力。故且不論性善或性惡的爭論，古人對於人性，或者說生理層面本性的差異，認為是非常小的。換句話說，如果不考慮風俗、地理、乃至文化等後天因素的影響，則不論是 1941 年 2 月 16 日出生在平壤的朝鮮族男童金正日，或是 1946 年 6 月 14 日在紐約皇后區川普家呱呱墜地的唐納，在人性的層面，並沒有太大的差別。

　　隨著時序邁入近代，人類對於心理方面的研究，也更加透澈深入。越來越多的資料顯示，即使是男女兩性間，出生時受到遺傳影響而產生的差異也很有限；更多的性別差異，其實是後天的社會化過程塑造出來的。從唐納．川普與金正日身上，驗證這個觀點，再明顯不過。兩人年齡相仿，但成人後意識型態卻大相逕庭，最後甚至分別成為兩大對抗集團的首腦。其中金正日雖然已逝，

* 淡江大學國際事務與戰略研究所博士候選人

1　BBC, "Zuckerberg: Facebook is in 'arms race' with Russia", *BBC*, April 11, 2018, http://www.bbc.com/news/world-us-canada-43719784.

卻有子金正恩繼承衣鉢。兩方均握有可對成千上萬人造成生命威脅的武器相互叫陣，凸顯了不論性善性惡，後天環境的影響力，確實驚人。就此而言，古、今之間，對於人類成長基本歷程的觀察，並沒有太大變異。

另一方面，自十九世紀以來，對普奧、普法、乃至第一、二次世界大戰產生重要影響的《戰爭論》，始終是西方軍事思想主流。作者克勞塞維茲被尊為西方兵聖，對各國建軍方向產生很大影響。但從現實層面觀察，部分他對戰爭型態的觀察，與當代現況已不盡相符。依據統計，自 1945 年以來，雖然國際社會的國家數持續增加、但跨國衝突卻大幅減少。[2] 事實上，回顧過往二個世紀，全球 83% 的衝突屬於非正規作戰，預料隨著全球化的發展，此種趨勢將更加明顯。[3] 顯然，戰爭已經不是當今國際社會用以解決政治問題的主要工具；至於「戰爭是一種推進到其最高限度的暴力行為」[4]，更有值得辯證的空間。

相對於此，全球最大社群媒體、臉書負責人祖克柏，2018 年初在美國國會的聽證中，則將與俄羅斯駭客在網路世界的較量，稱為「軍備競賽」，宛若「看不見的硝煙」。對照「劍橋分析」事件，公關公司運用數千萬個資，用於影響包括美國在內的各國選舉，顯示運用資訊、而非「暴力行為」貫徹政治意志，已然可行。雖然美國「通俄門」的調查，仍然真相未明，但不論結果如何，回顧歷史，在數十年冷戰對抗中，前蘇聯即使擁有數量龐大的核子武器，亦難撼動美利堅合眾國分毫，最終落得分崩離析的下場；而 21 世紀的今天，國力規模遠不如前蘇聯的俄羅斯，僅以臉書等社群媒體為武器、消息為彈丸，即展現了直取白宮的力量，[5] 怎不令人驚異、又具有什麼樣的代表意義？本研究嘗試從網絡社會發展、資訊流大幅成長角度切入，並借用美軍「多領域戰鬥」「資訊環境作戰」等概念，對現代戰爭理論，做初探式思索。

2　Bruno Tertrais, "The Demise of Ares: The End of War as We know it?" *The Washington Quarterly*, July 1, 2012, pp. 7-22, https://www.frstrategie.org/web/documents/publications/autres/2012/2012-tertrais-twq-demise-ares.pdf.

3　David J. Kilcullen, "The City as a System: Future Conflict and Urban Resilience," *The Fletcher Forum World Affairs*, July 26, 2012, pp. 19-39, https://static1.squarespace.com/static/579fc2ad725e253a86230610/t/57ec7faf5016e1636a22a067/1475116977946/Kilcullen.pdf.

4　克勞塞維茲著，鈕先鍾譯，《戰爭論精華》（*A Short Guide to Clausewitz on War*）（台北：麥田，1996 年），頁 58。

5　〈資訊爭鬥 新世紀認知戰成形〉，青年日報，2018 年 3 月 24 日，版 10。

貳、網絡社會

在三字經問世近 800 年後，隨著科技的快速發展，人類已由二戰後的工業社會，邁入網絡社會（network society）。依據 Manuel Castells，網絡社會係指「社會結構或活動受資訊科技與網路的結構所影響的社會」。[6] 在其觀點中，網際網路等新科技發展，帶動傳播媒介、交通工具等器物的長足進步，已經對社會中的生產、企業組織、人際溝通與互動方式；乃至於空間中移動的方式、對時間的認知方式等造成影響，並因此引發社會結構的劇烈改變。

舉例而言，傳統社會極易受地理空間的限制。但時至今日，人們可以利用高速交通工具，進行遠距離移動。如同高鐵的通車，讓台北高雄變成一日生活圈。再如利用各種資訊工具如網路、個人通訊器材等，現代人也可不受地理距離限制，傳遞資訊。台灣地區 line 等通訊軟體盛行，為顯例之一；在全球性範圍，即時通訊軟體的使用量近年超出社群媒體，[7] 則是另一個例證。

事實上，依據學者 Berney 的研究，可以被稱為網絡社會者，必須具備成熟的網絡通訊科技，以及資訊管理／流通科技，用以構成基礎建設，以利管理愈來愈多的社會、政治與經濟實務；同時，網絡社會再生產與體制化，也是特徵之一。Berney 進一步說明，網絡主要由三項元素構成，即節點（nodes）、紐帶（ties）與流量（flow）。其中的節點可以是個別的朋友、單一電腦或一家公司；而紐帶則可能是通信、電纜或合約；至於流量則可能是閒聊、資料或金錢。「網絡」描繪的是一種結構式的狀態，其中各個端點（節點）透過相互連結（紐帶）而產生連繫關係。一個節點可能與一個或多個節點相連；流量可能是單向、雙向或多向；其建構出的網絡可以是中心化（centralized）或非中心化、多中心的（multicenter）。[8]

由前述定義思考網絡社會，可以發現在類似結構中，資訊與知識的傳遞，

6　Manuel Castells 著，夏鑄九、王志弘等譯，《網絡社會之崛起》（*The Rise of the Network Society*）（台北：唐山，2000 年），頁 520。

7　Mckitterick Will, "Messaging apps are now bigger than social networks." *Bussiness Insider*, April 23, 2016, http://www.businessinsider.com/the-messaging-app-report-2016-4-23.

8　Darin Barney 著，黃守義譯，《網絡社會的概念：科技、經濟、政治與認同》（*The Network Society*）（台北：韋伯，2012），頁 31-32。

因各式網絡的綿密構連，脫離了時空限制、也不再受限於線性擴散，而是透過點與點之間、跳躍連結進行傳遞，進而形成一種點對點的網狀資訊傳遞。除了高速的訊息傳遞與交通運輸外，其在結構上具有節點與中繼站的特徵。相關特性導致社會組織、權力或文化等，產生相當大的改變，而形成網絡社會。

Manuel Castells 在其《資訊時代：經濟，社會與文化》系列鉅作中，描述了此種人類歷史上的重要變化，所產生的影響，認為以資訊為中心的技術革命，已經改變了人類思考、生產、消費、貿易、管理、溝通、戰爭的方式。在網絡社會，傳統時間與空間的限制，已經被打破。人類對於時間與空間的經驗，遭到「無時間的時間」（timeless time）及「流量的空間」（space of flows）所置換，[9]人類依附空間及時間生存的經驗，因而可能發生極大改變，不得不改變看世界的方式。在新通訊系統中，區域性時間已被抹除；流量空間取代地理空間，使得不同地點象徵的文化、歷史與地理意義，將被解放。

舉例而言，近年興起的 World Run 慈善路跑，在台灣即相當受歡迎。參與的跑者來自全球各大洲，即使所處時區不同、地理界限形成的季節不同，但卻以格林威治時間為準，同時在不同地點開跑。[10]透過即時視訊，跑者可以感受到，在全世界有成千上萬跑者，與自己同時從事路跑活動，而形成一種跨越式的經驗。這種與傳統時空定義不同的時間與空間，匯聚成一種全球化、由整合式媒體建構的文化，雖然混入了國際上不同文化的元素（如台北 World Run 中的亞洲風、義大利的歐陸色彩），但性質上已經不以地域為依據且超現實。

就此 Manuel Castells 進一步指出，一種動態的全球經濟，已經在地球各處建構起來，將全世界有價格或價值的人及活動連結在一起。其間，虛擬真實的文化，圍繞著交流頻仍的視聽宇宙被建構起來，滲透到每處精神表徵和溝通傳播中，以電子超文本整合文化的豐盛性。「流量空間支配了地方空間，無時間性的時間廢除了工業年代的時鐘時間。」[11]此一思維的現實驗證，同樣可自

---

9 Manuel Castells 著，夏鑄九、黃慧琦等譯，《千禧年之終結》（*The End of Millennium*）（台北：唐山，2001 年），頁 1-2。

10 Red Bull, "About: Wing for life world run 2018", https://www.wingsforlifeworldrun.com/tw/zh/about/about/.

11 Manuel Castells 著，夏鑄九、黃慧琦等譯，《千禧年之終結》（*The End of Millennium*），頁 1-2。

World Run 路跑活動中獲得支持。該項活動自 2014 年起舉辦，連續 4 年間已經吸引了來自 193 個國家的 441,021 人參加，總路跑里程數達 4,276,640 公里、所募得捐款總計達 2060 萬歐元。[12]

參、跨界資訊流

此種打破時區、地域流動的趨勢，在資料流方面尤其明顯。研究[13]顯示，自 1980 年至 2014 年間，歷經 20 年的快速成長，雖然跨界貿易及財務流動有所上下波動；但自 2005 至 2014 年間，唯獨跨界頻寬，已經持續成長了 45 倍。此種包括了商業、訊息、搜尋、影音、通訊以及跨國公司的資訊流量，至 2019 年將再成長 9 倍。換句話說，跨界資訊流（data flows）的快速成長，是近十年間的新興現象，我們正跨入人類歷史上的另一個新階段。

此一含括資訊、研究、通信、交易、影音、公司內部往來資料，大幅成長的跨國資訊流，對於全球經濟，究竟產生了何種影響？McKinsey 的研究進一步指出，過去十年，全球貨物流動、外國投資及訊息流對全球 GDP 成長的貢獻，較之於毫無流動的狀況，至少成長了約 10%。僅以 2014 年計，產值即約 7.8 兆美金；其中資訊流的貢獻約 2.8 兆，比全球貨物、貿易流動的貢獻值還高。分析認為，全球規模的數位平台減少跨國界溝通、傳輸的成本，降低了進入全球市場的最低門檻。新的競爭者可能從任何角落快速崛起，增加了產業的競爭壓力。「21 世紀的全球化，越來越傾向以資料與資訊的流通做為定義」[14]。

尤其重要的是，相關發展已經改變新時代的經濟模式。儘管具規模的大型跨國公司，仍然具有優勢，但卻受到價格競爭、創新研發的挑戰；發展中國家的新興競爭者，則有效縮短雙方距離。此外，3D 列印等新科技，降低了實體貨物流動的重要性、進而改變生產方式；對抗網路犯罪以保護資訊流，成為極重要的環節；社群媒體創造了全球社群，但也便利網絡極端份子串連……整體

---

12 Red Bull, "About: Wing for life world run 2018".

13 McKinsey Global, *Digital Globalization: The new Era of Global Flows* (San Francisco: McKinsey Global Institute, 2016).

14 McKinsey Global, *Digital Globalization: The new Era of Global Flows*, p. 1.

而言,「今日的全球化更為複雜、節奏快速,但連結度大幅成長」[15]。

其間,除了人際連結,物聯網(internet of things)的快速發展,讓機器與機器、乃至機器與人類的連結,也更為密切,並進而促成更多資訊的產生。全球網路巨擘 Google 的兩位靈魂人物 Eric Schmidt 及 Jared Cohen 在合著的「數位新時代」一書中指出,預估至 2025 年,多數人透過手持裝備,即可獲得所有資訊。而數據資料本身就是一種工具,可以將許多工作簡化,在日常生活中提供更多無縫接合形式,讓人類對時間的運用更有效率。兩人也發現,雖然文化差異、時區不同等因素,可能讓相隔兩地者,未必能完全連結,但運用相同平台、在了解彼此的情況下交涉,將成為家常便飯。[16] 重要的是,此種密切連結,將產生驚人巨量數據,既可能讓不同個體的詳細面貌,完整公開呈現、甚至侵犯傳統觀念的隱私;所有個別資料的匯集,又可能對現實世界,造成重大影響。

Eric Schmidt 及 Jared Cohen 指出,由於資訊豐富,大眾得以擁有更好的參考依據;相形之下,宗教、文化、種族的神聖化,就更需要費力傳播論點,才能受到青睞。非洲巫醫的權威可能受到挑戰,葉門的童媳陋俗有機會終結…不過從另一個角度來看,也因為大量、具有跨文化隔閡能力的資訊,得以廣泛流通,大眾對現實狀況的認知,將更容易受到影響。隨著愈來愈多人連結上網,不但政府要想操控國民將更加困難,[17] 傳統媒體的角色與定義,也都必須改寫。當任何人都可以透過手機等行動裝置,將突發事件向全球直播時,傳統媒體怎麼可能有即時「新」聞?當各式「爆料」平台不斷「解密」各種不為人知的消息時,媒體如何用「獨家」平衡必然的千篇一律?媒體有沒有能力協助大眾,判斷訊息的基本真偽?還是只能推波助瀾,在二手傳播中沉淪?類此問題值得關注,原因不在於媒體角色,而是其內容足以影響使用者之思想、生活,乃至使用者所處之更廣大生活方式與社會關係。[18]

---

15 Ibid., p. 2.
16 Eric Schmidt 與 Jared Cohen 合著,吳家恆、藍美貞、楊之瑜、鍾玉玨合譯,《數位新時代》(*The New Digital Age*)(台北:遠流,2013 年),頁 18。
17 同上註,頁 45。
18 Hodkinson Paul 著,黃元鵬、吳佳綺譯,《媒介、文化與社會》(*Media, Culture and Society*)(台北:韋伯,2013 年),頁 11。

肆、多領域戰鬥

如果人類從事戰爭的方式,其實是複製自其商業模式的話,那麼數位全球化網絡的緊密連結、大量的資料流動,將對戰爭產生何種影響?下個世代的戰爭,只是以機器人代替真人衝鋒陷陣,還是將產生一種完全不同的面貌?美軍近年提出的多領域戰鬥(Multi-Domain Battle, MDB)概念,提供了一個反思的起點。

依據美軍定義,所謂「多領域戰鬥」係在既有軍種與聯合作戰基礎上,將兵種協同方法擴大到實體領域外之抽象領域而形成之戰力,如網路空間(cyberspace)、電磁頻譜(electromagnetic spectrum, EMS)、資訊環境(information environment)以及戰爭之認知面向(cognitive dimension)[19](US Army TRADOC, 2017)。多領域戰鬥是聯合兵種作戰的延伸,必須整合陸地、水面(下)、空中、電子戰、太空、網路及政府機構等獨立作戰單位,以建構加乘的作戰效能,「進而在所望時間與空間,貫穿敵人防衛體系,在一個以上的作戰領域中,開創領域優勢之窗,以利在敵人整體防衛體系內機動用兵」[20]。

回顧歷史,可以發現美軍發展「多領域戰鬥」概念,有其回應安全環境變化的應然。冷戰時期,美軍即因應威脅型態變化,發展跨軍種聯合作戰概念,先有冷戰後期,因應指向西歐之優勢華約集團裝甲部隊,而提出之「空地整體作戰」(Air-Land Battle)概念;其後因應敵手反介入/區域拒止作戰概念之成熟,美軍又發展「空海一體戰」與「全球公域介入與機動聯合概念」,期適應世代革新之戰場需求。[21]其間,空地作戰的效能,在沙漠風暴作戰等行動中,已經獲得驗證。但其前提是美軍必須於空、海、太空乃至網路等領域,保持絕對優勢。事實上,在近代史的大部分時間,美軍真正能受到挑戰的領域,即為地面戰場,[22]但時至今日,情況已經產生重大變化。

---

19 US Army Training and Doctrine Command, *"Multi-Domain Battle: Combined Arms for the 21st Century."* February 24, 2017, http://www.tradoc.army.mil/MultiDomainBattle/docs/MDB_WhitePaper.pdf.

20 David Perkins & James Holmes, "Multidomain Battle: Converging Concepts Toward a Joint Solution" *Joint Force Quarterly, January* 10, 2018, pp. 54-57.

21 翁予恆,〈淺析美國陸軍多領域作戰概念〉,青年日報,2017年4月8日,版10。

22 David Perkins, "Multi-Domain Battle: Joint Combined Arms Concept for the 21st Century", Assciation of the United Stated Army, November 14, 2016. https://www.ausa.org/articles/multi-domain-battle-joint-combined-arms.

　　首先，隨著冷戰結束、美軍戰略層級勢均力敵的對手消失。為了分享和平紅利，部分國防資源已被轉用至其它領域。1990 年代幾場規模有限的戰事中，精準遠距打擊系統是「紅花」，地面部隊僅被視為「綠葉」。其次，稍晚登場的全球反恐作戰，在曠日費時的 15 年後，逐漸轉變為綏靖作戰。由於面對不對稱及恐怖主義威脅，戰力防護是裝備整備與部隊訓練的當務之急；相對的，美軍潛在敵手則不斷觀摩分析美軍戰術戰法，在空中、地面、海上、太空及網路空間全力發展，如今已經在很多層面足堪與美軍分庭抗禮、甚至有所超前。依美陸軍訓準部司令 Perkins 與美空軍作戰司令部司令 Holmes 上將的說法，未來美軍的敵人，將擁有強大的整體防衛戰力，防空、遠距火力，以及更精準的情監偵系統、攻勢與守勢資訊戰力及電子戰、網路戰力等。「美軍欲確保賡作戰全程，對所有領域保有全面主宰性優勢，已經不可能」[23]。

　　多領域戰鬥概念的提升，就是充分利用美軍在既有人力素質與訓練上的優勢，將陸、海、空、太空、網路、電磁頻譜、資訊環境、戰爭認知等所有領域戰力，統整至可行的最低層級聯戰部隊，以短暫創造、運用「機會之窗」，陷敵於多重困境，奪取、維持及發揮主動權。此種聯合部隊有能力嚇阻敵人侵略、限制敵行動自由，確保己方進出無虞，贏得未來勝利。

　　由美軍對多領域戰鬥概念的詮釋，可以發現除了因應敵我雙方相對能力變化，而發展的作戰能力調整；最大的變化，是在原有的陸、海、空、太空、網路等五大聯合軍種確認的領域外，新增電磁頻譜、資訊環境、戰爭認知等新的對抗領域。其中的資訊環境，依美軍定義，主要包括網路設備等實體、訊息及使用者之認知層面；與最後一項戰爭認知領域，有某種程度的重疊。兩者所指涉的具體目標對象，其實都是人因領域（human dimensions）。

伍、資訊環境作戰

　　從這個角度切入，美海軍陸戰隊的發展速度，顯然又更為快速，並提出「空地特遣隊資訊環境作戰」（Marine Air Ground Task Force Information

---

23　David Perkins & James Holmes, "Multidomain Battle: Converging Concepts Toward a Joint Solution", p. 50.

Environment Operations）運用概念，預劃於 2025 年新編一支資訊大隊（MEF information Group, MIG），指揮官編階上校，並下轄作戰中心（Combat Operations Center, MIG COC），整合所有資訊作戰。[24]

依據美海軍陸戰隊說明，「空地特遣隊資訊環境作戰」係指：整合計畫，運用海軍陸戰隊空地特遣隊、美海軍、聯合及跨部門資訊能力、資源與活動，強化海軍陸戰隊單一戰場概念，並提供或支援防衛、攻擊、擴張（戰果）等做為，以在對抗性的資訊環境中運作、戰鬥並獲勝。

預劃至 2025 年，該大隊將完成編組與訓練，併用資訊與火力、機動的力量，在實體與認知層面奪得先機。其任務包括：協調、整合及運用資訊環境作戰能力，以確保在資訊環境中，海陸空地特遣隊指揮官協力友軍機動、阻絕敵軍行動自由之能力；並在支援海陸空地特遣隊作戰中，提供通信、情報、支援兵力聯絡及執法等能力，核心能力如圖 1。

分析美海軍陸戰隊所提資訊大隊七大核心能力，分別涵括了傳統通資電技術（確保指管暨關鍵系統、攻擊、利用網路系統）、研製具影響力內容（告知國內外閱聽眾、影響外國目標對象、欺騙外國目標對象），以及整合兩大領域後，產生的資訊環境戰場覺知（情搜）與控制資訊戰能力、資源需求，與近年資通產業發展之數位匯流趨勢，若合符節。

一般商業流域的數位匯流，主要指涉的是四種傳統上相對獨立的產業融合過程，分別是 IT 產業 (Information Technology)、電信產業 (Telecommunication)、消費性電子產業 (Consumer Electronics)、和娛樂產業 (Entertainment)，是一種依據市場需求而產生的跨產業融合。其中，數位科技 (digital technologies) 與內容的數位化 (digitalized content) 均為關鍵（維基百科 2015）。[25] 將前述美陸戰隊即將成立之資訊大隊核心能力，對照數位匯流概念，即可發現其理則基本相通，主要重點在於將可資運用的新資訊科技，與可對目標對象產生影響的內容，

---

24 US Marine Corps, "MCFC 5-5 MAGTF Information Environment Operations Concept of Employment," US Marine Corps Concepts & Programs, July 28, 2017, https://marinecorpsconceptsandprograms.com/concepts/mcfc-5-5-magtf-information-environment-operations-concept-employment.

25 〈數位匯流〉，維基百科，檢索於 2015 年 10 月 14 日，https://zh.wikipedia.org/wiki/%E6%95%B8%E4%BD%8D%E5%8C%AF%E6%B5%81。

有效結合。析言之，商業上跨產業融合的動力，是因應市場需求、也就是「人」的需求；多領域戰鬥強調「人因領域」，要求的也是如何透過科學化的方法，對我、敵、友方人員之認知領域產生影響，以堅定我方戰志、打擊敵人士氣、爭取友盟支持。

提供資訊環境戰場覺知	• 針對資訊環境之實體、資訊、認知面向進行分類，以辨識挑戰、機會以利空地特遣隊。
攻擊、利用網路、系統及資訊	• 在獲得授權後，利用或攻擊敵人之網路、系統、訊號及資訊，以利空地特遣隊。
告知國內、國際閱聽眾	• 告知國內及國際閱聽眾，以建立理解、爭取支持。
影響外國目標閱聽眾	• 獲得授權後，影響選定之外國閱聽眾，改變其決策及行為以利作戰目標。
欺騙外國目標閱聽眾	• 在獲得授權後，採取混淆、誤導、資源錯置及拖延行動等，以誤導敵決策者，揭露敵強度、傾向及未來意圖，並保護特遣隊之能力、整備、立場、意圖。
控制資訊作戰能力、資源及活動	• 指揮管制所屬海陸、海軍及聯合資訊資產，強化特遣隊在資訊環境之作業能力。

圖 1：資訊大隊核心功能

資料來源：US Marine Corps, "MCFC 5-5 MAGTF Information Environment Operations Concept of Employment," US Marine Corps Concepts & Programs, July 28, 2017, https://marinecorpsconceptsandprograms.com/concepts/mcfc-5-5-magtf-information-environment-operations-concept-employment.

陸、新世代戰爭

「多領域戰鬥」與「資訊環境」作戰概念的發展，都象徵了現代軍隊回應外在環境變化，必須在作戰方式上不得不然的調整。但如果從更宏觀的戰爭理論角度思考，因應人類社會型態的劇烈變化，尤其是邁入數位科技快速發展、全面滲透的網絡社會後，人類緊密連結的程度，與全球資訊流的生產與流動，產生了相互加乘的效應，勢將全面改寫戰爭的面貌。

一、戰爭暴力本質的變化

雖然很難在正式準則中，發現相關定義，但美國軍方近年來相當程度地使用「動能」（kinetic）與「非動能」（non-kinetic），形容不同作戰能力可能產生的效果。其中資訊與電子，相較於火砲、戰機等，都被歸屬於具備非動能效果的作戰能力。雖然在傳統上，在陸、海、空、太空等有形領域的戰果，通常顯而易見；對目標施予摧毀、消滅等動能效力，也比較易於評估，但現代戰爭中的非動能效力，勢將日益明顯。

造成此種轉變的主要原因，正是日益密切連結的新型態社會。著名歷史學家 Harari 分析近代人類社會中，戰爭已經不是常態的主要原因時，即指出戰爭成本大幅上升、利潤下降，而和平卻來越划算、全球政治文化亦愈趨愛好和平等，都是重要原因。[26] 其中尤其關鍵的，是國際間的網絡日漸緊密，使得多數國家都不再能全然獨立，突發戰爭的可能也就減少。

以同樣理則，分析未來戰爭，實體摧毀等暴力成分，亦將日益減少。因為密切連結的網絡社會，不論在政治、經濟、貿易或文化等面向，都將是「你中有我、我中有你」，甚至難分你我。戰爭做為迫使他人改變意志的手段，勢將更聚焦於認知等人因領域，非動能效果相關能力將更形重要。

二、戰爭不確定性減低

傳統戰爭中，「戰爭之霧」或戰場上的不確定性，永遠不可能被移除。但

---

26 Harari Noah Yuval 著，林俊宏譯，《人類大歷史：從野獸到扮演上帝》（*A Brief History of Humankind*）（台北：遠見天下，2014 年），頁 415-420。

隨著科技的發展，從 1812 年克勞塞維茨戰爭論出版，到二百年後的全球反恐戰爭，軍隊掌握戰場覺知的能力，就已經產生翻天覆地的變化。

時至今日，巨量、跨界的資訊流，正成幾何級數成長，讓人類不斷登上資訊流通的新高峰。當個體隱私都將因而受到威脅時，牽涉大規模人類活動的戰場，勢將更加透明化。尤其是對於旗鼓相當的對手，在實體技術層面遮蔽己身、而又力求透視敵手的能量差距不大，唯一的變數，即在於決策者如何詮釋所獲資訊、形成自己相信的認知。

三、大量資訊對認知改變的影響

也由於處於緊密連結、資訊大量且深度流通的環境，欲改變目標體之認知、態度乃至形塑行為，難度也將降低。尤其是該等資訊包括了客觀環境、主觀形勢，以及論述文本時，都將讓「不戰而屈人之兵」從傳統的戰爭典範，回復成為具體可及的目標。

析言之，包括戰車、船艦、戰機等動能戰力，必須與資訊、文宣、心戰、新聞等「非動能戰力」相結合，乘機造勢、「形塑」有利的資訊環境。在此一戰場上，駭客高手必須與文宣專家、心戰能人乃至媒體公關整合，於埋木馬、放病毒同時，也必須生產數位化、具影響力的內容 - 可能是軟語溫馨，也可能酸言嘲諷；可能是以假亂真的圖像，也可能是感人肺腑的影片…這些「非動能火力」，將對人類認知領域產生影響，進而改變敵手的決策。

由此引申，本研究認為，多領域戰鬥強調之陸、海、空、太空、網路、電磁頻譜、資訊環境、戰爭認知等八個面向，在人類社會益趨緊密連結、跨域資訊流持續快速成長的情況下，將產生本質上的改變。戰爭認知做為人因領域的首要因子，將由爭鬥領域之一、轉變為主要的目標。「奪三軍氣」「迷將軍心」是獲取戰爭勝利的必經通道。讓敵人深信自己已遭擊敗、而非在實體上消滅對方不但可行，也將是戰爭中的常態。

Taking a Neorealist Approach on Sino-Japanese Relations 2012-2016, Lessons for Taiwan's Relation with China

Fernando Mauro Sambrani Cavalieri *

I. Introduction

One of the great changes we are experiencing in our times is the unstoppable rise of China that threatens put an end to almost 500 years of Western domination. Chinese numbers are truly stunning; China is the 2nd biggest economy on Earth as well as being the main exporting countries on Earth not to mention it has the biggest reserves too. This economic power enabled China to improve its military capacities and consequently the role it can play in international relations. Speaking of it, the Chinese desire of playing a bigger role not only comes as a consequence of an immense hard power, but also as a desire to reshape the current international order. The current order was not established by China nor the rules of it, therefore following the neo-realist theory this sum of an international order where China cannot use its full potential plus the domestic conditions of an increasing hard power can drive the foreign policy into a more assertive way.

The flipside of the coin to the rise of China is the current status of Japan. Too early to say that the country is in a bad situation, yet it is true that the country of the Rising Sun is far from its peak in the middle'80. Even though that the Japanese economy is growing a little bit, it is not what it once was, what is more its size is way behind China. Nevertheless, Japan has been improving its military capacities by acquiring new equipment. Also, it has been able to partially review its Constitution allowing the so called collective defense. From a Japanese perspective, the changing international order that is slowly being challenged by China as it wants to redefine it, putting pressure on the Japanese policy makers to counter-balance it.

* Graduate Student, International Master's Program in Asia-Pacific studies, National Chengchi University

The aim of this paper, is to analyze Sino-Japanese relations from 2012-2016 from a neo-realist approach, to understand how Japan has been counter-balancing the rise of China, after understanding I will try to extract implications and policy recommendations for Taiwan when dealing with China.

First of all, I believe that this topic that I will research can provide useful advice as well as recommendations for Taiwanese policy makers for dealing with China. Also it is interesting to study and analyze that despite Japan is far from its heyday, it can still counterbalance China's rise. I think that Taiwan may learn some lessons from here and apply them according to the Taiwanese reality, Japan as Taiwan has a much smaller hard power than China so dealing with this giant has to be carefully planned.

II. Paper Methodology

First of all, I will read available literature on neo-realism to have an insight, give a definition of it and explain why It is appropriate for this analysis.

By reading available literature, I will give an explanation of what does the rise of China means. How it can be said that China is Rising? Which are the consequences for the International System and the Regions.

Thirdly by using Japanese Government reports as well as available literature I will describe the Japanese counter-balance to China and its outcomes.

Finally, after establishing the consequences of the Japanese counter-balance I will give the policy implications for Taiwan according to its own national reality.

III. Neo-Realism

Following John Mearsheimer thought, the International order can be described as anarchic, meaning that there is no central authority to rule the States. No State is safe from the aggression of other state as all States have at least the potential to do some damage with its military force. In an event of an aggression, there is no one to ask for help, the State can only rely on itself, in other words States have to self-help themselves if they want to survive. Being powerful enough means being able to survive, therefore States seek power. This quest for power is triggered by the International System, not by evil human nature. States will increase its economic and military power as much as they can to have access to a better position in the International Order.[1]

But how much amount of power can should a State seek to obtain? That depends on the perspective we prefer. From a defensive approach having too much power can be actually dangerous as other states can feel it as a threat, their reaction can be from seeking a balance to putting an end to the threat. From an offensive approach States should have as much power as possible in order to become hegemons. By being a hegemon the surrounding states are less likely to challenge them.[2]

As I previously stated, the way foreign policy is conducted by a country does not entirely respond to the position it has in the international system, rather it also depends on its material conditions. A country may consider that the position that it has in the system is unfair, that according to its view it should be greater, but that view by itself cannot start a *revisionist* foreign policy. Apart from that feeling of unfairness that triggers the revisionist desires, a country requires material conditions, its hard power a robust economy and a strong army. Without them any desire of change will remain nothing more than a desire. To be more precise, a solid economy is the base of any potential revisionist foreign policy, without it is doomed.

To sum up, material conditions as well the awareness and confidence of the country's elite on them are vital to response to the systemic pressure, that response will try to reduce the threat or seek a better position in the international system.

---

1 John J. Mearsheimer, "Structural Realism," in Tim Dunne, ed., *International Relations Theories: Discipline and Diversity* (New York, USA: Oxford University Press, 2007), pp. 72-74.
2 Ibid., pp. 75-78.

In the recent years, China has been experiencing an impressive growth of its economy becoming the second largest economy on earth.[3] This robust economy has allowed China to have the 2nd biggest budget on military defense[4], in other words Chinese hard power has become extremely powerful, the question is what does China wants with its power. Will it seek to secure its position from any rival or will it seek to become a hegemon? Personally, I agree with the offensive realism view, China is increasing its hard power in order to become a hegemon in the long distance future. It will try to emulate the what the United States did during the 19th Century expelling European powers from its sphere of influence, only difference is that China will try to expel Americans from Asia.[5] As a consequence of Chinese targets on international relations, neighboring countries have been actively trying to increase its hard power in order to balance Chinese growing power.

The case of Japan is a more complex situation, the deterioration of the economy (even though today the economy is slowly growing it is far away from its glorious past)[6] has brought some issues, nevertheless the country keeps on upgrading its military equipment.[7] The reason for this big military spending in an economic situation far from good may be explained by the feeling of threat that Japanese elites are feeling, Chinese threat. Japan perceives that the increasing Chinese hard power might put in a bad situation. Actions taken by China in Japanese territory are a major concern, providing that they are successful they can deprive Japan from natural resources and its territorial integrity. In that event, Japanese status in the international order would be seriously diminished, making Japan a minor role State. Therefore, Japanese increase in its hard power as well

---

3 "The World Bank In China," The World Bank Website, April 19, 2018, http://www.worldbank.org/en/country/china/overview.

4 Brad Lendon, "China boost military spending 8% amidst ambitious modernization drive," *CNN*, March 6, 2018, https://edition.cnn.com/2018/03/04/asia/chinese-military-budget-intl/index.html.

5 John J. Mearsheimer, "Structural Realism," in Tim Dunne, ed., *International Relations Theories: Discipline and Diversity*, pp. 83-84.

6 "Japan", World Bank Website, https://data.worldbank.org/country/Japan.

7 Tim Kelly, Nobuhiro Kubo, "Japan Approves Record Defense Spending that Favors US made equipment," Reuters, December 22, 2017, https://www.reuters.com/article/us-japan-defence/japan-approves-record-defense-spending-that-favors-u-s-made-equipment-idUSKBN1EG081.

as in its bilateral relation with China from 2012 to 2016 can be seen as a way trying to preserve the status quo. What I mean, is that Japan is trying to increase its power and deal with China not in a way of seeking hegemony or diminishing Chinese place in the international order, but as a way of keeping its current position and influence in the world.

A rising power that is confident about itself and might soon challenge the existing international and at the same time a declining major regional power that see in its neighbor a potential threat surely is not the best combination. But, what do I mean when I say that China is a rising power?

IV. China's Rise

To answer the question that I have previously made, it is necessary first to make a short trip back in time. When the People's Republic of China was established in 1949, with Mao Zedong as it supreme leader, the country shut itself, having few ties around the World. Mao firmly believed in self-reliance, so nothing was needed from overseas, making no point on participating in international organizations or making alliances, being North Korea the only ally China had for a long time. China did not participate in international organizations, as the traditional approach was skeptic of them.[8] In the economic realm the country went through an accelerated socialist transition until communism was finally applied. The disastrous policies of the great leap forward as well as the cultural revolution made Chinese miserable, while weakening the country. During that time, the living standards of the Chinese people were among the lowest in the world, the economy was weak and small in size.

Regarding the military, despite of its large number of soldiers, around 5,000,000 the People's Liberation Army proved to be vulnerable and unreliable when facing a strong opponent like Vietnam. What it was meant to be fast operation to "Punish the Vietnamese" was in reality a disaster leaving 150,000

---

8 Bates Gill, *Rising Star: China's New Security Diplomacy* (Washington, USA: Brookings Institution Press, 2007), pp. 21-29.

casualties for the Chinese after a few months of combat. In short, the country that Mao left, soon to be commanded by Deng Xiaoping, was weak and vulnerable while having a miserable economy with even more miserable people.

Having Deng Xiaoping as the most powerful man in China, meant difficult but necessary reforms. Strongly believing that a major war world will not happen, the PLA stop being a priority what meant firing 1,000,000 million soldiers as well as taking out the PLA from the list of priorities. Deng could devote himself to reforming the country's economy, attracting foreign direct investment by giving incentives as well as making use of China's cheap labor force.

The success of this reforms are well known, the economy grew in an unstoppable way, surpassing in size and speed previous "Asian economic miracles". From 2010 China become the 2nd largest economy in the World, being on top of Japan that used to be number two for more than 40 years. By 2014, China made another milestone when it became the first economy in PPP terms. Today, the Chinese GDP is around 11.2 trillion USD dollars an impressive contrast to the situation in the mid 1970 when the GDP was below 180 billion USD dollars. Not only does China have the second biggest economy size, it also has the biggest foreign reserve, being up to 3.134 trillion USD dollars.[9]

The reforms that started in the late '70 made China an economic powerhouse. Deng's emphasis on economic transformation by foreign investment meant that thousands of factories relocated in China. From steel to toys, from cars to clothes China was the place for manufacturing any kind of product at a very low cost. The label "Made in China" become famous worldwide, Chinese exports were available globally.

The Asian giant quickly begun trading with almost all countries on Earth, as not only did China sell its industrial products it also required raw materials for making them. As Chinese economy keep growing the thirst for raw material grew larger as more products needed to be manufactured, slowly China started to be

---

9 "China," World Bank Website, https://data.worldbank.org/country/China.

the first or second largest trading partner for many countries.[10]

Chinese economic might has enabled the country to have a loud voice on international affairs. Economic ties can influence countries foreign policies, China is well aware of this, trying to boost its image by making others countries rich. For being this generous, China gains much needed political influence that China is seeking. Its powerful economic links can eventually compel countries to become closer to China.[11]

Since the last years of the 20th Century, China begun spending considerable amounts on the modernization of its troops. Today it is ranked second after the USA when it comes to military budget, what is around 131 billion US dollars. It is hard to tell if this is the real number, as many reports have argued that Chinese military spending goes beyond that number. During these years China has been focusing in transforming the People's Liberation Army by acquiring new military equipment, improve soldiers training in order to make them highly professionals as well as having a new mindset for the PLA. A mindset according to China's new political aspirations. Having the PLA modernized will make harder a possible Taiwan independence, as well as giving China a greater role in East Asia as major military power.[12]

On the international relations field, Deng Xiaoping originally strategy was keeping a low profile, in order to focus in the economic development of the country. Basically consisted on no contradicting the biggest power, the USA, discarding any type of leading role, seeking no expansion as well as seeking no hegemony. This policy stopped being applied, on the late 2000' and slowly being replaced for a more assertive one, called "Striving for achievement" by Chinese scholars. This new policy is focus on trying to make China a leader in the

---

10 "Is China the World's top trader?" China Power, January 25, 2016, https://chinapower.csis.org/trade-partner/.

11 Phillip C. Saunders, "China's Role in Asia Attractive or Assertive," in David L. Shambaugh, ed., *International Relations of Asia* (Maryland, USA: Rowman and Littlefield, 2014), pp. 147-172; Yan Xuetong, "From Keeping a Low Profile to Striving for Achievement," *The Chinese Journal of International Politics*, Vol. 7, No. 2, June 2014, pp. 153-184.

12 Phillip C. Saunders, "China's Role in Asia Attractive or Assertive," pp. 147-172.

International Order, that leadership will be obtained by making countries enjoy economic prosperity. China will seek expansion, sizing territories from unfriendly countries, like Japan is part of the grand strategy that will make China the new hegemon in the future.[13]

Coming back to the original question, what do I understand for Rise of China? In my own view, the consequence of an extraordinary economic transformation that not only changed Chinese people's lives but also how China see its position that it has in the international order. Chinese elites have become aware of their economic might as well as its growing military capacities. This awareness that Chinese elites have might lead to some changes in the international order, as they seek to become by 2050 the main superpower on Earth. By that year China should also have reunified with Taiwan, putting an end to the humiliation of having its territory separated. To sum up, China's Rise is the realization by the Chinese that the current international order is not good enough for them, that now that China's hard power is strong changes should be made on it according to them.

V. Japan's Perception of China through Time

The great transformations that China has been undergoing in the last years were closely watched by one of its closest neighbors, Japan. Before introducing the Japanese policy to counter balance China's Rise, it is best to know how Japan has seen China since 2012 until 2016. In order to do so I will use the Defense of Japan (DOJ). DOJ is a white paper written annually by the Japanese Ministry of Defense. Written with an everyday language can be understood by anyone. In this paper, the reader will find the official views of the Japanese government regarding security issues, international affairs as well as Japanese defense policy.[14]

---

13 Yan Xuetong, "From Keeping a Low Profile to Striving for Achievement," *The Chinese Journal of International Politics*, pp. 153-184.

14 Takashi Yamazaki, "Japan's Geopolitical Vision and Practices on the Indian Ocean," paper presented at *22nd International Political Science Association World Congress* (Madrid, Spain: International Political Science Association, July 12, 2012), p. 3.

Acknowledging Chinese economic might the 2012 DOJ mentions Chinese economic power, adding that it has allowed China a bigger spending in its military budget. The DOJ calls for a bigger participation of China in the international order, as it has the economic means to do so. Regarding the Chinese military, the DOJ makes a big critique to the Chinese authorities saying that the military budget that is shown does not reflect the total amount of money spent. What is more the DOJ also critics China's modernization of the its defense capacities as it is does not have a clear purpose for doing it. In this white paper, the Japanese Government believes that, the modernization of the Chinese navy has more than one purpose: being a deterrent to any possible Taiwanese independence, securing sea lanes, having access to the second chains of Islands in the Western Pacific as well as having a material back up for the legal claims that China has in the South China Sea. Finally, it is worth to mention that the DOJ reports and condemns Chinese espionage ships near Japanese waters, as well as reporting the entrance of Chinese fishing vessels near the Senkaku Islands. [15]

One great change introduced by the 2013 DOJ is acknowledging China's greater involvement in peace operations, by stating that the communist country has sent personnel to United Nations peace operations as well as the vessels China has dispatched to combat piracy on Africa. Nevertheless, the paper encourages China to follow the international norms, being a responsible power and settle disputes with others countries in a reasonable way. The DOJ still mentions previous incidents where Chinese fisheries law enforcement ships and fishing boats entered Japanese waters. [16]

The 2014 DOJ continues with the idea of a more influential China as a consequence of its larger hard power, the Japanese government keeps asking China to be a responsible power. What is new in this edition, is that the DOJ has a subtitle that talks about violation of Japanese air space by Chinese aircraft

---

15 Japan Ministry of Defense, *Defense or Japan Report*, Vol. 36, 2012, pp. 26-47, http://www.mod.go.jp/e/publ/w_paper/2012.html.
16 Japan Ministry of Defense, *Defense of Japan Report*, Vol. 37, 2013, pp. 30-49, http://www.mod.go.jp/e/publ/w_paper/2013.html.

in addition to the violation of Japanese waters by Chinese vessels. The DOJ is deeply concerned about these issues and it is fearful of a confrontation as a consequence of a mistake done by a Chinese vessel or plane.[17]

On 2015, the DOJ makes a skeptical reference to Chinese peaceful rise saying that while China is saying this in reality, when it comes to deal with disagreements with other countries, is using a non-peaceful bullying method. What is more the DOJ reports Chinese activities in South China Sea, condemning the building of islands in South China Sea by China. Furthermore, the reports criticize Chinese violation of Japanese waters and air space, reporting that bigger ships are entering Japanese waters. Nevertheless, the Japanese government is pleased to know that China is taking some small steps to prevent a major confrontation like signing the Code for Unplanned encounters at Sea with the United States and Japan. What is more China begun to have consultations with Japan to implement the Maritime and Air Communication Mechanism (MACM).[18]

The DOJ of 2016 deepens it concerns over Chinese activities in Japanese waters, especially because on 2015 a Chinese navy ship entered Japanese waters, also the report says that fishing vessels are now carrying cannons. Chinese ambition is not limited to Japanese waters; the reports warns the acceleration of the construction of Chinese islands in South China Sea and the assertive policy that China has been carrying out against claimants in the area. What is more, the report has serious concerns of Chinese moves in Africa Djibouti where the Chinese navy has a port as well as Chinese activities in the Indian Ocean where China has been conducting operations. Nevertheless, Japan remarks that China keeps the consultations for implementing the MACM.[19]

---

17 Japan Ministry of Defense, *Defense of Japan Report*, Vol. 38, 2014, pp. 32-52, http://www.mod.go.jp/e/publ/w_paper/2014.html.

18 Japan Ministry of Defense, *Defense of Japan Report*, Vol. 39, 2015, pp. 33-56, http://www.mod.go.jp/e/publ/w_paper/2015.html.

19 Japan Ministry of Defense, *Defense of Japan Report*, Vol. 40, 2016, pp. 41-70, http://www.mod.go.jp/e/publ/w_paper/2016.html.

From a neorealist approach, Japan understand that the Chinese giant increase of its hard power, booming economy and robust military, which keep growing in size every year at a fast speed are the forces behind its assertive foreign policy. However, since 2014 Japan is well aware that an open confrontation with China will not bring any kind of benefit for both parties, after having a tough line on China during 2012 and 2013, that is why it suggest China should engage with the international community, by being a responsible power. Moreover, Japan is pleased when China tries to dialogue for the bilateral friction due to the Chinese activities in Japanese water and air space. In line with this neorealist approach, Japan knows that strengthening its army and its economy is vital in order to keep its place in the international order and balance against China.

VI. Counter Balance Policy

Perhaps, in the near future, 2012 will be remembered as one of the toughest years' in the bilateral relationship of China-Japan. In the previous year's Japan already acknowledged China's growing material power while at the same time asking China to be a responsible power. What is more Japan has been reporting illegal Chinese activities on Japanese waters.[20] It can be said that this continuous enlargement of Chinese hard power made Japan anxious as a consequence of Chinese actions that were previously mentioned, therefore fearing a change of the status quo in the bilateral relationship and the international system Japan took a controversial decision, nationalizing Senkaku Islands. The islands claimed by Taiwan and China are under Japanese control since the Americans returned Okinawa prefecture to Japan in the early 1970'. Japan does not recognize the claims made by Taiwan and China, stating that the Senkaku islands are Japanese territory. Until 2012 the islands were owned by a Japanese citizen.

The official Statement made by the Japanese Government stated that it just acquired the islands from a private citizen. According to the press the main reason for doing this was to prevent the far right Japanese politician, Shintaro

---

20 Japan Ministry of Defense, *Defense or Japan Report*, Vol. 35, 2011, pp. 72-92, http://www.mod.go.jp/e/publ/w_paper/2011.html.

Ishihara from buying them[21]. Not only did the Japanese government prevented Ishihara from buying them, it was also a political message to the Chinese. The message can be understood as despite that China is increasing its hard power and it has been able to become a major power, at the same time that Japan is far from its heyday, Japan will not allow a change in the status quo in spite of Chinese pressures.

Not only Japanese actions did not stop Chinese violation of Japanese space, even worse it triggered a national outrage across China. Demonstrations and lootings against Japanese property in China took place across the Chinese territory.[22] Ordinary Chinese citizens were protesting heavily against the Japanese " bully", that offended and humiliated China.[23]Not only did the Chinese people reacted against the Japanese measure, Chinese government protested greatly and designed a counter offensive, which included renaming the claimed territory and creating new borders.[24] Chinese reaction can be understood in the sense that as a rising power with a strong hard power it cannot be intimidated by actions that tend to create a roof for its aspirations, what is more China in its rising position has to show to the World its real power.

As it was previously mentioned, rather than deescalating the tensions the Japanese nationalization of the Senkaku islands made things worse. Starting 2013, Chinese vessels entered Japanese waters much more often than they used to do it in the past and stayed for longer periods of time, staying up to 28 hours in one occasion.[25] During that year harsh words were exchanged from both sides

---

21 Jane Perlez, "China Accuses Japan of Stealing Disputed Islands," *New York Times*, September 11, 2012, https://www.nytimes.com/2012/09/12/world/asia/china-accuses-japan-of-stealing-disputed-islands.html.

22 Patti Waldemir, Rahul Jacob, Michiyo Nakamoto, "Anti-Japan Protests Spread in China," *Financial Times*, September 16, 2012, https://www.ft.com/content/3d69959e-ffce-11e1-a30e-00144feabdc0.

23 Ian Johnson, Thom Shanker, "Beijing Mixes Messages Over Anti-Japan Protests," *New York Times*, September 16, 2012, https://www.nytimes.com/2012/09/17/world/asia/anti-japanese-protests-over-disputed-islands-continue-in-china.html.

24 Ministry of Foreign Affairs of Japan, *Diplomatic Bluebook*, Vol. 42, 2013, pp. 10-13, http://www.mofa.go.jp/policy/other/bluebook/2014/html/index.html.

25 Ministry of Foreign Affairs of Japan, *Diplomatic Bluebook*, Vol. 43, 2014, p. 12, http://www.mofa.go.jp/fp/pp/page22e_000566.html.

and the idea of a coming war was in the head of decision makers.[26] The tough policy of counter balancing China raise without any other soft measure proved to inadequate, instead of making China more cautious it triggered a hard response. To make things worse on that same year China established an air-defense identification zone on the territory that claims from Japan, which covers the Senkaku Islands. The Chinese government requested that any aircraft that flies through that space has to identify itself as well as keep contact with Chinese authorities while it is in there. The consequence for refusing is facing emergency defensive measures.[27] After this announcement was made, the Japanese reaction was to condemn the Chinese action, indicating that this was of no benefit to the bilateral situation.[28]

Even though 2013, was a bad year for the bilateral relationship, it is important to mention that Japan tried to cool the situation by proposing a joint Maritime Communication Mechanism with China with the idea of avoiding a confrontation in case of some incident take part in the sea. From here onwards it can be seen that despite trying to keep the status quo, Japan is also willing to cooperate with China when possible. It is a wise decision as hard power gap keeps increasing despite Japanese policies.

Situation started to look brighter since 2014, both countries slowly realized the importance of keeping a healthy bilateral relationship while having their differences regarding territorial issues.[29] High Officials exchanges were reassumed, the foreign ministers from both countries met and paved the way for the Summit between the countries leaders in that year. The outcome of this

26 J. Michael Cole, "Japan Explores War Scenarios with China," *The Diplomat*, January 9, 2013, https:// thediplomat.com/2013/01/japan-explores-war-scenarios-with-china/; Jeff Kingston, "War of Words heats up China, Japan Tensions," *CNN*, November 1, 2013, https://edition.cnn.com/2013/10/30/opinion/japan-china-spat-kingston/index.html.

27 "China establishes 'air-defence zone' over East China Sea," *BBC News*, November 23, 2013, http://www. bbc.com/news/world-asia-25062525.

28 Japan, Ministry of Foreign Affairs, *Diplomatic Bluebook*, Vol. 43, 2014, p. 12.

29 Shanon Tiezzi, "A China-Japan Breakthrough: A Primer on Their 4 Point Consensus," *The Diplomat*, November 7, 2014, https://thediplomat.com/2014/11/a-china-japan-breakthrough-a-primer-on-their-4-point-consensus/.

substantial progress was the 4-point consensus: "Toward the improvement of the Japan-China relations, quiet discussions have been held between the Governments of Japan and China. Both sides have come to share views on the following points:

1. Both sides confirmed that they would observe the principles and spirit of the four basic documents between Japan and China and that they would continue to develop a mutually beneficial relationship based on common strategic interests.

2. Both sides shared some recognition that, following the spirit of squarely facing history and advancing toward the future, they would overcome political difficulties that affect their bilateral relations.

3. Both sides recognized that they had different views as to the emergence of tense situations in recent years in the waters of the East China Sea, including those around the Senkaku Islands, and shared the view that, through dialogue and consultation, they would prevent the deterioration of the situation, establish a crisis management mechanism and avert the rise of unforeseen circumstances.

4. Both sides shared the view that, by utilizing various multilateral and bilateral channels, they would gradually resume dialogue in political, diplomatic and security fields and make an effort to build a political relationship of mutual trust."[30]

Out of this 4 points I would like to highlight the last three. The second one, tries to make Japan acknowledge its Imperial Past and the consequences it had on China. Forgetting history or rewriting is not acceptable, by understanding the past both countries should not repeat old mistakes and work together for a better future. The third point, acknowledges the bilateral issue of Senkaku island and tries to bring a peaceful resolution to it. Both countries have realized that despite its hard power gap, a confrontation or a solution that moves away from the peaceful resolution brings no benefit to no one. Finally, the last point emphasis

---

30 Ministry of Foreign Affairs of Japan, Regarding Discussions Toward Improving Japan-China Relations, Japan-China Relations, November 7, 2014, http://www.mofa.go.jp/a_o/c_m1/cn/page4e_000150.html.

the use of dialogue to build mutual trust, which is key to avoid confrontation and conflicts.

The four-points consensus was agreed by Prime Minister Abe and President Xi defusing the previously investable bilateral relationship. What is more, from the Japanese perspective even though the fear of a potential change in the status quo still remains Japan welcomes China, and highlight the importance of both countries for keeping the peace and stability of the region. [31] Keeping a neorealist approach, Japan has understood that while keeping the status quo in the international order is desirable, the country can make its strategy according to its own material power taking into account the rival's material power. In other words, it is important to keep things the way they are, but in this case as the power gap between Japan and China keeps getting bigger makes sense that Japan seeks to cooperate with China in order to retain its position.

Having a reactivation of high officials exchanges as well as signing the four-points consensus on 2014, made that things went better on 2015. After almost three years both countries could have finance talks, which is a key aspect of the relationship, Japan is China second biggest partner where China is Japan biggest partner. Officials from both sides pledge for a better communication and cooperation between the two countries.[32]

On the security realm, the countries had it first talk in almost four years. Even though the Maritime Communication Mechanism could not be signed, it was a great progress that the countries could have this type of meeting.[33]

That warm relation made possible a new meeting between Abe and Xi as well as Abe and Chinese Primer Minister Li Keqiang for further improvement of the bilateral relationship. Even though, disagreement regarding Senkaku Islands

31 Ministry of Foreign Affairs of Japan, *Diplomatic Bluebook*, Vol. 44, 2015, pp. 27-28, http://www.mod.go.jp/e/publ/w_paper/2015.html.
32 Shanon Tiezzi, "The Latest Sign of a China-Japan Thaw," *The Diplomat*, June 11, 2015, https://thediplomat.com/2015/06/the-latest-sign-of-a-china-japan-thaw-2/.
33 Shanon Tiezzi, "Japan, China Test the water with Restarted Security Talks," *The Diplomat*, March 20, 2015, https://thediplomat.com/2015/03/japan-china-test-the-water-with-restarted-security-talks/.

still persists, Japan as well as China are willing to solve this issue peacefully. A sharp contrast from 2012 and 2013 where the idea of an incoming war was imminent.

As I previously stated Japan become more interested in having this friendlier approach as the tough line on China made things worse as well as the disparity of material power in the event of a confrontation can make a deep change in Japanese position in the international order.

The last year that is going to be analyzed in this paper, 2016, showed that some trends continue. Tensions and reports of Chinese illegal activities on Japanese waters continued, nevertheless both parties tried their best to resolve this issue in a peaceful way, avoiding a diplomatic escalation. It is interesting to mention that both parties have been continuously improving their military capacities with China spending in 2016 around 147 Billion US dollars for its military budget, even though the Japanese government claims that this number is largely behind the real number, while Japan spent roughly a quarter of the Chinese budget around 41.6 Billion US Dollars.[34] Both government are highly suspicious of each other military expenditure, fearing that it will be used against them. It is an interesting paradox, the stronger you get the more insecure you become, as you believe your rivals will try to match your capacities. Therefore, it is extremely important for both sides to keep the dialogue in order to prevent a major confrontation.

Regarding the high official exchanges, they continued the trend of meetings, Primer Minister Abe met with Premier Li Keqiang, later he also met with Xi Jiping when he went to China as part of the G-20 Summit.[35] These interesting meetings will have a positive outcome for 2017, when Abe visited by surprise the Chinese embassy in Tokyo a gesture that China appreciated, while talks begun for a possible visit of Xi Jiping to Japan during 2018.

---

34 Japan Ministry of Defense, *Defense or Japan Report*, Vol. 40, 2016, pp. 41-70.
35 Ministry of Foreign Affairs of Japan,," *Diplomatic Bluebook*, Vol. 46, 2017, pp. 31-32, http://www.mofa. go.jp/files/000290287.pdf.

VII. Conclusion

In this brief paper, i found out that constrains of the international order play a central role in the Japan-China relation. For Japan, it has acknowledged the great power gap between China and itself. Japan clearly understands now that the economy size difference is immense between the two countries, what in return has made possible for China spending almost 4 times more than Japan in its military. This great spending in the military has made Japan feel very anxious about it, greatly worrying of a potential revisionist China that seeks to change the status quo, making Japan an insignificant player that also has to lose part of its territory. It should not be a surprise then that not only did Japan improve its hard power, but also initially, showed a hard line on China by nationalizing the Senkaku Islands as well as considering war on China. First message to China was quite clear, do not pressure us, do not engage on illegal activities on our territory as we will sharply respond. In other words, respect the status quo, yes we understand that you are a rising power, but you have to respect the current international order. As things look pretty ugly, Japan was becoming aware that just having a tough line on China as well as potential war could be terrible for both sides, that is why Japan offered an olive branch to China. Even though it is not accurate to say that the bilateral relationship is friendly, it can certainly be said that it is better than it used to be. The high level officials exchange the 4-point consensus are signals of a better relationship.

My prediction for the future is that, it is highly possible that both parties keep improving its hard power as both need it for its ambitious goal on the international order, consequently the situation of Senkaku Islands can be tense at sometimes. Nevertheless, I see as remote the idea of war, both parties have realized how destructive it could be and how much damage it can inflict to the region. What is more, both Japan and China are making efforts not to clash in there. Finally, I will add that there is a possibility of a better bilateral relationship, but the suspicious of the other party will remain.

So what can Taiwan learn from this case? Well, keeping the neorealist approach the importance of having a strong hard power in order to keep to date with the constraints of the international order. Investing in improving your hard power it is a wise decision as it can make think twice before trying to change the status quo to whoever that is willing to do so.

Secondly, Taiwan should also learn that having a hard line on China can be utterly risky, especially, when observing the great power gap between China and Taiwan. Having this great difference in economic size as well as military equipment, makes Taiwanese choices very few. Personally, I think Taiwan should try to keep the status quo, by keeping a low profile while having a limited economic engagement with China. At the same time, is vital for Taiwan to improve its hard power, reinforcing its military is a must. A well-armed Taiwan, might be a hard target for any Chinese attack making the chances of an actual attack being considerably reduced. At the same time, a Taiwan that does keep a low profile, while not irritating Beijing has even fewer chances of being attacked.

International order constraints, driven by a power hungry State can be frightening to its members, especially when this power hungry state has a considerable hard power. Fear not, have a calm policy while at the same time improve your material power. When this power hungry state, understands that your hard power is big enough to inflict a great damage it will probably think greatly before taking an ambitious move that tries to change the status quo and replace it with one created by it rules.

台灣國家安全戰略

台灣海洋戰略的檢視與前瞻：
總體戰略的觀點

施正權 *

壹、前言

海洋是人類共同擁有的天然資產，也提供了食物與生計。由於陸上資源日益匱乏，占地球表面積 71% 的海洋，遂逐漸成為全球競爭的焦點。[1] 為滿足人類生存與發展所需，「走向海洋」似乎已是未來的趨勢。為此，世界各大陸的濱海國家與海島國家紛紛制定新的海洋政策，擴大對海洋的開發與利用與安全維護。

但是，海洋亦是複雜的力量競爭場域。基於對本國利益的考量與主權的維護，各國海洋政策大都採取排他性的規畫，難免在海洋交錯之際產生爭端。這些爭端並未因《聯合國海洋法公約》（United Nations Convention on the Law of the Sea, UNCLOS）問世而止歇，反而因國際政治的妥協性所訂定的模糊條文，讓各國有自行解釋的空間，讓海域爭端持續存在。

在此種情勢下，為滿足本身利益以及維護海洋方面的權利與權力，國家必須要綜合運用國家力量。而這種力量的運用，又難免與其他國家產生衝撞，因此有讓衝突上升的可能。為了避免誤判情勢、綜合運用國家力量來維護海洋利益，以戰略思維探究海洋利益的維護就有其必要性，因此海洋戰略成為各國必須要審慎思考的問題。

然而，學界對於海洋戰略內涵論述莫衷一是，大部分的論者認為海洋戰略只是針對海洋安全的回應，因此僅將海洋戰略視為維護海洋安全的戰略，探究的面向是海洋方面的衝突與國家軍事力量在強制方面的運用，而忽略了海洋合作的可能性，未將海洋戰略思維擴及於海洋外交、經濟、環境保育的面向。此

173

一將海洋戰略侷限於運用海洋軍事權力控制海洋、維護利益的傳統戰略思維，相對於當前「走向海洋」的大趨勢，實在有必要重新檢視。

法國戰略家薄富爾將軍（Andre Beaufre,1902-1975）[2]指出，不同社會背景的國家，往往會有不同的行動選擇。例如，以農業為基礎的國家會重視土地、人口與牲畜；工業型國家則重視原物料的獲得與貨品的市場；貿易型的國家重視市場、國際貨幣穩定及貿易交通航道的安全。這些說明了各個國家會依照其國家環境的特性，採取不同的行動來面對威脅。[3]台灣四面環海，是典型的海洋社會，近年也以「海洋國家」自居，自然必須要關注來自於海洋方面的問題。

台灣地處亞洲主要航道的關鍵位置，可箝制韓國、日本、中國大陸北方各港往南向印度洋航道，因此在地緣上成為各國關注的焦點。同時，因為與經濟海域重疊，台灣與許多國家有海界劃分的紛爭，海洋環境並非風平浪靜。但是，以我國現有國家總體力量而言，台灣在東北亞並非強國，過去基於國際現實的限制，對於海洋事務採取保守態度，也因此讓自身的行動自由受到諸多限制。如果台灣真的想成為「海洋國家」，除了在海洋政策的擘畫與開放上必須要加緊腳步，對於執行政策的海洋戰略也必須要深入思考。

但是，海洋議題具有其複雜性，涉及國家權力運用與經濟、環保、漁業、交通、科技等各方面，因此在擬定海洋戰略時，要考慮到每一個面向，以免顧此失彼。而擬定每一個面向的戰略時，也需要考慮彼此之間的相互協調配合，才能採取一致的行動，達成共同的目標。因此在各個領域的戰略之上應有一個整體的思維，才指導各個戰略分工與行動。也由於海洋戰略所應關注問題的面向非僅限於軍事方面，所以在規劃海洋戰略時，應該將所有的國家力量放進權力工具箱之中，包含最暴力與最溫和的。這樣的戰略思維，是薄富爾的總體戰略（total strategy）思想的核心概念。所以，採取總體戰略的思維，綜合運用外交、經濟、軍事等各種工具來思考海洋戰略，在深化海洋永續經營與發展的同

---

2 薄富爾將軍（Andre Beaufre）出生於 1902 年，是法國的知名戰略家，曾參與二次大戰與蘇黎士運河作戰等，官拜上將退伍後開始著作，主要的著作有戰略序論（An Introduction to Strategy）、嚇阻與戰略（Deterrence and Strategy）與行動戰略（Strategy of Action）等，詳見鈕先鍾著〈薄富爾的戰略思想〉，收錄於薄富爾著，鈕先鍾譯，《戰略緒論》（*An Introduction to Strategy*），（台北：麥田，1996 年），頁 180-181。

3 André Beaufre, translated for R. H. Barry, *Strategy of Action* (New York: Frederick A Praeger, 1967), p. 31.

時，或許能讓台灣成為真正遨遊四海的「海洋國家」。基此，本文將以總體戰略的觀點來檢視與前瞻我國的海洋戰略。

貳、海洋戰略與總體戰略的概念界定

為使讀者理解為何海洋戰略必須要具備總體戰略（total strategy）的思維，在進行分析之前，自有必要將總體戰略與海洋戰略的概念與以界定，並說明兩者之間的關聯性。

一、總體戰略

「總體戰略」為薄富爾重要的戰略思想之一。解讀薄富爾的總體戰略，可歸納出三個總體特性。第一是戰略運用領域的總體性。薄富爾認為戰略在運用時須加以分項，基於物質條件不同，在每一個特殊領域內應有一個適用的戰略。而不同形式的戰略之間雖有差異，但卻又是相互依賴。須將這種差異有明確的了解，才能結合成為一套有協調的行動，並指向同一個目標。[4]

第二是戰略層級劃分的總體性。薄富爾認為，戰略是有層級性，他以金字塔為喻，將戰略劃分為三個層級。最高層次的是總體戰略，意指在政府（最高權威）階層的戰略，是在政府的控制之下，其任務即為決定如何指導「總體戰爭」，亦應替每一種特殊分項戰略指定目標。並決定政治、經濟、外交、軍事等方面應如何配合協調。其次是分類戰略（overall strategy），意指在軍事、政治、經濟、外交等領域都應該有各自的戰略，其任務是在某一特殊領域內，分配工作並協調各種不同的活動。[5] 在分類戰略下還應有運作戰略（operational strategy），意指每一個領域下的次領域都應該有自己的分支戰略。[6]

第三是戰略工具選擇的總體性。薄富爾指出：「戰略必須要有一整套的工具，包括物質與精神在內，其範圍從核子轟炸起一直到宣傳或貿易協定。」[7] 意味著國家權力的工具箱應放進各種可用的工具，包含最暴力到最溫和的。這三

---

4 薄富爾著，鈕先鍾譯，《戰略緒論》（*An Introduction to Strategy*），頁38。.
5 同上註，頁38-41。
6 同上註，頁40。
7 同上註，頁29。.

個總體特性，勾勒出薄富爾總體戰略的內涵。

薄富爾曾定義戰略是「二個相對意志使用力量以解決彼此爭端的辯證法藝 術」（the art of the dialectic of two opposing will using force to resolve their dispute），這個定義亦具有總體戰略的意涵。此一定義意味著戰略的目的在於解決「爭端」，不只是運用在「戰爭」；其次在談到「力量」（force）這個名詞時，所指涉的並非只是軍事力量，而是包含國家所有的力量；而軍事力量的運用也並非只有在戰爭中，而是包含在平時以軍事力量存在所產生威脅，與所有「非軍事力量」結合產生的施壓作用。[8]

基於前述的戰略思維，薄富爾將總體戰略的行動模區分為「直接模式」（direct mode）與「間接模式」（indirect mode）。[9]以軍事因素為主要角色的類型稱之為「直接模式的總體戰略行動」（total strategy action in direct mode）。以軍事武力扮演次要的角色，有時甚至只是輔助的角色的類型稱為「間接模式的總體戰略行動」（total strategy action in indirect mode）。換言之，「直接模式的總體戰略行動」的基礎是嘗試使用軍事武力的「威脅」以達成一個「決定」（decision），並將之視為主要的行動手段。[10]「間接模式的總體戰略行動」則以「非軍事方法」來達成預期主要目標的結果。[11]換言之，戰略可以用來處理「戰爭」，但也可以用來處理「平時」的爭執。這種在平時使用力量來進行意志辯證的情境，正是當代大國爭霸出現的場景。而總體戰略的思維提供了我們面對此一情境處理複雜的國際事務時的戰略思考。

薄富爾指出，過去在政治、經濟、軍事等這些國家的活動中，除了軍事之外並未存在「戰略」的觀念，但在這些領域內，戰略的觀念卻又幾乎每天都被運用，只是未曾以「戰略」為名。[12]在過去，古典戰略將行動的定義往往限制在軍事行動。但是，在今日軍事權力的效用已日益降低，因為國家會選擇以武

---

8 André Beaufre, translated for R. H. Barry. *Strategy of Action*, p. 103.

9 這種戰略「模式」（modes）的分類的原則是以軍事力量在整體行動中扮演的角色來區分，其運用的手段範圍包括從最陰險的手段到最暴力的行動。Andre Beaufre, translated for R. H. Barry. *Strategy of Action*, p. 103.

10 André Beaufre, translated for R. H. Barry. *Strategy of Action*, p. 103.

11 Ibid., p. 112.

12 薄富爾著，鈕先鍾譯，《戰略緒論》（*An Introduction to Strategy*），頁 39-40。

力來解決彼此紛爭的可能性已大幅下降，即使需要使用軍事力量來解決紛爭，通常也不會是主要的工具。[13] 薄富爾強調，當代戰略概念的延伸已涵蓋了所有的「強制」現象，因為不論採用何種強度的強制過程（不論是否為軍事），都必須要合併在單一的思考體系之中。[14] 因此，薄富爾的總體戰略思想或許可以提供我們在比冷戰時期更為複雜的當代一種行動的指導。

值得注意的是，薄富爾不但強調總體戰略有其層次，且是須由上而下來進行指導，不可本末倒置。薄富爾指出，「總體戰略」應該由政府的首長親自主持。[15] 意味著總體戰略不應該是由分項戰略的規劃者來進行，而是由政府的決策者來規劃。為了避免混淆，在最高決策單位的總體戰略規劃，薄富爾用「最高政策」（high policy）來表示。「最高政策」是選擇所要達成的目標，而次一階層的「總體戰略」則是選擇能夠實現政策目標的手段。[16] 簡單的說，最高政策的工作是選擇目標，而總體戰略是要依照目標來選擇執行的手段。此一概念將「政策」與「戰略」的任務做了清楚的分野。

政策應該指導戰略是人類歷經兩次大戰後的經驗教訓。歷史告訴我們，如果放任軍人以戰爭目標來左右政策的決策，其結果通常是帶來悲劇性的發展。薄富爾也一直強調政策與戰略是不可混淆的。總體戰略的意涵，是希望降低軍事戰略的獨特性與嚴密性，讓軍事戰略能夠服從由政治所支配並由政府官員制定的「全面戰略」，[17] 這樣才符合總體戰略的思維。李德哈特（B.H. Liddell Hart, 1895-1970）亦曾指出：「通常一個有限目標的戰爭政策，一定會產生一個有限目標的戰略，只有獲得政府的批准之後，軍事指揮官才可以去追求一個決定性的目標，而只有政府才有權決定何種目標是值得追求的。」[18] 此即說明只有政府才能夠為戰略決定目標，也就是由政策來決定戰略，如此才不至於產生「尾巴搖狗」的情況。

13　André Beaufre, translated for R. H. Barry. *Strategy of Action*, p. 16.
14　Ibid., p. 20.
15　薄富爾著，鈕先鍾譯，《戰略緒論》（*An Introduction to Strategy*），頁39。
16　André Beaufre. translated for R. H. Barry. *Strategy of Action*, p. 22.
17　Ibid., p. 20.
18　李德哈特，鈕先鍾譯，《戰略論》（台北：麥田，1996年），頁404。

二、海洋戰略

　　海洋戰略是戰略思想在海洋事務上的運用，是海洋政策的行動指導。[19] 但學界對海洋戰略的定義並無統一之見解。例如，中國大陸學者劉繼賢、徐錫康指出海洋戰略是「國家籌畫與指導海洋方面各種事務的方略」。[20] 德國學者哈曼（Gerd Hamam）則認為海洋戰略是「對政治行動的指令，以期在符合海洋政策前提下實現一國的海上利益，是一國戰時、危機時期及和平時期的總體戰略的一部份。」[21] 雖然這些定義不盡相同，但基於以上學者的定義，我們可以認知到海洋戰略的制定是國家為了執行海洋政策、爭取海洋利益的一種行動指導。然而，由於學界對於海洋戰略缺乏明確的定義，在檢閱現有對於海洋戰略研究的文獻之後，即發現許多研究經常將「海洋政策」與「海洋戰略」混用，這將使讀者感到困惑。因此，在論述海洋戰略之前，也必須要先將「海洋政策」與「海洋戰略」分別加以界定。

　　所謂「海洋政策」是一個國家在海洋事務的最高指導。俄羅斯的海洋文件內指出，「海洋政策是一國在海岸、內海水域、領海、專屬經濟區、大陸架和公海內為實現國家利益而由國家確立的目標、任務、方向和措施。」[22] 換言之，海洋政策是一個國家追求海洋利益的政策，是國家政策在海洋方面的重點。循此思路，可以認知到海洋政策應該服從於國家政策。而薄富爾認為國家政策是政府為達成國家目標的手段，因此我們可以說「海洋政策是政府為達成國家在海洋方面國家目標的手段」，而「海洋戰略」則是國家執行海洋政策的戰略，是國家為達成海洋政策目標而進行選擇工具的指導、計畫與行動。簡單的說海洋政策是目標與行動路線的選擇，海洋戰略是達成目標的工具與行動模式的選擇。[23]

19 需先說明，本文對於戰略的界定並非指傳統上狹義的「戰爭指導」，而是將戰略視為是「思想、計畫、行動」三位一體的行動指導。
20 劉繼賢、徐錫康主編，《海洋戰略環境與對策研究》（北京：解放軍，1996 年），頁 3。
21 喬爾根・舒爾茲等編，鞠海龍、吳艷譯，《亞洲海洋戰略》（北京：人民，2014 年），頁 16。
22 劉建永、張波主編，〈2020 年前俄聯邦海洋學說〉，《世界大國海洋戰略概覽》（南京，南京大學，2015 年），頁 54。
23 分析各國的海洋政策的手段可從溫和到暴力，而海洋戰略的工具從非軍事到軍事，範圍相當的廣泛如同光譜的兩極。國家若是選擇暴力，則軍事力量則是必要工具。反之，國家若希望和平解決海洋爭端，則外交努力、經濟制裁、輿論施壓、心理施壓則是主要的工具。由此可見政策選擇對於戰略選擇的重要性。

　　鈕先鍾曾指出，政策是國家行動方向的概括構想，戰略是執行此一構想的技巧。[24] 簡單的說，政策是指導，戰略是執行。政策在上，戰略在下，不可以混淆。基此分析，我們可認知到海洋政策與海洋戰略兩者是有上、下關係的；但是，必須明白，政策與戰略兩者是相互作用，而非僅是上下關係而已。而儘管戰略亦有指導、計畫與行動，但是一切都為了行動。因此，我們在研究海洋戰略時，釐清政策、戰略、行動是非常重要的，否則在論述上層層疊疊，根本難以區分那些應該事務是政策單位的權責，那些是執行單位的權責。

　　另一個在探討海洋戰略時常見的問題是將「海洋戰略」與「海權戰略」、「海軍戰略」的概念混淆。大部份研究者在討論海洋戰略時還是以軍事為主，特別是海軍力量的運用，也因此容易與海權戰略、海軍戰略混淆。實際上，「海洋戰略」與「海權戰略」、「海軍戰略」的定義與運用的層次並不相同。正如前述「海洋戰略」是國家海洋政策的行動指導，其範圍應該包含國家在政治、經濟、環保、軍事方面政策的具體實踐，而「海權戰略」是指國家為了海上利益而運用海上力量以控制海洋的戰略。[25]「海軍戰略」則是在「海權戰略」指導下的運用海軍維護海上利益的軍事戰略。[26] 三者顯然有所不同。

　　海洋戰略與海權戰略、海軍戰略的混用，導致學者在論述海洋戰略時，實際上是在談論海軍戰略。例如哈曼指出，海軍戰略應被引申為「有計畫的、精心考慮的、集中的實現一個國的政治、經濟、軍事、科技和人力資源，以確保

---

24 為釐清政策與戰略的關係，鈕先鍾曾以航海做為比喻來說明。國家如同一艘船，在國際政治的海洋中航行；國家利益是大致的航向，國家目標是具體的目標，國家政策是船隻保持的航線，國家戰略是航行學術，也就是保持既定航線航向目標的技巧。參照鈕先鍾著，《大戰略漫談》（台北：華欣，1977 年），頁 85-86。

25 Alastair Cooper 定義海權理論是「為了國家的利益而控制海洋」的理論，參照張國城《東亞海權論》（新北市，遠足，2013），頁 39。由此可知海權的概念是為了海洋利益而控制海洋，這必然會牽涉到國家海上力量的運用，筆者認為可以比較清楚的界定海權戰略是國家為了海上利益而運用海上力量以控制海洋的戰略。

26 1908 年德語百科辭典就海軍戰略的定義是「海軍戰略是關於海上戰爭的戰略，如通過在和平時期建立海軍艦隊基地來保證獲得勝利。制定海軍戰略的目的是在戰時及平時支撐並增強一國的海權，包括維海戰提供全方位的準備工作，以及海戰作戰指揮，即以一國的海軍力量為依據，為該國的海上行動提供必要的命令和指示，只在征服或擊退的敵軍」。哈曼（Gerd Hamam）指出戰略可以劃分為兩個層次，第一個層次是總體戰略，包含政治目標。第二層次是軍事戰略，包含軍事力量和資源使用方式的綜合規劃的所有原則的總合。軍事戰略主要依據各種軍事力量對地理、經濟、社會、政治和科技等方面的因素進行合理的考慮，海軍戰略是軍事戰略的組成部分。參照喬爾根 • 舒爾茲等編，鞠海龍、吳艷譯，《亞洲海洋戰略》，頁 15-16。

該國的海上長期目標得以實現」[27]此一說法很難釐清「海洋戰略」和「海軍戰略」究竟有何差異性，若希望一國的海軍能夠確保國家在海洋方面的政治、經濟、軍事、科技上的長期目標，未免對於海軍抱有過度的期盼。

　　研究海洋戰略的盲區，在於誤以為「戰略」一詞是指針對軍事而言，因此每論及海洋戰略時，即易於涉及海洋上的軍事風險與軍事衝突、海權、海軍戰略等面向。[28]例如漢斯‧佛蘭克（Hans Farnk）在討論海洋戰略時首先討論的即是海洋安全政策及未來的衝突形勢、雖也有討論到海洋合作，但是係以預防衝突戰略的安全政策為核心，明顯的是以討論海洋安全問題為核心來討論海洋戰略。[29]哈曼在討論海洋戰略時，是以「制海權」為核心，意味的維護海洋利益有時必須要使用到「強制手段」來加以實現。[30]然而，薄富爾在《戰略緒論》中所強調，即使是戰爭，也絕對不是純粹的軍事現象，而是一種具有總體特性本質的現象。薄富爾指出國家採取「強制行動」時，雖意味需要採取軍事行動，但實際上，軍事行動在強制過程中往往不具有主導作用，而是與外交、經濟等國家力量相互配合。[31]因此，將戰略思維運用在海洋議題上，更有助於解決當代複雜海洋問題的思考；而運用總體戰略的思維來分析海洋戰略，可避免過於偏重軍事議題的缺失。

　　目前雖有學者在討論海洋戰略具總體觀念者雖有，但是不多，這也是筆者近年來陸續蒐集文獻希望加以補實的地方。例如，哈曼認為海洋戰略比海軍戰略更為廣泛，是對政治行動的指令，以期在符合海洋政策前提下實現一國的海上利益，是一國戰時、危機時期及和平時期的總體戰略的一部份。[32]此一分類方法接近總體戰略的層次劃分，但實際上是將海軍戰略劃分為戰時、危機時期及和平時期而已，無助於釐清海洋戰略的總體性。若是依照薄富爾對戰略的分

---

27 參照喬爾根‧舒爾茲等編，鞠海龍、吳艷譯，《亞洲海洋戰略》，頁 15-16。
28 例如喬爾根‧舒爾茲主編的《亞洲海洋戰略》一書，其內容不論是討論海洋政策還是海洋戰略的文章，大多是以討論各國面對海洋爭端、海洋威脅的海洋力量運用為主，甚少討論合作的面向。
29 喬爾根‧舒爾茲等編，鞠海龍、吳艷譯，《亞洲海洋戰略》，頁 8-14。
30 同上註，頁 14-15。
31 薄富爾認為戰爭是一種具總體本質的現象，需要國內政策、對外政策、經濟與軍事行動結合在一起，相互重疊。他指出總體性的特徵從十字軍東征到百年戰爭，從宗教戰爭到十七和十八世紀的「孩子手套戰爭」，從拿破崙戰爭到美國南北戰爭，一直二十世紀兩次世界大戰，與標誌著去殖民化過程的各種活動的歷史中，都可找到。André Beaufre, translated for R. H. Barry. *Strategy of Action*, p. 29.
32 參照喬爾根‧舒爾茲等編，鞠海龍、吳艷譯，《亞洲海洋戰略》，頁 16。

類，海洋戰略應屬於分類戰略的層次，而在海洋方面的海權戰略則是屬於運作戰略。[33] 至於海軍戰略則是在海權戰略之下的戰略行動層次的指導。如此一來，我們就可清楚的知道海洋戰略與海權戰略、海軍戰略與海軍戰略之間的關聯性與總體性。

筆者一向主張海洋戰略是不同於海權戰略或是海軍戰略。海洋戰略雖包含海權戰略以及海軍戰略，但不等同於海權戰略或是海軍戰略。海洋戰略的指涉應該更為廣泛，意指以維護國家海洋利益為重心的戰略思維，將海洋視為是國家利益目標的戰略。這種利益不只是保護國家不受他國侵害，而是更進一步的在海洋永續發展、資源開發上著眼，是一種真正體認海洋重要性而必須要以戰略行動來達成國家海洋政策、維護國家海洋利益的思維。換言之，海洋戰略除了涉及與其他國家的海洋權利爭執的國家權力運用之外，也應包含將海洋開發與海洋維護包含在內，才能產生完整維護國家海洋利益的行動指導。因此筆者則曾將海洋戰略區分為三個主要面向，一是國家權力的拓展，是屬於國家間政治霸權發展的觀點；其次，是以國家安全為主體的海權戰略的建構，也就是對於國家主權的維護的海洋安全戰略；再次，是以海洋為主體的經濟與資源開發，是屬於海洋發展戰略的主要內涵。[34] 這三個面向並非各自獨立的戰略面向，而是一個總體宏觀的戰略，符合總體戰略的旨趣。

參、我國的海洋政策的檢視

海洋給台灣帶來財富，也是台灣生存發展的重要資產，因此必須要有完整的海洋政策維護自身利益與保護這些可貴的資產。以下試就我國的海洋環境、海洋政策發展歷程，以及蔡英文政府的海洋政策逐一探討，並進一步提出我國海洋政策的反思，作為前瞻我國海洋戰略的基礎。

---

33 薄富爾指出總體戰略之下是全面戰略（Overall Strategy），包含政治、經濟、軍事、心理等面向的戰略，說明軍事戰略只是其中一項。

34 施正權，〈走出冷戰鎖國思維，發展海洋戰略迎向海洋〉，《海洋新風潮：海洋休閒文化新思維》（高雄市：高雄市海洋局，2010 年），頁 40。

一、台灣的海洋環境

台灣既然需要良好的海洋政策，就必須要先理解自身所處的海洋環境，並依據所面對的環境與資源進行配置。台灣的生態、歷史、文化、政治與經濟都與海洋息息相關。而就地理位置言，台灣位處西太平洋海上交通門戶，扼東亞南北往來樞紐，東邊濱臨世界最大海洋－太平洋，西邊有世界最大陸地－亞洲大陸。[35] 就海洋運輸航線而論，台灣西南側扼有台灣海峽、巴士海峽及北側東海等國際海上重要通道；東北亞國家的大部份能源運輸都必須經由這條黃金水道。東側則是西太平洋南北海運航線的要衝，更是東亞的戰略要域。[36] 台灣地理位置是《孫子兵法》所稱的爭地（我得亦利、彼得亦利）、交地（我可以往、彼可以來）、衢地（諸侯之地三屬，先至而得天下之眾者），[37] 所以是東亞諸國不敢忽視，卻又不容他國覬覦之地。

地理位置的優勢是台灣的機遇，卻也是一種風險。尤其是與周邊國家海域有重疊與爭議的情況，卻迫於國際現實讓台灣的處境困難。我國專屬經濟海域與日本、菲律賓、中國等海域多所重疊，在東海有釣魚台主權爭端，在南海海域劃界與島嶼主權爭端的紛擾，對我國海洋開發活動造成嚴重影響。[38] 這些因為歷史因素、地緣政治、對國際海洋法公約解釋不同造成的紛爭，涉及中國大陸與日本這東亞兩大強權與美國之間的合縱連橫，也涉及到東南亞諸國與華人社會之間的恩恩怨怨。不論成因如何，都是台灣目前面臨的海洋風險。

中國大陸對台灣地理位置的重視，是台灣在地緣政治上的另一個壓力來源。中國大陸學者認為，台灣是中國在亞太地區戰略企圖的「重要關鍵」。[39]

---

35 行政院海洋事務推動小組，〈海洋政策 - 政策綱領〉，海巡署，2014 年 3 月 12 日，http://www.cga.gov.tw/GipOpen/wSite/ct?xItem=69299&ctNode=7497&mp=cmaa。

36 胡思聰，《我國應強化海洋政策堅衛海疆》，國政評論，2012 年 7 月 5 日，http://www.npf.org.tw/1/10997。

37 孫子，〈孫子・九地十一〉。陽明先生手批，《武經七書》（台北，陸軍指揮參謀大學，1966），頁 129-130。

38 南海諸島係由我國最早發現、命名、使用並納入領土版圖。第二次世界大戰結束之後，中華民國自日本手中收復南海諸島。1952 年 4 月 28 日生效的《舊金山和約》及同日簽署的《中日和約》及其他相關國際法律文件，確認原由日本占領的南海島礁均應回歸中華民國，其後數十年間，中華民國擁有並有效管理南海諸島的事實，亦被外國政府及國際組織所承認。參照條約法律司，〈南海和平倡議〉，外交部，104 年 5 月 26 日，http://www.mofa.gov.tw/News_Content.aspx?n=604CBAA3DB3DDA11&sms=695940。

39 北航大戰略問題研究中心教授張文木在 2016 年 1 月一篇「從整體上把握中國海洋安全」的文章中

對中國大陸而言，台灣的樞紐地位有助於中國向北解決釣魚台、向南解決南海爭議，亦可間接控制麻六甲海峽附近海域，以及向東構成阻擋美國對大陸本土威脅的屏障等四個方面的安全問題，強化中國大陸的海洋領土防衛與專屬經濟區的保護。[40] 由此可知中國大陸不會輕易放棄台灣。

台灣海洋經濟亦受到地緣經濟整合的影響，而受到限制。其中最大的挑戰是來自於中國大陸的「21 世紀海上絲綢之路」（以下簡稱海上絲綢之路）的挑戰。中國大陸的「海上絲綢之路」是一種海洋經濟戰略，沿此一海洋路線的國家，似乎將此一規劃視為機會。但是，亦不乏國家將之視為挑戰，諸如越南、印度等國都對中國的意圖抱持警覺。[41] 對台灣而言，自蔡英文政府就任後，兩岸關係中斷，台灣將中國的任何舉措都視為具有敵意的行動，「海上絲綢之路」構想難免就成為一種具有威脅性質的統戰作為。為因應兩岸變局，民進黨政府提出「新南向政策」以轉移對中國大陸的經濟依賴。但新南向政策與中國大陸的海上絲路有重疊的部分，可能會變成一種經濟戰略上的競爭。雖然蔡政府指出新南向政策與海上絲路並無競爭關係，但實質上不可能沒有競爭關係。[42] 也因此增添了我國海洋戰略的變數。

除了地緣政治上的紛爭，海洋權利的競爭也是紛爭所在。由於台灣與周邊各國的海域重疊嚴重，各國對於各自認定的經濟海域，均採取強勢的排他性作為，導致台灣遠海作業漁船過去經常成為各國獵捕的對象。不論是在釣魚台海域與沖之鳥礁海域被日本海上保安廳驅趕，或是在南海海域被印尼、菲律賓、

指出，當前東中國海及南中國海出現的許多問題，「其根源在於台灣問題的牽制」，如果完成中國統一，台灣島和海南島可以對中國東南經濟黃金地帶形成一個寬闊的拱衛海區，要解決南中國海的問題也會相對容易得多。張文木認為：「中國從地緣政治角度而言，台灣是中國海洋安全戰略的關鍵所在，如果台灣能回歸祖國」，中國的有效安全邊界就能推至西太平洋的深海海域，中國的核潛艇就能發揮反擊作用，中國的航母建設也能繼續向前邁進，中國的經濟建設成果才得以獲得保障，因此今後的國防新常態應該包括台灣回歸祖國的安排」。參照〈有中國學者認台灣是中國海洋戰略關鍵〉，美國之音中文網，2016 年 3 月 14 日，https://www.voacantonese.com/a/china-taiwan-geopolitics-20160311/3234830.html。

40 王俊評，〈從習近平的海洋戰略說起：美中亞太戰略再布局〉，聯合新聞網，2015 年 11 月 9 日，https://global.udn.com/global_vision/story/8663/1303169。

41 〈缺席中國東盟首秀 越南到底怕什麼〉，多維新聞，2017 年 11 月 1 日，http://news.dwnews.com/global/news/2017-11-01/60021055.html。

42 為推動新南向政策，蔡英文總統接受印度等 6 國媒體聯訪時表示，新南向政策的用意，不是在這個區域內提出一個政治宣示，不是與大陸競爭，而是強調台灣做為本區域的成員，以自身優勢促進互惠互利發展。〈蔡：新南向不是與大陸競爭〉，中國時報，2017 年 5 月 6 日，http://www.chinatimes.com/newspapers/20170506000348-260118。

越南捕捉，都讓台灣的漁民心酸。台灣附近海域海洋生物約佔全球物種的十分之一；就漁獲量而言，我國排名世界第 20 位，而遠洋漁業則為世界六大遠洋漁業國之一。[43] 但是因為海洋環境惡化，生態遭到破壞，漁業資源過度開發，除了嚴重影響漁民生計，也牽絆了台灣向海洋發展的腳步，也讓海洋興國淪為口號。[44]

二、我國海洋政策發展

海洋政策攸關海洋戰略的規劃，在討論我國的海洋戰略的問題前，必須要先了解我國的海洋政策。

過去中國在明清時代都曾有「片板不得下海」的禁海規定，將海洋當成阻體，隔絕外國文化、經濟、宗教的影響。此一傳統影響中國經營海洋的觀念，海洋事務成為國家政策中的枝節，導致國家自我封閉，未能善用海洋的孔道、經濟、資源的功能，海洋遂成為少部分人的專利。而台灣因為 1949 年之後的兩岸情勢發展，亦將海洋當成是外敵入侵的可能路徑，在戒嚴令的要求下，對於海洋亦採取嚴格的管制措施，導致民眾雖居住於四面環海的島嶼，但「親海性」薄弱。過去除了發展貿易與漁業所需的港口建設受到重視之外，其他的海洋事務，例如海洋生態保育、海洋觀光等未受到應有的重視。

此一趨勢隨著台灣解嚴後有所改變。1996 年總統候選人彭明敏喊出「海洋國家」口號，開啟政黨將海洋政策作為競選議題開始，台灣展開海洋政策的擘劃。1998 年的《中華民國專屬經濟海域及大陸礁層法》相應而生，成為我國最早和海洋保育事務相關的行政法規。[45] 2000 年陳水扁總統執政後，「海洋立國」成了響亮的口號。但形之於文字、正式公布的海洋政策，主要僅限於行政院研究發展考核委員會於 2001 年公布的《海洋白皮書》和 2006 年的《海洋政策白皮書》。[46] 行政院在 2004 年設立「行政院海洋事務推動委員會」（簡稱海

---

43 行政院海洋事務推動小組，〈海洋願景〉，海巡署，2014 年 3 月 12 日，http://www.cga.gov.tw/GipOpen/wSite/ct?xItem=69298&ctNode=7496&mp=cmaa。

44 李武忠，《面對爭端台灣需要總體海洋戰略》，東森新聞，2012 年 9 月 6 日。

45 〈海洋國家口號 20 年，海洋專責主管機關《海洋委員會》終於上路〉，農學堂，2015 年 6 月 16 日，https://www.newsmarket.com.tw/blog/71500/。

46 邱文彥，〈台灣海洋政策與管理：如何讓「海洋立國」不只是響亮的口號？〉，關鍵評論，https://www.thenewslens.com/article/18322。

推會），作為跨部會的平台。[47] 2004 年 10 月 13 日政府核定發布「海洋政策綱領」，作為海洋施政的根基。[48] 從歷年來政府在海洋事務各個不同領域採行具體施政作為，可看出「海洋立國」已是我國發展的新藍圖。[49] 海洋政策的轉向也讓海洋事務逐漸受到重視。

惟行政院於 2006 年再修訂發行《海洋政策白皮書》後，迄今未再提出新的海洋政策宣示文件，顯見我國在海洋政策上不夠積極。教育部雖於 2007 年公布了《海洋教育白皮書》，作為我國推動海洋教育的參據；可惜的是，我國海洋教育的推動仍然困難重重，加上人才出路受限，「教、考、用」整合的策略成效依然有限，使我國海洋人才業出現明顯的斷層，海洋實力的推進面臨巨大挑戰。[50]

為了整合我國各部會資源，馬英九總統提出「藍色革命、海洋興國」的政策藍圖，並在 2011 年由行政院提交《海洋委員會》草案，經立法院於 2015 年 6 月 16 日三讀通過相關組織法，負責擘畫海洋總體政策。其下設置的海巡署、海洋保育署、國家海洋研究院，是我國朝向設立海洋專責機構努力的實踐。

三、檢視蔡英文政府的海洋政策

2018 年 4 月 28 日，立法後延宕兩年成立的海洋委員會終於掛牌，而讓蔡英文政府自豪之處居然是「海洋委員會是第一個在南部成立的中央部會」。[51]

---

47 〈海洋國家口號 20 號 20 年，海洋專責主管機關《海洋委員會》終於上路〉，農學堂。
48 我國海洋政策綱領中有九項政策主張，包括了：一、確認我國是海洋國家，海洋是我國的資產，體認我們國家的生存發展依賴海洋。二、享有與履行國際海洋法賦與國家在海洋上的權利與義務，並響應國際社會倡議之永續發展理念。三、重行認識國家發展中之海洋元素，尊重原住民族海洋與智慧，並建立符合國家權益之海洋觀。四、調查國家海洋資產，瞭解社會對海洋之需求，掌握海洋活動本質，規劃國家海洋發展。五、採行永續海洋生態及世代正義的觀點，建立海洋環境保護、海洋生物資源養殖及合理利用海洋之海洋管理體制。六、強化海洋執法量能，以創造穩定之海洋法律秩序與安全之海洋環境。七、創造有利之政策與實務環境，實質鼓勵海洋事業發展。八、推動以國家發展為導向之海洋科學研究，引導各級水產、海事、海洋教育發展，以利海洋人才之培育。九、提供安全、穩定之海洋環境，鼓勵民眾親近海洋，培養海洋意識與文化。行政院海洋事務推動小組，〈海洋政策 - 政策綱領〉。
49 行政院海洋事務推動小組，〈海洋願景〉。
50 邱文彥指出國內海洋相關系所（如造船系）逐漸更名、海洋院校學生就業情形不佳、大專畢業生升學轉入其他非海洋系所而逐漸流失、實習船不敷需求、商船上多為外籍船員，以及公務船船員年齡過大等情況而言，我國海洋人才業已出現明顯的斷層。邱文彥，〈台灣海洋政策與管理：如何讓「海洋立國」不只是響亮的口號？〉。
51 行政院組織法於 2010 年修正，即明訂行政院轄下 37 個部會將整併為 29 個，並成立新單位「海洋

但實際上筆者認為此一布局未考慮到海洋事務需要各部會配合，未來將可能導致協調時的困難。將海洋委員會放在高雄，遇有需與各部會協調、配合的海洋事務，各部會勢必只能派代表到高雄參與協調，重大事項的決議勢必無法決斷，讓海洋事務的決定需耗費更多的時間。而海洋委員會主委每周需前往台北參與行政院會，南北往返奔波，如何在專注於海洋事務的統合？如此配置對於海洋事務沒有幫助，卻凸顯了民進黨政府心中只有南北平衡與資源分配的考量，沒有將海洋事務視為優先重點，也忽略了海洋事務的複雜性。且台灣既然自稱為是海洋國家，不論是新北市還是基隆、台中、高雄都是台灣海洋都市，何須將海洋委員會刻意放在高雄？除了增加資源的浪費、延宕決策的時間外，這樣的配置，實難以顯示政府對海洋事務的重視。且海洋委員會雖已式揭牌成立，但媒體報導相關單位的業務根本尚未移交至海委會，甚至無具體時間表，[52] 要能夠全面掌控台灣的海洋事務，恐怕還有一段磨合期。

事實上，檢視蔡英文政府的海洋政策似乎仍停留在口號與宣傳的階段，卻乏具體實踐的行動指導。2015 年，時任民進黨黨主席的蔡英文在出席第二屆台灣海洋產業研討會時指出，在海洋事務的推動有幾項重要的工作要去開展、完成：包括了第一、要迅速完成整合性的海域利用規劃。第二、推動海岸海洋的資源復育、保育及利用。第三、積極開發海峽風力及黑潮能源。第四、全力發展海洋觀光休閒產業。[53] 而在 2018 年，蔡英文總統出席「海洋委員會首長布達暨揭牌典禮」時強調，海洋是台灣最重要的出路，並表示「我們是海洋國家，海洋就寫在台灣人的 DNA 裡面。」[54] 蔡總統並提示「健全海洋法制，做好生態保育工作」、「配合政策，推動海洋產業」及「強化海洋研究能量，培育海洋

委員會」，海洋委員會組織法也早在 2015 年 6 月就已完成立法，當時行政院長毛治國核定施行日期為 2016 年 7 月 4 日。但因政黨輪替，民進黨立委及黨內高層有不同意見，行政院前院長林全在 2016 年 6 月 30 日廢止施行日期，2017 年敲定海委會於 2018 年 4 月成立，並考量高雄是台灣最重要商港與漁港，以及顯示政府重視北中南均衡發展，決定海委會設於高雄。〈拖了 8 年！海巡署降編納入海洋委員會今高雄掛牌〉，聯合新聞網，2018 年 4 月 28 日，https://udn.com/news/story/6656/3112280。

52 〈海委會今揭牌 業務移交卻無具體時間表〉，自由時報，2018 年 4 月 28 日，http://news.ltn.com.tw/news/politics/breakingnews/2409756。

53 〈出席第二屆台灣海洋產業研討會 蔡英文：推動永續 有競爭力海洋產業〉，民主進步黨新聞中心，2015 年 7 月 30 日，https://www.dpp.org.tw/media/contents/6699。

54 〈經歷兩次政黨輪替、三任總統，台灣第一個專管海洋的「海委會」成立〉，關鍵評論網，2018 年 4 月 28 日，https://www.thenewslens.com/article/94562。

人才」三個努力方向，期勉與會人員要立足台灣、航向海洋，讓「臺灣的海洋事務蒸蒸日上」。[55] 這兩次的宣示應可視為是蔡英文政府的海洋政策。但是這些口號式的政策宣示，並未對於台灣面臨的周邊海域爭端、海洋安全維護、台灣在國際漁業作業面臨的困境等問題做出具體作為的說明。此亦顯示蔡英文政府並未打算將這些問題劃歸給「海洋委員會」來處理。

此外，行政院長賴清德在行政院院會聽取海巡署「行政院海岸巡防署的回顧與展望」報告後表示「台灣是海島國家，海洋治理攸關國家發展，為彰顯政府對海洋事務的重視，成立海洋專責機關『海洋委員會』，將原本分散在各部會掌理的海洋事務予以系統性統合。」[56] 亦顯示民進黨政府對於海洋事務的錯誤認知。海洋事務的複雜性涉及到國防、內政、外交、經濟、科技、環保、漁業、交通等各部會之事務，無法交給單一部會來處理，必需由各部會來共同努力，因此係以「委員會」的型態來處理海洋事務，其原因正在於此。我國的海洋委員會依組織法的規定「設置委員十七人至十九人，由行政院院長派兼或聘兼之。」但並未明文規定這些委員的派任單位應由行政院各部會主管擔任，來提高委員會的地位。

而且，蔡英文政府就任至今兩年的時間，除了在會議場合的宣示外，似乎也並未見到對於海洋政策的具體政策文件、白皮書。檢視總統府網站的焦點議題，可以發現當前總統府重視的議題包含創新經濟、年金改革、司法改革、轉型正義、原民正義、人權、兩岸、國防、外交、前瞻基礎建設、長照 2.0 等 11 項，[57] 但卻並沒有關於海洋事務、海洋發展、海洋戰略等議題。

此外，依行政院網站所載，當前行政院重要政策包含內政及國土、外交國防、法務、教育文化及科技、財政經濟、交通建設、農業環保、衛生福利勞動、綜合行政等九項，亦無「海洋事務」。有關海洋方面的政策說明於內政及國土、外交國防兩項之中共有三篇政策說明，分別為〈展開臺日海洋事務合作對話，確保臺灣漁民權益〉（民國 105-11-16）、〈經略南海、永保太平—收復南海

---

55 〈出席海洋委員會揭牌儀式　總統籲立足臺灣，航向海洋〉，中華民國總統府網站，107年4月28日，https://www.president.gov.tw/News/23292。

56 〈拖了8年！海巡署降編納入海洋委員會今高雄掛牌〉，聯合新聞網。

57 中華民國總統府網站，〈關鍵議題〉，https://www.president.gov.tw/Default.aspx#。

諸島 70 周年紀念〉（民國 105-12-08）、〈捍衛南海主權，維護漁民權益〉（民國 105-08-08）。至於財政經濟、交通建設、農業環保政策、中，則無任何有關海洋事務的政策說明。[58] 顯示蔡英文政府對於海洋政策的議題並沒有列為優先、重要政策。

四、我國海洋政策的反思

海洋政策涉及的問題包括天然資源開發、漁業資源保育、海洋作業管理、海洋權益保護等面向。但從陳水扁、馬英九、蔡英文等歷任總統的海洋政策來看，台灣現今的海洋政策的走向，似乎正走向海洋保育、海域開發、海洋管理方面，而忽略了軍事與外交對於海洋政策的重要性，因此我們可以發現許多關心海洋事務的先輩在推動海洋政策之際，對於海洋專責機構的設置不遺餘力，卻往往忽略了海洋事務絕非單純的「海洋專責單位」可以處理的。設置海洋專責機構，卻可能變成許多真正的核心問題無法處理，例如，涉及到軍事安全與海洋外交的工作。此種對於海洋事務的狹隘視野，源自於文人本位主義的觀念，實不利於我國海洋政策的整體發展。

就海洋事務的廣度來看，台灣以對外貿易為主要的經濟命脈，但是當論及海洋事務時，卻往往只將視野侷限於周邊海域，很少願意去談論海洋運輸安全問題。這可能是受限於台灣的外交環境所限，卻也是政府應該要設法突破的困境。筆者認為如果政府要談論海洋事務，就要放眼世界，看台灣在國際海域遭遇的問題。台灣的遠洋漁船遠赴太平洋、南海、印度洋、北極洋等各地作業時，會遭遇海盜、颶風、颱風、海上碰撞、越界作業等等問題。政府在考量海洋政策時，是如何去思考這些問題的處理？如何對我國海上從業人員進行安全維護？如果欠缺這種總體宏觀的海洋政策與海洋戰略思維，台灣充其量只是一個以陸面海，只顧海域的島國，而非自稱的「海洋國家」。質言之，我們的海洋政策只是「島嶼的海域政策」，而非真正的海洋政策。

就海洋政策的總體性來看，新成立的海洋委員會的分工與職權，說明我國的海洋政策，大部分集中在海洋開發、利用、保育、環保、管理等思維，似

---

58 中華民國行政院網站，〈重要政策〉，https://www.ey.gov.tw/Page/2124AB8A95F79A75。

乎未將屬於國家海洋軍事戰略納入，甚至於排斥軍人參與海洋事務。例如海洋委員會組織法修訂過程中，部分人士對於海洋委員會下轄海巡署具有軍人身分者，竟然會以「軍人干政」的理由，反對讓軍職人員於海洋委員會海洋保育署中任職，而非以職務專長不相干為考量。[59] 此等狹隘的海洋思維，令人感到失望。我們必須要明白，總體海洋政策必須要考慮的絕對不會是只有海洋保育、海洋開發、海洋管理等海洋發展戰略的面向，亦必須要考量到海洋安全維護的問題，尤其是海上維權力量的建構。狹隘地將海洋委員會用來處理海洋發展問題，將會讓海洋問題的處理變成各管各段。特別是這種思維多半來自於組織內部的本位主義，文人與警察、軍人之間的權力爭鬥。將組織利益置於國家利益之上的偏頗心態，無助於國家海洋利益的維護。

借鑑於美國的「海洋委員會」成員，我們可反思台灣海洋政策的狹隘。[60] 美國的「海洋委員會」委員包含國防部長、情報機構首長，這些委員可指派該機構的人員履行其在委員會的職能，被指派的人員除了是由總統任命的官員外，亦包含陸軍或海軍的將官，[61] 顯示美國在處理海洋事務上的軍文並重。美國「海洋委員會」的組成，凸顯了複雜的海洋問題並非單一機構可以完全處理，並須是國家總體力量的投入。

就政策與戰略的劃分來看，台灣〈海洋政策綱領〉的內涵即呈現出台灣在海洋政策與海洋戰略的混淆。〈海洋政策綱領〉中明列我國海洋政策的目標是「掌握國際發展趨勢，增進海洋國際合作，強化海洋政策法制，健全海洋行政體制，以維護海洋權益，確保國家發展。」政策既然是指導，雖然是掌握大方向，但是僅以空洞的「維護海洋利益、確保國家發展」為政策的目標，難以顯示政府在海洋方面的戰略思維。〈海洋政策綱領〉同時強調要透過「強化海域執法功能、健全海域交通秩序、提昇海事安全服務、充實海域維安能量」的

---

59 〈3 千連署增設「海洋保育署」籲海巡軍警退出〉，自由時報，2015 年 5 月 19 日，http://news.ltn.com.tw/news/life/breakingnews/1321816。

60 美國的海洋委員會是協調機構，納入了國務卿、國防部長、財政部長、農業部長、運輸部長、能源部長、國土安全部、司法部長、環境署長，甚至於中央情報局、參謀首長聯席會議主席等人參與，幾乎是美國政府的行政部門均納入其中。李景光、張士洋、閻季惠主編，《世界主要國家和地區海洋戰略與政策選編》（北京：海洋，2016 年），頁 3。

61 李景光、張士洋、閻季惠主編，《世界主要國家和地區海洋戰略與政策選編》，頁 4。

策略來到達維護海上安全之目標。」[62] 這些具體的作為是海洋政策之下各個部門的行動綱領，應該是屬於戰略層次的東西，卻放在了政策綱領之中。其他如「掌握國際發展趨勢，增進海洋國際合作，強化海洋政策法制，健全海洋行政體制」，究竟是戰略，還是政策，也是模糊不清。再者，〈政策綱領〉中提及「強化海域執法功能、健全海域交通秩序、提昇海事安全服務、充實海域維安能量」[63] 就其文字的表述應該是海洋戰略的內涵，但卻沒有將海洋防衛與海軍力量的運用納入，顯然認為這是國防戰略的責任，而非海洋戰略應有的內涵。此一問題也顯示台灣的海洋戰略卻乏總體思維，難以達成維護海洋權益的政策目標。

肆、台灣海洋戰略的前瞻

誠如前述，台灣的海洋政策明顯地欠缺乏總體性，並且在政策與戰略上有所混淆，實在有必要進一步釐清海洋政策的實質內涵。但政策是由政府所提出，筆者除提出反思之外，實難為海洋政策提出處方，畢竟這是「海洋委員會」的職責；然而，基於維護國家的海洋利益與海洋安全，仍試就戰略層次提出建言。筆者認為，台灣需要提出具有總體性質的海洋戰略，以因應複雜的海洋環境問題。基此，以下透過總體戰略的觀點，從外交、經濟、保育、軍事等四個面向，概述台灣應有的海洋戰略思維，前瞻台灣成為海洋國家發展的未來。

一、透過外交戰略拓展海洋空間

著實而論，台灣當前所處的國際環境並不容樂觀，在海洋方面遭受的壓迫亦不容忽視。面對目前國家之處境，如何發揚台灣的海洋地緣優勢，更是成為台灣突破國家發展瓶頸與國際困境之新思維。[64] 因此台灣在海洋外交上必須要保持彈性，不可過於僵化。對固有海域疆界的捍衛，政府更應拿出堅定的態度與作為，不應讓周邊國家一再的侵佔掠奪，而竟噤聲不言。[65]

為了解決台灣面臨的海域紛爭，前總統馬英九於任內致力於以和平手段處

---

62 行政院海洋事務推動小組，〈海洋政策 - 政策綱領〉。
63 同上註。
64 行政院海洋事務推動小組，〈海洋願景〉。
65 胡思聰，《我國應強化海洋政策堅衛海疆》。

理海域爭端來避免衝突，這是一種「間接戰略」的思維。此一戰略一方面聲張中華民國海洋權力，一方面透過外交手段穩定海域和平。島嶼主權歸屬與海域劃界雖然是涉及國際法的爭端，但在實務層面卻是一個難以解決的政治議題，也因為如此，就需要有更具智慧的解決手段。「自我克制」與「建立合作開發機制」就成為一個思考的重要方向。[66]

馬英九任內對區域內海域安全的具體行動，首先是積極改善與中國大陸關係，使台海緊張局勢趨緩，讓兩岸關係得以持續和平發展。其次是於 2012 年 8 月針對東海海域紛爭及釣魚臺主權問題，提出「東海和平倡議」，一方面緩和東海緊張局勢，另一方面促成中華民國與日本於 2013 年 4 月簽署《臺日漁業協議》，解決兩國四十年之漁權爭議。[67] 我國並於 2015 年 5 月 26 日發表「南海和平倡議」，[68] 將「主權在我、擱置爭議、和平互惠、共同開發、全面規劃、分區開發、自由航行、飛越自由」做為我國的南海政策。[69] 並於 2015 年 11 月 5 日與菲律賓簽署「台菲漁業協定」。[70] 這些行動有效的緩和了區域的緊張形勢，讓台灣從麻煩製造者轉變為和平促進者。

我國要融入區域體系，亦可藉由積極參加人道救援與維和任務來取得道德正當性。[71] 例如，我國長久以來在東沙島以及太平島上所投入的人力與物力資源，正可展現台灣的南海存在的價值。特別是當前南海周邊國家對於非傳統安

---

66 王冠雄〈東海南海爭端 我無法置身事外〉，國政評論，2015 年 8 月 12 日，http://www.npf.org.tw/1/15302。

67 條約法律司，〈南海和平倡議〉，外交部，2015 年 5 月 26 日，http://www.mofa.gov.tw/News_Content.aspx?n=604CBAA3DB3DDA11&sms=695940。

68 「南海和平倡議」內容包括了：一、自我克制，維持南海區域和平穩定，避免採取任何升高緊張情勢之單邊措施。二、尊重包括聯合國憲章及聯合國海洋法公約在內之相關國際法原則與精神，透過對話協商，以和平方式解決爭端，共同維護南海地區海、空域航行及飛越自由與安全。三、將區域內各當事方納入任何有助南海和平與繁榮的體制與措施，如協商建立海合作機制或訂定行為規範。四、擱置主權爭議，建立南海區域資源開發合作機制，整體規劃、分區開發南海資源。五、就南海環境保護、科學研究、打擊海上犯罪、人道援助與災害救援等非傳統安全議題建立協調及合作機制。條約法律司，〈南海和平倡議〉，外交部，2015 年 5 月 26 日。

69 周晨晰，宋連海，〈南海爭端對我國海權發展之影響〉，《海軍學術雙月刊》，第 50 卷第 1 期，http://navy.mnd.gov.tw/MediaRoom/Paper_Info.aspx?ID=30099&AID=57。

70 台菲漁業協定於 2015 年 11 月 5 日簽署，目前已執行的避免使用武力、執法前相互通報及扣押逮捕後儘速釋放等三項共識法制化，並於 2015 年 12 月 5 日生效。參照〈台菲漁業協定簽了 12 月 5 日生效〉，中國時報，2015 年 11 月 20 日，http://www.chinatimes.com/newspapers/20151120001488-260102。

71 周晨晰，宋連海，〈南海爭端對我國海權發展之影響〉。

全事務機制有所需求，包括了海洋科學研究、海上搜救與人道救援、海洋環境保護、漁業資源永續、共同打擊海上犯罪行為，包括海盜、人口販運、毒品走私等，都需要區域合作來達成。由政府所發布的相關聲明中，可以見到我國在處理南海爭端過程裡具有能力、能量與意願，更可以期待因為我國之參與，而可以創造一個 SAFE（科研、救助、漁業資源管理、環境保護）的南海區域和平環境。[72] 而在蔡英文政府時期，前台灣省長宋楚瑜代表政府參與亞太經濟合作會議（APEC）與東協領袖進行非正式對話時，即表達台灣希望與東協各國深化在教育、資通訊、醫療、農業、基礎建設與災害防救等方面合作與交流。[73] 此種採取和平手段來宣示主權，表達積極參與合作立場的作法，符合間接戰略的思維，也有助於擴大台灣的海洋空間。

二、透過經濟戰略建構海洋連結

我國是以「出口導向」為主的海島經濟國家，在海洋經濟上亦須具有戰略層次的規劃。台灣在地理型態上是標準的海島，軍事上依賴海洋為屏障，經濟上也仰賴海上貿易交通。再者，可用土地面積及資源有限，能源及民生物資須仰賴海上航運。所以，應該將環繞四周的海洋視為延伸出去的藍色領土，予以有效地經營。例如，利用台灣四通八達的地理位置，一方面吸引海洋國家利用台灣作為轉運中心，二方面與主要貿易航線國家取得連繫，運用這些國家的港口作為台灣遠洋航運的後勤支援，以保障台灣航運的安全，並疏通台灣的經貿動脈，凡此都是戰略層次的問題。

同時，海上貿易線的安全與穩定同樣攸關台灣的國家安全，特別是蔡英文政府提出的「新南向政策」，也需要一個穩定的海上交通，如此一來，中國大陸的「海上絲綢之路」與台灣的海上貿易線便產生了關聯性甚至是重疊性。「新南向政策」與「海上絲綢之路」兩者都離不開海洋，兩者都需要海上貿易，兩者也都需要安全保障。在此一情境下，選擇衝突還是合作，即為兩岸的決策者必須要審慎思考的問題。特別是居於弱勢且受限於國際現實的台灣，在戰略選擇上就要特別的小心。

---

72 王冠雄，〈東海南海爭端 我無法置身事外〉。
73 〈APEC 與東協領袖對話 宋楚瑜盼深化合作〉，中央社，2017 年 11 月 11 日，http://www.cna.com.tw/news/firstnews/201711115003-1.aspx。

　　台灣可藉由參與經貿整合著手，與各國密切的經貿關係，並設法加入區域整合，讓彼此經貿體系能夠休戚相關、禍福與共的緊密結合，進而在安全議題上產生一定程度合作的需要與空間，方可避免進一步升高衝突。[74] 此種思維並非只有台灣會才有，區域國家也會希望透過經貿合作來降低衝突。例如是中國大陸提出的「海上絲綢之路」，將中國大陸與東協之間透過再一次的經濟整合，將關係提升到「全面戰略夥伴」關係，從而保持南海情勢的穩定。[75]

三、透過發展戰略永續經營海洋

　　台灣的海洋發展戰略首先需展現對自然環境的尊重，採取兼容並蓄的「海陸平衡」思維。將海洋經濟發展視為台灣經濟發展的引擎，提出海洋產業發展政策，積極開發海洋漁業、交通運輸、石油天然氣、休閒旅遊、高階造船等產業，並完善海洋服務體系，增加海洋科技創新研發的經費，積極培育從事海洋科技的高階技術人員，提高科技對海洋經濟發展的貢獻率，並積極投入海洋環境保護與魚類資源的保育工作，以確保台灣的海洋經濟可以永續經營與發展。[76] 唯有政府正視海洋產業，才能鼓勵民間企業投資海洋經濟相關產業，擴大海洋經濟的範圍。

　　此外，要成為真正的「海洋國家」，台灣應強化海洋資源保護，更應積極增進海洋國家特色，讓台灣成為真正與海洋相生共榮、海洋資源與生態生生不息永續發展的「海洋國家」。[77] 海洋的永續經營是海洋國家的根基。海洋既是人類共同資產，海洋保育不能僅依賴政府，必須全民共同擔負護衛海洋的責任，因此，培養與提升國民海洋意識是海洋國家最根本的工作。換言之，台灣在活絡海洋經濟的同時，必須要設法提高國民的海洋意識，讓全民重視海洋，以強化海洋保育的觀念，建立陸海連動的認識，從日常生活做起，例如‑避免日常廢棄物、家庭與工業廢水汙染海洋，讓沿岸海域的珊瑚礁群得以復育，並成為各種魚類的安全覓食區，才能讓海洋的永續經營成為可能。[78]

74　周晨晰，宋連海，〈南海爭端對我國海權發展之影響〉。
75　王冠雄，〈大陸「21 世紀海上絲綢之路」對臺灣海洋經濟的挑戰與機會〉，對外關係協會，http://www.afr.org.tw/product_detail.php?lang=tw&id=10。
76　李武忠，《面對爭端台灣需要總體海洋戰略》，東森新聞，2012 年 9 月 6 日。
77　胡思聰，《我國應強化海洋政策堅衛海疆》。
78　邱文彥，〈台灣海洋政策與管理：如何讓「海洋立國」不只是響亮的口號？〉。

四、透過軍事戰略展現海洋實力

　　台灣位處交爭之地，為東亞交通要衝，極具戰略價值，難免成為各國亟欲拉攏與控制之地，也因此與周邊國家在海洋畫界問題上有所爭執，必須要有全面性的海洋戰略，才能避免成為各國覬覦與打壓的對象。本文雖主張維護海洋安全不一定要使用軍事力量，但軍事力量也是總體戰略工具不可或缺的一環，因此海洋軍事戰略與海軍軍事實力的建構仍有其必要性。

　　過去台灣缺乏海洋意識，因此海軍在國家戰略與軍事戰略中淪為配角，對於海軍建軍的目標亦缺乏長遠與完整的規劃。例如海軍軍艦採購往往並非以滿足戰備任務為優先考量，而是以政治外交目的為著眼，因此海軍的艦艇不一定符合台灣的戰略需求。但是台灣有許多港口，十分依賴海外貿易，只有海軍能確保港口及海上交通線的暢通，要防止敵方的封鎖損傷經濟，並維持長久的生存發展，就必須重視海軍建設。[79]要保持海上交通線的暢通與維護遠洋作業船隻安全，除了強化海巡署遠洋巡護能力外，我們也需要重視海軍。

　　就我國海軍戰略的規劃而論，似乎仍是以海域保衛為主軸，於是近岸防衛成為海軍的主要任務，同時並肩負協同海巡署執行海上護漁的任務；但是台灣既然自認為海洋國家，除了海洋永續經營與管理的戰略規劃外，也應該延伸海軍的任務半徑，將遠洋護衛納入建軍規畫，才能夠確保海上運輸線的安全。區域國家如日本、韓國都派有海軍在印度洋巡護，唯獨台灣沒有，這並非我國海軍不願意或沒也能力，實在是受限於國際現實的限制，讓海軍失去了行動自由。

　　台灣想要獲得海洋行動自由，就必須要認識遠洋巡護的重要性。台灣或許無需建立龐大的遠洋艦隊，但是至少要能夠具備滿足至印度洋巡護的戰力。[80]目前政府已將國艦國造列為政策目標，雖然發生慶富造船利用獵雷艦採購套利

---

79 周晨晰，宋連海，〈南海爭端對我國海權發展之影響〉。

80 這必須在增建遠洋補給艦上多加著墨。筆者認為過去我們需要靠他國的保護是因為能力不足與國際情勢不允許，但唯有獨立自主的護衛能力才能贏得他國的尊重，若能配合外交戰略上爭取幾個港口作為補給點，亦可降低補給艦的壓力，但是外交現實往往讓區域國家不願提供港口給我國海軍，所以唯有自主發展更大型的遠航補給艦，才能夠滿足作戰需求。特別是要能在不靠港的前提下滿足前往中東至非洲的一個航次的作戰需求的補給量，足以支撐一個小型支艦隊的救援任務所需，以宣示政府對於本國籍航運船艦的護衛的決心。我國商業船隻的製造能力已獲世界肯定，遊艇業更是全球第六名，且中船已有為國軍建造 17,000 噸級的武夷艦之經驗，增建遠洋補給艦並非難事。

的弊案，但不應因此而改變政策。就長遠看，國艦國造才能將自主權掌握在手中，依照自己的建軍規畫來興建船隻，而不需受制於他國，這才是海洋國家應有的建軍思維。

而在海域安全維護上，台灣仍必須要具備足夠的反擊能力與宣示保衛海疆的決心，才能贏得區域國家的尊重。誠然以和平解決紛爭已是趨勢，但試問台灣周邊有那一個國家放棄軍事防衛的準備？正所謂「能戰才能和」，海洋戰略的規劃不能一廂情願，在透過國際法和平解決爭端，在運用執法機構維護權益同時，我們不可忘卻海軍力量的建設與軍事反擊的準備。

此外，對於孤懸海外的領土，台灣亦應將之視為走向遠洋的戰略支點，積極投入防衛與戰略設施的建設。例如，在南海海域，應思考強化南海諸島的軍事建設與宣示作戰的決心，對區域國家產生嚇阻的能力。[81] 建構海洋力量並非意味著我們要與周邊國家以武力解決爭端，但唯有讓區域國家知道我們的決心，才能真正的產生嚇阻的力量。我國目前在南海海域中實際行使管轄權的島礁有東沙島、太平島與中洲礁，這些島礁給予我國在南海事務的參與，有了可以著力的依據。[82]

台灣過去在海南諸島的經營已有一定的成效。雖在李登輝總統任內撤回太平島上的海軍陸戰隊，改由海巡署駐守太平島，[83] 讓台灣失去有力的防衛力量。但陳水扁總統在太平島修建一條可供 C130 運輸機起降的跑道；馬英九總統任內則由陸戰隊代訓海巡署人員來強化防務。馬英九時期並開始興建新的碼頭，於 2015 年年底完工啟用。新的碼頭可讓海軍與海巡署的船艦可以靠泊進行行政卸載，提高補給效率。馬英九任內也持續強化太平島的軍事整備，並進行「衛

---

81 王冠雄，〈東海南海爭端 我無法置身事外〉。
82 在這些島嶼中，東沙島可朝海、空軍的中繼基地方向規劃建設，其任務就是以監控中共海、空軍動向，並建有適合大型船艦靠泊的深港，以及足夠戰機起降與整補的機場，並具備維修、補給與醫療能量，平時可提供周邊海域救援使用，應急時可供海、空軍中繼補給。而太平島屬於我國實際控制的國土的最南端，周邊均是與我國有主權爭議的國家，因此太平島亦須建構獨立支持防衛作戰的能力。周晨晰，宋連海，〈南海爭端對我國海權發展之影響〉。
83 在李登輝任內，最重要行動就是 2000 年 1 月撤回了防守南沙太平島近半世紀的海軍陸戰隊官兵，改由「海岸巡防署」隊員駐守太平島。當時政府認為此舉既可以讓「宣示主權」的意涵不變，「以警（海巡）代軍」更可以率先表態降低區域緊張。〈八年任期中，馬英九政府怎麼經營南海？〉，端傳媒，2016 年 7 月 12 日，https://theinitium.com/article/20160712-taiwan-South-China-Sea/。

疆演習」以展現護衛國土的決心。[84] 蔡英文總統就任後也在南海進行「南援三號」海上救援演練，以宣示政府捍衛南海和平安全的決心，這些對台灣海洋能力的展現具有正面的意義。

伍、代結論—台灣海洋戰略的行動模式

總結上述的分析可知，海洋資源是台灣持續繁榮發展主要命脈之一，有關海洋資源之開發、養護與管理，海洋空間使用及海洋環境保護，對台灣未來發展有相當大的影響。因此，以戰略思考來維護海洋永續發展與維護自身利益，才能夠真正的維護台灣的安全與利益。

海洋政策與海洋戰略是相互為用的，一個國家若重視海洋，無論其將海洋視為是生存發展的空間，還是商業運輸的載體、資源的產地、漁業的作業空間，其政府必然有一個海洋政策作為其戰略規劃的指導。過去中國、日本都曾有「禁海」的政策，自然會封閉海洋，將海洋做為阻隔外國的阻體。由此顯示「政策」具有決定性的作用，決定了運用海洋的方向。海洋戰略是海洋政策的具體實踐，國家的洋政策若是決定要「禁海」，海洋戰略必然會走向封閉海洋的努力；相對地，海洋政策若決定開放海洋，海洋戰略必然會邁向開拓發展的道路。是以，海洋政策的優劣將決定海洋戰略的成敗。政府在海洋政策的擘畫上更積極，台灣的海洋戰略才能夠更為清晰明確。

誠如前述，海洋戰略是具有總體特性的行動指導。而所謂的總體特性，意味著行動戰略的工具必非只有「軍事行動」一種而已，而是必須要將「軍事」與「非軍事」組合起來運用。實質上，戰略是一種國家力量運用的規劃，但力量並非只有軍事力量。因此，論述海洋戰略就不能只討論軍事面向，而應該是包括海洋經濟、海洋生態、海洋安全、海洋防護等議題的反應，以國家外交、經濟、環保、漁事、軍事部門的整體力量，以保護國家的海洋利益，這才是總體戰略思想在海洋事務上的體現。

---

84 詳請參閱〈八年任期中，馬英九政府怎麼經營南海？〉，端傳媒。

　　至於台灣應該有甚麼樣的海洋戰略，筆者則認為，台灣應該採用間接模式的總體戰略行動。這種總體戰略的思維，是在評估國家實力與可能威脅下做出的決策。台灣現有的實力不足以同時跟周邊海洋爭議國家進行軍事對抗，也無需一味地採取軍事對抗，而是應該審時度勢地採取務實的彈性外交，例如透過擱置爭議的作法。此外，台灣應更進一步與周邊國家進行海洋事務的合作，共同開發與利用海洋資源，共同維護海洋安全，以及共同維護海洋的永續發展。

　　或許有人會質疑這是一廂情願的做法，但卻是小國最好的選擇。台灣與大陸之間的合作關係，是展開與區域國家合作的門鑰。台灣與日本之間的釣魚台爭端，涉及於漁業資源的開發運用，過去很難與日本達成漁業協議，但馬英九政府時期與大陸改善關係之際，台灣便成為日本必須要拉攏的對象。此一角色的轉換，讓台灣與日本簽訂了漁業協定，保障台灣漁民的安全。在簽訂漁業協定時或許會對釣魚台的主權問題擱置，而傷害台灣對釣魚台的主權聲索，但至少也爭取到維護漁民作業安全的利益。台灣無需成為區域的刺蝟，但是也不能被視為軟腳蝦，唯有堅強的實力與堅定的意志，讓區域內國家理解台灣維護自身利益的決心，才能在海洋利益談判時，提高自身的籌碼。

　　歸結言之，台灣在海洋政策上並不缺乏宣言、白皮書和行動綱領，缺的是行動。[85] 我們必須要有具體行動來實踐自己作為海洋國家的宣示，不是光憑海洋國家的口號就能達成海洋國家的目標。台灣若想在海洋戰略上確保自身的行動自由，就必須評估自身的實力大小，選擇適切的行動模式。很明顯地，台灣現在護衛海洋的實力仍有待加強，經濟實力不若從前。若想透過海洋資源來豐厚自身的經濟實力，自然必須要透過間接模式的總體戰略行動，妥善的運用自身的地理位置與區域國家之間的利益矛盾來獲取台灣的利益。要言之，以非衝突、非對抗的方式來達成戰略目標，這才是真正具有間接、總體、行動意涵的海洋戰略。

---

85 邱文彥，〈台灣海洋政策與管理：如何讓「海洋立國」不只是響亮的口號？〉。

展望 2020 年台灣的國防戰略

沈明室 *

壹、前言

　　一般而言，在官方版的國防報告書出爐之後，國防戰略或是國防政策[1]議題討論的空間並不大，有興趣討論者只要引用官方資訊，即可了解國防戰略與政策的相關內涵，而其形成過程及思維，不太會引人注意。不同國家在不同時期的國防戰略有期考量因素，如果當初制定國防戰略時的背景與戰略環境無任何重大變化，或所處戰略格局延續性發展，國防戰略應不會發生經常性的轉折。即使不同的戰略決策者上台，只在應該投入的資源與使用方法，其考量重點不同罷了。如台灣、日本及韓國在威脅格局未變下，防衛戰略延續，但優先性有所不同。

　　《中華民國106年國防報告書》在台灣國防戰略的篇章中，提及我國國防戰略時強調「面對當前及未來的戰略環境與安全挑戰，須以創新思維，務實推動改革，有效運用資源，以達成最適之建軍規劃與戰備整備。點出了戰略環境與安全挑戰、思維、資源運用、建軍與備戰的內涵。若以戰略強調計畫與規畫來看，此處雖未明言國防戰略具體內容為何？但其理則可歸納為「以創新思維，務實推動改革，有效運用資源，以達成最適之建軍規劃與戰備整備。」

　　新版國防報告書以此為基礎提出和以往國防報告書不同的五項國防戰略目標。現階段我國國防戰略目標包括：防衛國家安全、建制專業國軍、落實國防自主、維護人民福祉、促進區域穩定」。[2]在過去的國防報告書中，有關國防戰略目標敘述有所變化。如《中華民國93年國防報告書》列出國防政策目標有三項，

---

* 國防大學戰略研究所副教授兼所長

[1] 國防戰略與國防政策的差別不容易區分，克勞賽維茲曾言，政策指導戰略，好像必須要有政策才能有戰略。但事實上，克勞賽維茲所言，政策乃政治戰略之意，戰略則為單純軍事戰略的內容，故強調政治戰略指導軍事戰略。但如果以戰略及政策的基本定義，戰略乃布局與安排，政策泛指政府、機構、組織或個人為實現目標而訂立的計畫，包含一連串經過規劃和有組織的行動或活動。可以想見戰略就是指形成政策前的規劃與安排，政策則是具體行動計畫。

[2] 國防部編，《中華民國106年國防報告書》（台北：國防部，2017年），頁55。

分別為：預防戰爭、國土防衛、反恐制變。[3]這三個目標，簡單明瞭地指出我國國防政策目標為何，也說明如何達成這些目標。《中華民國98年國防報告書》的戰略基本目標增加「防範衝突、區域穩定」兩項，[4]將區域安全問題與責任納入了國防戰略層次的目標。[5]《中華民國106年國防報告書》則將過去延續的三大目標或是五大目標大幅修改，但僅維持促進區域穩定，其他都已改變。[6]

不同年代國防報告書在國防戰略目標的說法不同，是否代表國防戰略已經改變，還是施政重點與觀點不同，恐怕仁智互見，這也是研究國防戰略或政策的難題。基於此，本文對於2020年國防戰略的探討，並不從過去國防報告書的概念與架構去分析，希望從國防戰略的實際內涵，加以探討。有關戰略分析的架構，參考美國陸軍戰院的萊奇模式（Lykke Model）加以運用，因為萊奇（Arthur F. Lykke, Jr.）認為戰略定義為目標、方法與手段的平衡。[7]從戰略目標與方法手段的運用與平衡，可看出整個戰略的適切性。戰略通常由上而下區分為國家安全戰略、國防戰略、軍事戰略、軍種戰略、野戰戰略（戰區戰略）等，本文研究主題在國防戰略層次，探討我國在國防戰略層次上，在面對2020年戰略環境變化時，戰略內涵（包括目標、方法、手段及風險）將產生何種變化。

貳、2020年台灣的戰略環境

制定戰略或是在調整戰略之前，必須對國家戰略環境進行客觀評估，尤其是中遠程的發展，才能據以研判可能威脅，明確戰略方向。就台灣戰略環境而言，亞太區域互動與安全連帶的會影響台灣戰略的走向，本文針對未來亞太戰略環境的發展進行評估，[8]探究至2020年時，亞太戰略環境變化及對台灣戰略發展的影響。

---

3 國防部編，《中華民國93年國防報告書》（台北：國防部，2004年），頁61。
4 國防部編，《中華民國九十八年四年期國防總檢討》（台北：國防部，2009年），頁42。
5 應該列於外交戰略，或者從屬於國家安全戰略的外交面向。
6 另外三項如建制專業國軍、落實國防自主、維護人民福祉，比較類似過去國防報告書有關國防政策內容。參見國防部編，《中華民國98年四年期國防總檢討》（台北：國防部，2009年），頁41；國防部編，《中華民國102年四年期國防總檢討》（台北：國防部，2013年），頁22-23。
7 Arthur F. Lykke, Jr., "Toward an Understanding of Military Strategy," in Military Strategy: Theory and Application (Carlisle Barracks PA: U.S. Army War College, 1989), pp. 3-8.
8 制訂戰略前，國家利益的認定與排序，國家目標的確認是非常重要的。參見 "Guidelines for Srategy Formulaton," J. Boone Bartholomees, ed., *U.S. Army War College Guide to National Security Policy and Strategy* (Carlisle Barracks, PA:U.S. Army War College, 1989), pp. 3-8.

一、外部環境

（一）北韓的核武擴張與轉折

2012 年金正恩接任北韓領導人後，為鞏固其絕對權力，曾發起多次整肅反叛勢力的事件，甚至殺害姑丈及多名高層官員，引發內部動亂。在外交方面，金正恩宣稱停戰協定失效，使兩韓關係再度緊繃；其強行發展核武引發國際非議，因而與原本關係友好的中共出現分歧。北韓是否發生巨變，造成東北亞戰略情勢的變化，備受亞太國家關注。在完成七次核武試爆後，美日等國決心對北韓採取軍事行動，然而北韓態度驟然轉變，轉而拉攏南韓，一起組隊參加冬季奧運會，另又密訪中共，以放棄核武作為條件，與美國會談，改變過去朝鮮半島的戰略格局。川普與金正恩在越南河內第二次高峰會的失敗，也令外界質疑。北韓是否又在玩弄兩手策略，是可能徹底改變朝鮮半島的緊張情勢，都將牽動亞太格局，對台灣國防情勢與安全也造成影響。[9]

（二）中日東海主權爭端的持續

日本首相安倍第二次擔任首相後，加強西南海域島嶼的軍事部署；中共則派出大量船艦和戰機在釣魚台海域活動，以回應東海主權問題，兩國陷入對峙態勢。2014 年中共公布東海防空識別區，引起美國、日本、韓國的關切，使東海成為區域衝突與爭議的源頭。中共軍事武力擴張，在民族主義催化下，對於釣魚台主權抱持強硬的立場，直接挑戰日本主權界線，將東海油田衝突與釣魚台爭議連結在一起，成為向日本外交施壓的重要議題。中共在軍力擴大後，不斷透過海警船及海軍軍艦，測試及衝撞日本公務船，中共船艦穿越宮古島航行，強化對釣魚台軍事行動的準備。日本則因此建立海軍陸戰隊，擴大西南諸島的軍事部署，反制中共對釣魚台軍事部署與意圖。[10]

---

9 Eric Chan and Peter Loftus, "The Taiwan Strait After a Second Korean War," *The Diplomat*, January 10, 2018, https://thediplomat.com/2018/01/the-taiwan-strait-after-a-second-korean-war/.

10 Shawn Snow, "First Japanese Amphibious Combat Unit Activated Since WWII Welcomed by U.S. Marines," *Marine Corps Time*, April 10, 2018, https://www.marinecorpstimes.com/news/your-marine-corps/2018/04/10/first-japanese-amphibious-combat-unit-activated-since-wwii-welcomed-by-us-marines/.

（三）南海主權與航行自由爭議

南海問題起於油源的爭奪及領海劃界的主權之爭，現有中共、印尼、汶萊、馬來西亞、菲律賓、越南及我國在強調南海主權的聲張。其中，中共、越南及菲律賓三國間爭執最烈，甚至引發軍事武裝衝突。印尼雖未占據任何島嶼，卻宣稱該國專屬經濟區、大陸棚海域與南沙群島重疊。東南亞國家之所以搶占南海諸島，無非是覬覦南海的地緣戰略價值。

南海是中共東南面海洋的戰略要衝，有利於中共海洋權利的維護，掌控南海軍事價值須先掌控南海諸群島礁的主權。中共在確保主權的前提下，雖未與東協及南海國家發生正面的對立與衝突，卻在近期採取強硬態度，不僅填海造陸以擴大島礁範圍，在南海部署雷達及反艦飛彈，試降民航機與軍用運輸機，展現海空武力；[11] 這些舉動壓縮到周邊國家的航行自由，未來可能引起軒然大波。未來中共解決南海主權問題後，打破美、日兩國的島鏈封鎖戰略；而在中共自製航空母艦建造完成，準備成軍部署之際，南海情勢未來的演變尤其令人關注。

（四）台海兩岸關係的冷和

中共對台一貫政策就是以「和平統一、一國兩制」為目標，希望透過密切往來，建構兩岸「和平發展」框架，避免台灣逃脫「一個中國」原則。從 2016 年 5 月，台灣再度政黨輪替之後，中共以台灣尚未承認「九二共識」為由，中止兩岸兩會的互動，改變了過去兩岸交流熱絡的情況。中共領導人習近平對台政策仍堅持軟硬兩手策略，主要在反獨促統，以軍事武力恫嚇反對台獨，另以經貿文化手段促進統一。蔡總統強調尊重 1992 年兩岸兩會會談的歷史事實，主張兩岸應該共同珍惜與維護 1992 年後，20 多年來雙方交流、協商所累積形成的現狀與成果，並在既有政治基礎上，持續推動兩岸關係和平穩定發展。未來會依據中華民國憲法、兩岸人民關係條例及其他相關法律，處理兩岸事務。[12]

---

11 "China Deploys Cruise Missiles on South China Sea Outposts – Reports," *The Guardian*, May 3, 2018, https://www.theguardian.com/world/2018/may/03/china-deploys-cruise-missiles-on-south-china-sea-outposts-reports.

12 曾薏蘋，〈總統蔡英文：盼 19 大後兩岸思考互動新模式〉，中時電子報，2017 年 10 月 3 日，http://www.chinatimes.com/realtimenews/20171003002656-260407。

雖然這樣的說法強調維持現狀，但中共以未完成「九二共識」的答卷為由，暫停兩岸的制度性交往。但在十九大之後，習近平掌握大權，人事也大幅變動，兩岸關係的改變仍在動態調整中。尤其在 2019 年年初，習近平發表對台五點講話，希望與台灣對話討論「一國兩制」台灣方案，進一步增加對台壓力。到了 2020 年，兩岸雙方如仍處冷和情況下，將形成動態平衡，因而延續過去戰略態勢。

二、內部環境

（一）政治勢力消長

2020 年以前的兩岸關係大致因為台灣政治情勢的延續而無重大變化，但是在 2018 年地方選舉後，政治版圖變化可能影響總統選舉結果，也讓兩岸關係發展受到牽動。毫無疑問的，中共藉由兩岸密切互動，干預或介入台灣政治的程度會越來越高，尤其是重要的選舉。也不能排除中共將會運用各種政治、經濟、文化的影響力，主導台灣政治風向，或是藉銳實力及假新聞操作選舉過程。[13] 2018 年台灣九合一選舉，執政的民進黨重挫，政治版圖改變。如果 2020 年蔡總統繼續執政，且兩岸並無重大政策轉折，兩岸關係將會延續冷和現況。

如果 2020 年台灣產生政黨輪替，由中共所公開或暗中支持的對象取得政權，有可能與 2008 年政黨輪替一樣，因為國家安全政策與指導的改變，影響國防戰略發展。如果台灣兩岸政策改弦更張，且往和平促統方向發展，台灣與周邊國家關係，尤其是在軍事合作面向，會立即受到影響，連帶在軍購及小型聯合演習等，將會轉變。在國防戰略方面，只要中共不放棄對台使用武力，或者繼續以武力當作政治工具，壓迫台灣接受其政治條件，國防戰略不會因為兩岸關係發生重大變化。如果兩岸關係越來越密切，政治與經濟將成為優先議題，國防戰略議題會弱化，軍隊角色與規模裁減也會不斷地提出討論。尤其在兩岸情勢和緩下，軍隊功能受到質疑後，募兵也會受到影響。[14]

---

13 吳敏之，〈中國特色的「銳實力」套餐，正在滲透全世界〉，關鍵評論網，2018 年 3 月 15 日，https://www.thenewslens.com/article/91055。

14 此處所指的影響是指因為軍隊功能弱化，規模裁減，社會價值降低，導致年輕人入伍意願低落，必須裁減規模以配合募兵人數的減少。

（二）社會對國防的支持（全民國防）

軍隊與社會關係密不可分，當社會尚武風氣甚佳，許多軍隊本身的價值與捍衛國家安全的目的也會受到肯定。反言之，如果社會普遍認為國防並非最優先國家事務，國防政策得不到人民支持，整個國防戰略與施政也會受到影響。原則上，經過全民國防教育實施多年的結果，社會對於國防事務已有一定的了解，從而能支持國防建設。尤其當國內政治都能在國防戰略具備共識時，社會對國防支持不會有重大變化。但如果因為不同政黨在兩岸政策的差異，衍伸到對國防戰略內涵的分歧，或是國防政策的優位性降低，社會對國防事務的支持也是分歧的。

（三）兵役與動員制度的變化

兵役制度改為募兵制之後，召募人員可以長期服役，訓練層次及內涵可以更為提昇，但是根據國際經驗，召募人數是否足夠，以及對我國原有動員制度的影響，就成為值得關注的議題。2020 年時，仍須服義務役的役男除特殊個案外，所剩無幾，動員對象絕大多數均為受過四個月訓練役的役男。這與過去至少服役一年，或者具備「一專多能」的後備士兵而言，不論在選充及訓練上，都有重大差異，連帶也會影響國防戰略運用。

除非未來遇到戰爭等重大事件的影響，才有可能改變政策，否則募兵制應會持續實施。在此情況下，兵力規模遇到戰爭必然不足，只能仰賴動員制度來增加戰力。過去傳統防衛作戰，需要許多透過動員編成的守備旅，在責任地境內進行防衛作戰任務。如果選充動員對象都是四個月訓練役的後備軍人，整體動員制度必須隨之調整。防衛作戰架構下，作戰區與動員的責任與指揮隸屬關係也須加以調整。如果動員制度無法滿足傳統作戰所需，因應防衛作戰的戰略與部署，必然有所變動。

（四）國防科技能力

在國防自主政策下，[15] 2020 年有些武器系統與平台，已能服役部署，並發

---

15 因為變數太多，此處並不探討 2020 年可以獲得軍購品項及影響。

揮應有戰力。如台灣現有 144 架的 F-16A\B 性能提升為 F16V，至 2022 年將全部完成；[16] 空軍藍鵲高級教練機於 2020 年原型機首飛，2026 年完成 66 架高教機交機，[17] 屆時可大幅提昇空軍戰力。

在船艦製造方面，我國自製潛艦「造修合一」，估計將建造 6 至 8 艘潛艦，第一艘 2024 年就會下水。[18] 國艦國造的政策下，首艘配置海劍二防空飛彈第一批量產型的沱江級軍艦，除配備雄二反艦飛彈及雄三超音速反艦飛彈外，因強化防空作戰能力，沱江軍艦 2.0 版會配置中科院自行研發的艦射型天劍二型飛彈（海劍二飛彈系統），以及對空 3D 雷達。首艘防空型的沱江艦預計在 2020年下水成軍後，即可大幅提升艦隊的區域防空能力。[19]

在飛彈製造方面，根據媒體報導，我國有關中程戰術飛彈研發計有代號「雲峰專案」、由中科院研發改良型射程 2000 公里的中程地對地飛彈，以及正在量產、代號「戟隼專案」的雄二 E 增程型巡弋飛彈，射程達到 1000 公里以上。為了加強戰力強度，雄二 E 在 2019 年將量產 240 枚，2020 年開始量產雄二 E 增程型，提昇制海能力。[20] 在遠程反制武力完成部署後，可以部份取代海軍艦艇與空中戰機，增加重層嚇阻的縱深。

（五）戰術戰法變化

因為中共軍事能力改變，至 2020 年整個台澎防衛戰術戰法也應隨之調整。例如資訊戰或是反三戰都是戰爭發生前及作戰全程必須密切注意的戰法。過去強調戰力保存、制空、制海、反登陸以應對來自西方中國大陸來犯敵軍，最後則於台灣本島進行決戰。但因為中共軍改之後，除了三戰與資訊戰之外，戰略戰術飛彈的猝然攻擊、特戰部隊的占領要地會師及包圍作戰、航母戰鬥群的海

16 〈台灣明年起升級 F-16 未來 10 年不會獲得新戰機〉，大時代，2017 年 8 月 23 日，http://www.twgreatdaily.com/cat35/node1086846。

17 喻華德，〈國造藍鵲高教機 可能增加對地攻擊能力〉，中時電子報，2017 年 12 月 6 日，http://www.chinatimes.com/realtimenews/20171206005984-260417。

18 洪臣宏，〈台船明年初建造潛艦廠區首艘估 2024 年下水〉，自由時報，2018 年 1 月 23 日，http://news.ltn.com.tw/news/business/breakingnews/2320521。

19 王炯華，〈沱江艦配備海劍二 2020 年前將成軍下水保衛台海〉，蘋果日報，2018 年 1 月 8 日，https://tw.appledaily.com/new/realtime/20180108/1274211/。

20 〈台灣重啟軍備飛彈研發解封〉，亞洲週刊，第 32 卷第 6 期，2018 年 2 月 11 日，https://www.yzzk.com/cfm/special_list3.cfm?id=1517455306467。

域封鎖及顛倒正面作戰，都讓台灣既有戰術戰法面臨新的挑戰。另如中共隱形戰機的攻擊、配合太空衛星的指管通情運用特戰與空降的聯合運用，將使原有防衛作戰計畫必須相應調整。

參、2020 年中共軍事威脅的發展

根據習近平觀點，以及修改憲法延長國家主席的任期，在經歷兩個任期的 2022 年以後，習近平仍將掌握黨政軍權力，並且親自驗收建黨百年（2021 年）的施政成果。如果台灣仍舊堅持兩岸維持現狀的主張，中共對台昇高軍事威脅，迫使台灣接受「一個中國」政治框架的意圖與做法不會改變。中共對台軍事戰略以積極防禦為主軸，就台海現狀發展出特定作戰手段與方法。積極防禦運用在台海戰場與兩岸關係上，更凸顯其積極性與攻擊性的意涵，不只是一種守勢戰略，更成為共軍作戰與建設的戰略指導方針。[21]

中共積極防禦戰略在台海的運用其實質目的不在追求和平，而是希望以積極防禦中的積極手段，加強軍事鬥爭準備，提升軍隊現代化作戰能力，尤其是資訊化條件下作戰能力，做為以戰迫和，或爭取更大政治籌碼的主要憑藉。[22]除了運用三戰及銳實力等軟性作為誘使台灣屈服之外，國防戰略主要面對中共的軍事威脅，在 2020 年期間呈現出以下發展變化。

一、隱形戰機大量運用

在現代高科技戰爭中，隱形戰機在空中突襲、制空權爭奪及戰略目標轟炸扮演重要角色。中共隱形戰機以殲 20 最為成熟，已經部署空軍作戰部隊。殲 20 是中共自製的隱形戰機，2018 年 2 月列裝空軍作戰部隊，最大速度 2.8 馬赫，作戰半徑約 2,200 公里，幾乎覆蓋東海、台海、南海的空域。可以和殲 16、殲 10C 等多型戰機聯合展開實戰背景下的空戰訓練，擔負起不同作戰任務，也參與繞台巡航的任務。[23] 2020 年以後，隱形戰機數量增加，可以執行單一作戰任

---

21 姜普敏、臧士明，《當代中國的馬克思主義軍事理論》（北京：國防大學，1997 年），頁 205。

22 張萬年，《當代世界軍事與中國國防》（北京：軍事科學，2002 年 7 月），頁 204。

23 〈參與繞台？殲 20 首次展開海上方向實戰化軍事訓練〉，東網，2018 年 5 月 9 日，http://hk.on.cc/cn/bkn/cnt/news/20180509/bkncn-20180509160314461-0509_05011_001.html。

務。除了殲 20 之外，中共也研發殲 31 戰鬥機 [24] 及轟 20 隱形轟炸機 [25]，目前雖在研發階段，預估在 2020 年代，應該可以測試及列裝部隊服役，並且形成對外隱形空中作戰的主力。

二、兩個以上航母戰鬥群成軍與作戰

　　從中共海權觀及海軍發展來看，建構遠洋海軍的意圖，應該毫無懸念。目前中共僅有一艘從舊艦改良的訓練用航空母艦遼寧號，並擔負著訓練與作戰的任務。雖然其航行範圍僅在第一島鏈附近，最遠到南海海域，但在中共第一艘與遼寧號同型航空母艦自製完成下水，[26] 並完成所有測試服役後，將使遠洋作戰更具彈性，遠程兵力投射的運用更為頻繁。與航空母艦完成相關的航母打擊群的作戰準則發展，以及隨伴陸戰隊特遣隊的兵力遠程投射，也會成為常態。屆時台灣東部海域即可能面對中共兩個航母打擊群（含潛艦）的軍事威脅。[27] 據美國國防部在 2017 年 6 月初公布的「2017 年中國軍力報告」研判，[28] 中國後續自製航空母艦，可望在 2020 年具備初始作戰能力（Initial Operating Capability, IOC），2025 年具三艘航母戰鬥群，[29] 2030 時將有四艘航母及戰鬥群。[30] 如果中共參照美國處理區域危機的做法，派遣三至四個航母打擊群到台灣海峽及東部海域。過去台灣傳統制海作戰思維與戰法，也必須隨之調整。

---

24 〈罕見！中國兩架殲 31 同時試飛〉，聯合新聞網，2018 年 3 月 20 日，https://udn.com/news/story/7331/3041319。

25 許依晨，〈恐劍指台海！殲 20 近日首次海訓轟 20 將首飛〉，聯合新聞網，2018 年 5 月 10 日，https://udn.com/news/story/11323/3133726。

26 涂鉅旻，〈中國首艘自製航艦下水最快 2019 服役〉，自由時報，2017 年 4 月 27 日，http://news.ltn.com.tw/news/politics/paper/1097600。

27 王洪光，〈遼寧艦戰時應搶佔台灣東部陣位威懾美日〉，環球網，2017 年 1 月 16 日，http://news.sohu.com/20170116/n478805412.shtml。

28 Office of the Secretary of Defense, *Military and Security Developments Involving the People's Republic of China 2017*(Washington D.C.:U.S. DoD, 2017), pp. 50-65.

29 Liu Zhen, "China Aims for Nuclear-powered Aircraft Carrier by 2025," *South China Morning Post*, March 1, 2018, http://www.scmp.com/news/china/diplomacy-defence/article/2135151/china-aims-nuclear-powered-aircraft-carrier-2025.

30 Nyshka Chandran, "China's First Homegrown Aircraft Carrier Is A Sign of The Country's Military Aspirations," *CNBC*, April 25, 2018, https://www.cnbc.com/2018/04/25/chinas-homegrown-aircraft-carrier-type-001a-will-start-sea-trials.html.

三、特戰、空降與陸戰隊等全境快反兵力運用

美國處理區域性衝突時，經常使用特戰部隊、空降部隊與陸戰隊投入戰區，以快速兵力投射，並以少量精銳部隊快速掌握戰場狀況，或執行戰場綏靖與重建任務。目前中共陸軍合成旅部隊正在進行跨境演習及遠程投送訓練，逐步建立全境作戰能力。另外像陸戰旅與特戰旅的擴編，都已經建立隨時快速反應的作戰兵力。中共陸軍各集團軍都編有特戰旅，空降軍也改編為六個旅，陸戰隊則預計擴編為十萬人，這些部隊編成等於解放軍遠程投射主力。換言之，基本上如果中共透過空降突襲方式作戰，以其戰略空運能力，只能空投一個空降營，20 架運 20 往返 5 次可投送一個完整的空降旅。[31] 但如果中共在 2020 年擴增重型運輸機隊，其遠程投送能力更強。

四、遠距精準戰略飛彈擴增

中共過去對台海使用反介入戰略時，其所依賴的殺手鐧武器除了空中與海上武力之外，主要就是各型戰略戰術飛彈的運用，如反艦彈道飛彈、洲際彈道飛彈與戰術飛彈。在遠程兵力建構完成及航母打擊群建構完成後，海空武力能夠攜帶更多射程更遠、更精準的飛彈，中共也研發製造許多專門特定對象的彈道飛彈。如東風 21D 型飛彈專門用以反制船艦（航空母艦）使用、[32] 東風 26 型飛彈則是針對美國關島基地使用、[33] 東風 16 則是針對駐守琉球美軍及台灣。[34]

五、一體化聯合作戰的提昇

中共軍隊當務之急除了發展先進武器系統、強化部隊演習之外，重點在如何讓軍隊在指管通情系統整合下，提昇一體化聯合作戰效能。根據中共軍改的

---

31 〈要多少架運 20 才能一次完整空投一個空降旅？〉，每日頭條，2017 年 9 月 18 日，https://kknews.cc/zh-tw/military/yneypen.html。

32 Harry J. Kazianis, "Is China's "Carrier-Killer" Really a Threat to the U.S. Navy?" *The National Interest*, September 2, 2015, http://nationalinterest.org/blog/the-buzz/chinas-carrier-killer-really-threat-the-us-navy-13765.

33 楊俊斌，〈關島快遞陸秀東風 -26B 飛彈旅〉，中時電子報，2018 年 4 月 18 日，http://www.chinatimes.com/newspapers/20180418000062-260301。

34 中共火箭軍在 2015 年公開東風 -16 飛彈的部署訓練情況。東風 -16 射程 998 公里，可以涵蓋「第一島鏈」如打擊日本沖繩、菲律賓和台灣地區，尤其美國在日本沖繩的軍事基地，素有「沖繩快遞」之稱。參見陳孟孟，〈東風 -16 最新部署訓練曝光「沖繩快遞」可打美軍基地〉，今日新聞網，2017 年 2 月 9 日，https://www.ettoday.net/news/20170209/863338.htm。

內容，主要在適應一體化聯合作戰指揮要求，建立軍委、戰區兩級聯合作戰指揮體制，要求平戰一體、精幹高效的戰略戰役指揮體系」。在軍隊作戰指揮體系上，「按照聯合作戰、聯合指揮的要求，調整規範軍委聯指、各軍種、戰區聯指和戰區軍種的作戰指揮職能」。[35] 換言之，過去守勢傳統作戰形態已經無法因應快速的攻勢作戰需求，或已經具備現代化聯合作戰能力的軍隊。所以解放軍才會強調對網絡超廣域、無線超寬頻的追求，不僅是國家資訊基礎建設重點和經濟利益增長依賴，也成為建構資訊化軍隊、打贏資訊化戰爭的核心。中共有關集團軍所屬合成旅作戰的演習已經納入一體化聯合作戰概念，至 2020 年時，運用將更為成熟。

肆、2020 年台灣國防戰略的探討

一、國家利益與目標的變與不變

　　國家目標是制定戰略的基礎，但是國家利益優先順序則是制定國家目標的重要考量。國家利益多半是指能夠維持國家安全與發展的利益，或者是應該得到的利益，如經濟、安全、民主價值等。就一個國家而言，國家利益與目標與國家安全戰略息息相關，故將國家安全戰略視為追求國家利益所制定的目標，以及達成目標所運用的方法及資源合理配置，可稱為國家安全戰略。[36]

　　但是在國家安全戰略之下，不同領域或部會如何達成國家安全戰略目標，也須制訂本身的戰略。例如在國家（安全）戰略之下，如何在國防領域上達成，就成為國防戰略。當然在國防戰略目標下，如何達成的方法與資源分配，必須與之相平衡。2020 年台灣的國家利益首要仍在經濟發展與安全，但在面對逐漸走上軍事強國的中國，並以軍事、政治與經濟的綜合力量壓迫台灣之際，如何維持民主自由生活方式，確保獨立政治主權下，提昇經濟發展，創造人民福祉，應為 2020 年代的國家首要利益。

　　國家目標就是在營造維護國家首要利益的環境，並透過國家各種能力，如內政、外交、國防、經濟、文化等，避免各種安全威脅，以保障及提昇各種利益。

---

35 任春陽，〈未來一體化聯合作戰靠什麼？〉，人民網，2005 年 10 月 19 日，http://military.people.com.cn/BIG5/1078/3781946.html。
36 個人參考萊奇模式（Lykke Model）對於國家安全戰略的定義。

如蔡總統在國防安全研究院揭牌所言，國家將堅定捍衛國家，守護台灣，保護自由民主生活方式，不會容許這些基本價值受到威脅，這是政府不變的使命。[37] 政府也主張維持現狀是不變立場，也是台灣向區域、向世界的有效承諾。未來會善盡區域安全的責任，持續保持善意，維持一個穩定、一致、可預測的兩岸關係。[38]

為了防範中共軍事武力進犯，必須強化國防整備與戰力訓練。為避免在國際社會被孤立，強化與國際社會的連結，必須提昇務實外交，以自身經濟與文化實力，參與國際社會活動，獲得國際認同。為強化安全預防工作，維護區域穩定，應強化與印太國家的安全合作等。這些分項工作就是在國家安全戰略指導下，各部會所應執行的政策。

二、國防戰略目標

如上述，國防戰略應該是從國防層面思考如何去達成與落實國家安全戰略，在目標、方法與手段上進行的平衡與安排。我國現行國防戰略目標包括：防衛國家安全、建制專業國軍、落實國防自主、維護人民福祉、促進區域穩定」等五項。若就其性質而言，前三項屬於國防戰略的權責與範圍，但是第四項的維護人民福祉會與內政部權責混淆，第五項的促進區域穩定，則與外交部或是國安會的負責事務重疊。

因此，本文認為綜合考量 2020 年的內部與外部環境因素，中共軍事威脅狀況，國防戰略目標應該包含下列三點：以有效嚇阻的國防武力確保國家安全；以精實訓練與組織變革促進國軍專業化；以國防自主的效益提昇作戰持續力。分述如下：

（一）確保國防安全

如果國家安全戰略對於兩岸關係採取維持現狀的態度，除非中共內部發生重大情勢變化，或者必須刻意挑起兩岸戰爭，否則中共並無發動對台戰爭的合

---

37 〈出席國防安全研究院揭牌總統對我與多明尼加終止外交關係表達政府嚴正立場〉，總統府新聞稿，2018 年 5 月 1 日，https://www.president.gov.tw/NEWS/23304。

38 〈主持年終記者會總統重申持續各項改革帶領臺灣邁向高峰〉，總統府新聞稿，2017 年 12 月 29 日，https://www.president.gov.tw/News/21895。

法性與正義性。但是國防安全不能依賴敵方的善意，必須具備戰力堅強的武力，以嚇阻敵方的貿然進犯。軍隊的存在本來就是在捍衛國防安全，至 2020 年，國防武力的首要目標仍是在確保國防安全。

（二）促進國軍專業化

提昇軍隊作戰能力產生足夠嚇阻能力是每一個國家軍隊的目標，但是能力的提昇並非軍隊編制形成就可達到目的。應該透過精實訓練與國防組織變革，使軍隊透過專業能力的發展，維持軍隊運作於不墜，隨著時代的延伸而不斷精進。尤其是軍事教育與人才培育方面，除了擔任不同軍事職務所應具備的本質學能外，軍人武德的涵養，包含軍事倫理與領導能力的培養，都在專業訓練的範圍內。在國防組織變革方面，國軍歷經數次組織變革，都是認為敵情威脅和緩、國防支出過多前提下，逐步削減軍隊規模。實施募兵制後，如何在 2020 年前決定軍隊最適規模，以及因應高科技戰爭的軍隊組織非常重要，這是促進軍隊專業化的基礎。

（三）國防自主的效益提昇

受到軍購環境的影響，台灣許多高科技武器必須仰賴先進國家的出售或援助。這樣的作法雖能使台灣在極其艱難情況下，獲得先進武器系統，但是在軍售價格、後勤補保、技術合作、使用自主性等，都受到一些限制。尤其當先進武器系統的零附件與補保受制於人，瀕臨戰爭狀態之際，或在作戰期間將面臨作戰持續力的問題。未來如果要捍衛台灣主權，維持區域和平，不能單憑他國的承諾，必須透過自身國防能力的提升，才能確保國防安全。因此，如何擴大國防自主的能力與科技範圍，提昇國防科技成效，也是國防戰略的主要目標。

三、國防戰略的方法

（一）適合台海戰爭軍事戰略

對中共而言，解決台灣問題的理想軍事戰略是不戰而屈人之兵，透過強大軍力的威脅與脅迫，讓台灣屈服並接受中共政治條件。如果必須動用軍事武力，則可能採取美國入侵伊拉克的方式，希望在最短時間內完成軍事行動，以免陷

入戰爭泥淖。換言之，在攻勢作戰過程中，如能以高科技戰爭的方式，迅速有效的摧毀戰略目標，讓敵方屈服，達到政治目的。中共如果進犯台灣，同樣也透過快速有效的方式達到目的，並集結所有可以遠程投射兵力，在外力尚未介入的最快時間內達成戰略目標。

就守勢作戰一方而言，如果不能形成戰略嚇阻武力，就必須設法以本身的常規武力，反制、破壞與延遲敵人目標的達成。如果攻勢一方採取速決戰的方式，守勢一方就必須採取持久戰的方式。但是如果兵力規模過於懸殊，守勢一方如要採取持久戰就必須透過總動員制度，或是以不對稱作戰或劣勢戰略的方式，破壞敵人軍事目標的達成。在軍事戰略方面，台灣依據防衛固守、重層嚇阻的軍事戰略，以及濱海決勝，灘岸殲敵的作戰指導，日益傾向以遠距精準的飛彈武力逐次損耗中共犯台武力，以取代那些戰爭時期不易戰力保存的武器。選擇載台應基於戰力保存，維持關鍵戰力機動性的思考。至 2020 年，類似軍事戰略改變空間不大，必須注意中共軍隊在精準打擊、遠程投射、特種作戰能力的提昇，若無足夠遠程反擊與防護武力，必須思考持久戰的各種戰術戰法的運用。

（二）堅實的動員制度

一般國家因為常備部隊人數減少，如果遇到大型戰爭，則須依賴動員制度來加以彌補，美國也發現類似問題。[39] 由於募兵制已經無法回頭改為徵兵制，則必須從動員制度去發展，使過去訓練優異的後備軍人可以補充常備部隊的缺額及戰耗補充，或是投入後備部隊執行各種守備任務。如果執行不對稱作戰，應該以特戰部隊擔任動員部隊的基幹，擴大編成城市後備部隊與山地游擊部隊，進行側翼襲擾與牽制行動。一些高科技作戰部隊如資通電、飛彈、飛行等，則應該做好精準動員編管工作，定期進行長時間的召集訓練，或是融入常備部隊共同進行演習任務，使兵役制度變革產生的兵源問題透過動員解決。

---

39 美國眾議院軍事委員會日前通過 2019 財政年度國防授權法案（NDAA FY 2019），根據軍委會公布的草案內容，第 1243 條規定，美國國防部長應與相對應台灣部門協商，全面評估台灣軍力，特別是後備軍力，以提高台灣自我防衛能力。參見〈美眾院 60：1 通過「國防授權法」草案，要求全面強化台灣軍力〉，關鍵評論網，2018 年 5 月 11 日，https://www.thenewslens.com/article/95483。

（三）穩定的兵役與召募制度

除非發生重大變局或戰爭，兵役制度大致持續實施募兵制。因為少子化及社會風氣影響，使兵役制度變革造成召募的問題。人事乃執行軍中一切事務執行的基礎，如果人力不足或是人員編制與訓練不能結合，連帶影響戰力及任務遂行。[40] 兵役與召募又與社會支持有關，社會的支持度則與社會風氣、政治環境有關。如果兩岸情勢和緩，可能會募到一些以軍人為穩定職業的太平軍人，缺員的迫切性較不嚴重。

當兩岸情勢升高，甚至瀕臨戰爭狀態時，貪生怕死者不願入營，但充滿愛國情操者，卻可能願意投入救國行列。但是上述兩者都非常態，不能被動的依照兩岸情勢或是內部政治因素發展，解決兵源問題。而是要建立一個穩定的兵役與召募制度，讓每一個軍種不同層級職務的發展，都能獲得公正評價，獲得社會與人民的信賴，同時可以召募以軍隊為職志或職業的專業人員。

（四）持續永久軍備發展政策

在中共積極擴張軍備，以新型飛機船艦繞越台灣，進行文攻武嚇之際，台灣如能藉由台美合作獲得尖端武器研發製造能力，從而具備後勤維修能量，將使國軍戰力大幅提升，擴大戰爭持續力，增強自主作戰能力。最後，可以讓國防產業提升可以帶動國內經濟發展，以國防自主能量建構嚇阻性武力。[41]

針對軍隊所青睞的夢幻武器提出需求清單不難，只要將適用於台灣防衛作戰的美軍現役武器列出即可，就好像過去美國一次同樣出售八項武器系統一般。卻因為預算總額太高，台灣只能針對三項武器系統列出六千億預算，終因在野黨的抵制而功敗垂成。美國要台灣大膽提出軍購項目，其實並無十足把握可以全數賣給台灣，而是希望列出清單優先順序，提供台灣適切的高科技武器。但這做法，必須因永久性國防自主政策及軍事戰略的改變，而須有所調整。[42]

---

40 因此參謀作業的第一項就是人事作業，即所謂的參一。
41 社論，〈台美國防產業合作的里程碑〉，臺灣時報，2018 年 5 月 13 日，http://www.twtimes.com.tw/index.php?page=news&nid=715693。
42 社論，〈台灣需要那些軍購項目〉，臺灣時報，2018 年 2 月 4 日，http://www.twtimes.com.tw/index.php?page=news&nid=701298。

在國防產業發展上如能運用政策支援及與美國國防核心業者合作，籌建產業關鍵技術能量，補足航太船艦供應鏈產業缺口，為台美雙方國防產業共創雙贏。更重要的是，台灣正在追求的航太和船艦產業，都有技術門檻與附加價值高的特性，也是先進國家科技研發的重要指標。

四、國防戰略手段（資源）

（一）預算

任何優劣的國防戰略都需要足夠的國防預算來支持。根據媒體報導，面對中國大陸軍事威脅不斷升高，以及在國家全力推動國防自主的情況下，台灣的國防預算將在 2025 年提升至 3,800 億台幣。目前台灣每年國防預算約 100 億美金（約 3,300 億台幣），將增加至少 20%。增加的國防預算將優先投入新型飛彈、無人飛行載具、電子戰系統、戰鬥機，以及彈道飛彈防禦系統等軍事設備及武器購買或研發。[43]

前總統馬英九在競選總統期間，曾承諾當選後國防預算要占國內生產總額（GDP）的 3%，事實上卻逐年下降。蔡英文政府上台後，國防預算也未達 GDP 的 3%。2018 年實施募兵制後，人員維持費用將占大多數國防預算支出，勢必排擠武器採購與維護費用。如果要增加職業軍人的各種加給，或調升薪資，都或增加年度預算的需求。如果要研發先進的高科技武器或是要購置美國的先進武器系統，都需要增加國防預算。除了國防自主花費仍留在國內國防產業，既可提昇國防又能促進經濟發展，將產生經濟的循環。但如果要外購高科技武器，以及支應募兵制後薪資調整，國防預算必須大幅增加，除特別預算外，應該恢復至國內生產總額（GDP）的 3%。

（二）人力

軍隊到底需要多少人力才能捍衛台海安全，一直是一個人云亦云的量化數字。對守勢作戰一方而言，防衛作戰所需兵力是越多越好。但是在人力配置及預算負擔上，不可能無限制的滿足人力需求。依照國防報告書資料，目前軍隊

---

43 呂炯昌，〈建構不對稱戰力國防預算將增至 3,800 億〉，今日新聞，2018 年 1 月 13 日，https://www.nownews.com/news/20180113/2682448。

總員額約為 21 萬人，如果按照軍種及戰鬥、戰鬥支援、後勤支援等類別加以區分，其實投入第一線作戰兵力仍然有限。若在召募無法滿足下，應先以現有兵力總員額為準，進行依照作戰方法所調整的兵力結構。例如海空武力如何配置可以反制中共遠海航訓的威脅、發展不對稱作戰應該編組何種相對應的部隊編制、守備部隊人數及常後備比例等，都必須經過精準的計算與安排。更重要的是，應該避免專業人才流失，產生人力不足的狀況。

（三）政策與部會整合

依照國防法的邏輯，國防戰略與政策的核定屬於行政院的權責，[44] 國防部門僅提出政策建議。由行政院主導的主要目的是因為國防事務非常龐雜，涵括國防部以外的其他相關部會，如外交、經濟、文化、教育、交通等，並且融合在防衛動員會報的機制之內。如精神動員準備方案由教育部主管；人力動員準備方案由內政部主管；物資經濟動員準備方案由經濟部主管；財力動員準備方案由財政部主管；交通動員準備方案由交通部主管；衛生動員準備方案由行政院衛福部主管；科技動員準備方案由行政院國家科學委員會主管；軍事動員準備方案由國防部主管。[45] 按照這樣的分工，作戰時期的戰爭準備已經有了一套非常完整的法規與程序。

但是在平常時期，與國防有關事務也需要跨部會的合作，如資安作戰、戰場經營、重大演訓、年金改革、國防自主、全民國防教育、精神動員、兵役制度、救災應變、反恐等，更需要國防部與其他部會之間的整合。當國防需求殷切，但無法將預算或資源集中放在國防部門時，亦可由相關部會針對軍民通用科技、輔助措施等，編列預算支應，或以專案部會整合的方式，組成任務特遣隊，處理相關事務。另應定期也應該針對與國防有關的跨部會事務，進行檢討與修正，以擴大政策整合的效益。

---

44 國防法第 10 條規定，行政院制定國防政策，統合整體國力，督導所屬各機關辦理國防有關事務。
45 參見全民防衛動員法第 9 條。

五、可能面臨風險

（一）區域衝突與戰爭

任何戰略皆有風險存在，必須在戰略制定時將其排除或化解。從兩岸關係發展來看，如果持續冷和，兩岸現行戰略不會在短期產生重大的變化，但會依據軍事變革與重大武器系統的陸續服役，強化戰略戰術運用。但是如果發生區域衝突與戰爭，有可能台海受到牽動而爆發衝突或者因為中共企圖轉移內部政爭焦點，在部分衝突熱點挑起小型戰爭，連帶造成原有戰略情勢的變化。如果這些戰爭係周邊國家與中共間的衝突，我國無法消弭衝突下，只能做好戰爭擴大的應變防範準備。此種防範與準備不會改變原有國防戰略，但如果變成長期性衝突，或是台灣被捲入衝突，國防戰略與準備也會受到影響。

（二）兩岸內部政治重大變化

軍事衝突通常因為政治原因所產生，軍事武力則是達成政治目的所使用的工具。國與國之間的衝突可能因為領導人更替而昇高，也可能因為政黨輪替而和緩。或者因為執行高度爭議政策引起他國報復，另因為內部動亂，召引他國的入侵。2020 年中共政治發生動亂的可能性將會降低，習近平可能繼續掌握黨政軍大權，陸續完成軍隊改革，增強軍事武力，將之用於領土主權問題的解決。台灣內部政治情勢則以 2020 總統選舉為關鍵，如果兩岸持續冷和，但都以經濟發展為發展首要，國防與軍事衝突不致發生。但如果因為區域情勢或中共內部發生變化，使兩方誤認可以成功採取改變現狀的作為，則有可能造成衝突與戰爭。

（三）國防戰略的延續與變化

我國國防戰略基於敵情威脅與地理環境的設想，台澎防衛作戰思維與戰法大多為延續性。但隨著中共軍力變化，以及周邊情勢的發展，外來威脅的進犯方式也可能採取猝然突襲方式。共軍對於東海及南海所進行的遠海軍事演習，或是航空母艦逐漸跨越第一島鏈繞行台灣東部海域的例行性訓練航行，使台灣面對來自東方軍事威脅大增，整個作戰線都會受到影響。以往將台灣東岸為第

二線戰場的國防戰略概念，恐怕要立即加以調整，[46] 甚至要有逆向思維，周密準備來自不預期方向的攻擊行動。

如果國防戰略不知變通，或是未隨時空環境改變而調整，致使國防戰略的延續與變化，將造成作戰失敗的風險。作戰內外環境因素瞬息萬變，國防戰略也應該隨之調整。面對 2020 年國防戰略的風險，國防戰略與時俱進，又能避免錯誤非常重要。

伍、結語

如果依照習近平在十九大所說新三步走策略，2020 年只是近程目標，希望達成全面小康社會。[47] 在全面達成小康社會的同時，軍隊訓練與組織變革也仍在動態變化當中，尚未發展成穩地狀態，追求小康社會也不宜對外發動衝突。就中共而言，2020 年是非常重要的轉型與發展關鍵期，不論是國家體制或是軍隊體制與訓練的發展。中共不會輕率對外發動戰爭，但是對於領土主權爭議，因為面對民族主義的倡導，也絕不能示弱。

對台灣而言，同樣的，在民主多元環境下，任何單一激進主張都不會過半數，多數人仍主張維持現狀，但維持民主自由價值與生活方式，如此也不會挑釁中共，或者並無中共入侵的合法性藉口。若能因為雙方克制而維持兩岸動態穩定的和平狀態，都可以給雙方一段努力發展經濟、致力國防改革的空間。但問題是，如果兩岸不能建立互動溝通管道，即使能夠因為冷和，得到一段可以致力經濟發展與國防轉型的穩定期，兩岸之間的衝突因素仍然存在。如果以武力解決的選項仍存，透過國防戰略贏得戰爭的思考仍會持續存在，相關作戰準備及軍備武器發展也會持續，國防戰略也必須隨著戰略環境、目標、方法及手段的平衡而與時俱進。

---

46 沈明室，〈中國積極強化軍事訓練的對台意圖〉，臺灣時報，2013 年 2 月 9 日，http://taiwanus.net/news/news/2013/201302091705111284.htm。

47 〈楊偉民介紹新目標：全面建成小康社會〉，中國評論新聞，2017 年 10 月 26 日，http://hk.crntt.com/crn-webapp/touch/detail.jsp?coluid=7&kindid=0&docid=104857436。

法盾與矽盾：兩岸天平傾斜下台灣國家安全戰略的再審思

羅慶生 *

　　長期以來，中國大陸不排除以武力統一台灣的企圖一直是我國家安全上的最大威脅，且隨著大陸經濟與科技的快速增長，兩岸實力天秤愈益傾斜，威脅性愈來愈高。雖然習近平在〈中共十九大政治報告〉中，顯示「兩岸關係和平發展」仍為政策的主旋律，[1]但「武統論」在大陸網路社群的討論聲浪一直很高，若長期無法降溫甚至發展成全民共識，則中共也可能被迫回應民意而有動武作為。面對解放軍可能的武裝犯台，國防部在蔡政府執政後提出「防衛固守、重層嚇阻」之軍事戰略，強調「發揮重層嚇阻戰力，防衛國土安全，嚇阻敵不敢輕啟戰端」，[2]雖然這在國軍仍具有局部的質量優勢，有機會依靠非對稱戰力否定解放軍攻勢下有嚇阻力，卻終將因兩岸軍事差距的愈益擴大，而使防衛固守的嚇阻力下降。以往依靠質量優勢的軍事力量已不足以支撐台灣安全，有必要再審思我國家安全戰略。

　　所謂「再審思」，是以往已有相當多學者鑑於兩岸軍事力量即將失衡，已開始審思台灣安全戰略並提出非軍事力量的因應構想，然而卻缺乏共識。戰略是目標與手段的連結，本文將透過戰略研究途徑釐清此一問題，再運用安全研究學者 Buzan 所主張國家必須統合軍事以外的政治、經濟、社會與環保等政策手段，才有能力維持其獨立身分與健全功能的複合式安全戰略概念，[3]以「在確保我獨立身分與健全功能下維持兩岸和平」為目標，檢討戰略工具箱中可作為

---

* 淡江大學整合戰略與科技中心研究員

1 在該報告第三項「新時代中國特色社會主義思想和基本方略」中有關兩岸關係的描述是：必須堅持一個中國原則，堅持「九二共識」，推動兩岸關係和平發展，深化兩岸經濟合作和文化往來，推動兩岸同胞共同反對一切分裂國家的活動，共同為實現中華民族偉大復興而奮鬥。請參閱新華社，《習近平：決勝全面建成小康社會 奪取新時代中國特色社會主義偉大勝利—在中國共產黨第十九次全國代表大會上的報告》，三、新時代中國特色社會主義思想和基本方略，中華人民共和國中央人民政府官網，2017 年 10 月 27 日，http://www.gov.cn/zhuanti/2017-10/27/content_5234876.htm。

2 中國民國國防部，《國防報告書 106》，我國軍事戰略演進圖，頁 56，中國民國國防部官網，2018 年 12 月，https://www.mnd.gov.tw/NewUpload/ 歷年國防報告書網頁專區 / 歷年國防報告書專區 .files/ 國防報告書 -106/ 國防報告書 -106- 中文 .pdf。

3 Barry Buzan, *People, State and Fear: An Agenda for International Security Studies in the Post-cold War Era*, (Brighton: Harvester Wheatsheaf, 1991).

國家安全支撐的工具，分析其適用性，以建構複合式防盾，形成我國家安全的支柱。

壹、複合式安全戰略與台灣安全的「目標」

複合式安全戰略是冷戰後愈趨複雜的國際形勢下產物，在 1990 年代安全意涵擴展的大辯論後，戰略（安全）研究學者多同意，僅憑藉軍事能力無法確保國家安全。除 Buzan 提出必須統合軍事以外的政治、經濟、社會與環保等政策手段外，其他如國際關係學者 Rosecrance 與 Stein 也主張國家要考慮可支配的全部資源（不僅是軍事資源），以有效運用確保平和戰時的國家安全。[4] 相較以國防安全為核心的傳統思維，複合式安全戰略強調的是維護國家安全的手段，亦即可稱為「戰略工具箱」中的工具，除軍事之外，應包括所有國家可操作的資源。

在台灣，面對基礎國力遠較龐大的大陸，在以往軍事上擁有質量優勢時，也多以國防安全作為安全戰略的核心思考。然而當兩岸天平逐漸向大陸傾斜，學界也開始運用複合式安全戰略概念，探討其他的非軍事手段以維護國家安全。例如翁明賢在〈國家安全戰略研究典範的移轉 - 建構淡江戰略學派之芻議〉中闡述複合式安全的相關脈絡，並提出「雙重平衡」戰略架構，主張台灣必須採取「嵌入式」策略的設計，將台灣與美、日等「連橫」戰略設計下的國家緊密結合，並透過國際合作與開發能量，將台灣與「合縱」思考下的小國，嵌入台灣的整體國家利益中。[5] 前總統馬英九也在 2008 年總統競選期間提出台灣國家安全要奠基於：軟實力、軍事嚇阻、保證現狀、修補互信等 SMART 的概念，[6] 也屬於複合式安全戰略觀，即不再僅以武力作為國家安全保障的核心。

然而，即便各方都依據複合式概念提出台灣國家安全戰略的解方，戰略構想與所選用工具仍有相當大差距，並無共識可言。尤其是加入美日聯盟，亦或

---

4 Richard Rosecrance and Arthur A. Stein, eds., *The Domestic Bases of Grand Strategy* (Ithaca: Cornell University Press, 1993), p. 5.
5 翁明賢，〈國家安全戰略研究典範的轉移：建構淡江戰略學派之芻議〉，《台灣國際研究季刊》，第 6 卷第 3 期，2010 秋季號，頁 92-97。
6 轉引自翁明賢，〈國家安全戰略研究典範的轉移：建構淡江戰略學派之芻議〉，頁 86。

與大陸建立更友善關係，更有較大的差異性。這主要是因為戰略三要素－目標、手段、連結－中「目標」選定的差異，才導致「手段」選用的不同，因而讓「連結」的部分也出現不同邏輯。例如翁明賢「雙重平衡」戰略架構的目標是「讓台灣成為小國發展的領頭羊」，[7]馬英九 SMART 概念的目標則是「降低中國動武念頭」。[8]後者是透過連結大陸以建構海峽和平，前者則有「目標」亦或「手段」的爭議；或只能視為中間目標，而作為連結另一個「未來面對崛起中國時，美台具有相同戰略目標」[9]之目標的手段。

　　未明確指出「目標為何？」或刻意模糊，是台灣安全戰略缺乏共識的主要原因。研究者（或許不自覺的）受其統獨立場影響，在未明言的政治理想引導下所提出的戰略構想，通常政治性高於專業性，所設定之讀者也多為說服社會大眾而非提供政治菁英作為擬定國家政策的參考。雖然「戰略模糊」是個專業術語，但指的是「手段」而不是「目標」，否則將喪失方向而使戰略無意義。因此，在隱藏著未明言的，或統或獨的政治理想，而刻意模糊「目標」下，自然導致戰略構想的各說各話。統、獨是台灣高度爭議的政治議題，戰略研究者即便無需處理此爭議，也有必要釐清。

　　冷戰後西方戰略學界反思戰略研究的困境，共識是將戰略研究學科定位為安全研究的次領域，而在戰略研究之下的次領域則是軍事科學。[10]這不僅界定了戰略研究是以「手段與目標的連結」而處理安全問題，且區隔了其他的安全研究領域，設定只有可透過軍事手段處理的議題，才屬於戰略研究的範疇。[11]同時強調研究者只是針對特定議題，提出一系列可操作的信條、價值觀與主張，以導入決策者的決策選項；[12]戰略研究者與決策者扮演不同角色，前者負責理

7　翁明賢，〈國家安全戰略研究典範的轉移：建構淡江戰略學派之芻議〉，頁 97。
8　轉引自翁明賢，〈國家安全戰略研究典範的轉移：建構淡江戰略學派之芻議〉，頁 86。
9　翁明賢，〈國家安全戰略研究典範的轉移：建構淡江戰略學派之芻議〉，頁 97。
10　持此一立場的戰略研究學者包括 Betts, Baylis 與 Buzan 等人；請參閱 Richard K. Betts, "Is Strategy an Illusions?," pp. 7-9; John. Baylis, et al. *Strategy in the Contemporary World: An Introduction to Strategic Studies* (Oxford: Oxford University, 2002), pp. 11-12; Barry Buzan and Lene Hansen, *The Evolution of International Security Studies* (Cambridge: Cambridge University, 2009), pp. 1-3.
11　例如環保與少子化，雖然都將影響國安，但解決途徑沒有軍事選項，故非戰略研究範疇。
12　Friz Ermarth, "Contrast in American and Soviet Strategic Thought," *International Security*, Vol.3, No.4(1978), pp. 138-155.

性分析與方案制訂，後者負責決策與風險承擔。[13] 這表示戰略研究者是在客觀立場上，依據決策者所制定的政策目標，在可用軍事手段解決的議題上分析「手段與目標連結」的各項包含軍事在內的途徑，而後提出供決策者參考的選項。

至於所謂的「客觀立場」，因為戰略研究在國家中心、行為者理性、國際無政府狀態、國家追求權力與安全等基本假定上與現實主義學派相同而被歸類於現實主義學派，[14] 因而研究立場是採取國家利益為導向的現實主義思維。這表示，即便研究者有自己的政治理念，所謂「國家利益」的界定，也經常受政黨政治影響而有客觀上的爭議；但研究者仍應排除政黨（選舉）利益，以及個人政治理念的干擾，非理想主義的不受意識形態拘束而進行分析。

在此研究立場下，本文以「在確保我獨立身分與健全功能下維持兩岸和平」為目標，而提出國家安全戰略的政策建議。選擇「確保我獨立身分與健全功能」作為目標前提，是因為就統、獨的政治理想而言，若依據大陸《反分裂國家法》所揭示的統一與和平之連繫性邏輯，台灣若同意統一將無所謂動武的問題，而符合該法定義的分裂則必然動武；因而台灣的複合式安全戰略目標需以中性的政治理想為前提才有意義，否則若不是開始談判統一進程，就是全面進行備戰。

因此，所謂「獨立身分與健全功能」在不刻意界定條件下最接近介於統、獨之間的「維持現狀」概念。而「維持現狀」不僅是當前執政的蔡英文政府所揭示的兩岸政策，事實上從李登輝總統以降無論政黨如何輪替，此一政策主軸都未改變。這表示「維持現狀」符合台灣長期的主流民意，如此也處理了「界定國家利益」的問題：國家利益是由政治菁英基於主流民意所界定。國家安全目標必須有主流民意的政治基礎，國家安全戰略才有達到共識的可能。

---

13 陳偉華，〈戰略研究的批判與反思：典範的困境〉，《東吳政治學報》，第 27 卷第 4 期，2009 年，頁 31。

14 Ken Booth, "Strategy," in A. J. R. Groom, and Margot Light, eds., *Contemporary International Relations: A Guide to Theory* (London: Mansfield, 1994), p. 109.

貳、台灣安全的戰略工具箱

在「獨立身分與健全功能」前提下，本研究以「維持兩岸和平」作為國家安全戰略的目標。由於台灣已放棄「光復大陸」的政治目的，因而兩岸和平與否決定於中國大陸是否對台動武。

「動武」是個重要決定，取決於能力與意願此二同時成立的要素。因此，戰略工具與「維持兩岸和平」目標連結的邏輯，若不是能夠「否定大陸動武能力」，就應具有「降低大陸動武意願」的功能。在此兩條件下，本文列舉一般較常被提及的手段（工具），製作台灣複合式安全戰略工具箱（如表1）。戰略規劃受資源有限與環境因素的限制，同時也要受風險的檢驗；[15] 因此適用性高的戰略工具在連結目標的同時，應具有低風險、高能動性且避免資源排擠的效果。本文將依據前述因素進行分析，而後做出適用性的結論。

表1：台灣複合式安全戰略工具箱

項次	工具	與目標的連結	風險	環境與資源限制	適用性
1	戰術武力	防衛性嚇阻：發揮重層嚇阻戰力，防衛國土安全，嚇阻大陸不敢輕啟戰端。	兩岸軍力發展失衡，成功防衛機率降低後將逐漸喪失嚇阻力。	增加國防經費將排擠經濟與社福等財政支出。	高，但逐漸降低。
2	戰略武力	毀滅性嚇阻：發展核武或攻擊水庫、核電廠，以大量殺傷的可能性嚇阻大陸不敢輕啟戰端。	違反國際法且將導致台灣無法承受的毀滅性報復。	發展中長程導彈與核武將排擠傳統武力經費。	低，具台灣無法承受的風險。

---

15 國防部軍務局譯，美國海軍戰爭學院，《戰略與兵力規劃》，（台北：國防部軍務局，1998），頁21-28。

項次	工具	與目標的連結	風險	環境與資源限制	適用性
3	民主價值	以民主之生活方式吸引大陸民眾認同，降低中共動武正當性。	風險低，但效果易被大陸的民族主義風潮抵銷。	無	高，但不具決定性。
4	外部資源	尋求加入美、日同盟，以外部力量嚇阻大陸不敢輕啟戰端。	有引起大陸反彈的風險，台海衝突時不排除美國「向台灣說再見」。	受限於美中對峙國際戰略結構，缺乏能動性。	低，風險大於可能利益。
5	兩岸交流	透過互動與交流營造兩岸友善的康德無政府文化，降低大陸動武意願。	交流初期凸顯差異性而非形成同質性，可能產生反效果。	無	高，利益大於風險，但不具決定性。
6	一中憲法	堅持中華民國憲政體制，挑戰大陸動武合法性，降低其意願。	對國內政治有影響，但對外風險低。	無	高，利益大於風險。
7	貿易鏈結	強化台灣在全球互賴體系角色，營造大陸動武時「投鼠忌器」的效果。	有被捲入貿易戰風險。	無	高，利益大於風險。

一、戰術武力

選項 1 連結目標的邏輯，在於否定大陸動武能力，「防衛性嚇阻」的核心概念即是以成功的防衛，嚇阻敵不敢輕啟戰端。

防衛性嚇阻的嚇阻力來自軍事力的對比，邏輯上「台灣軍事力／大陸軍事力」若大於 1，將有 100% 的嚇阻力，若低於 1，則嚇阻力等比降低。

軍事力的增長決定於資源投入的多寡。相對台灣每年 3% 左右的經濟成長率無法支撐國防經費的穩定成長而使軍事力增長受限，大陸經濟成長率若繼續維持 6.5% 以上，將可支撐其每年 8% 左右的軍費增幅，如此解放軍事力將快速增長。

依據〈中共十九大政治報告〉中的政策規劃，解放軍將分別以 2020 年、2035 年與 2050 年為目標，三階段的建設成「基本實現機械化、資訊化」、「實現國防和軍隊現代化」與「世界一流軍隊」。[16] 若解放軍的「現代化」與「世界一流」是以美軍為參考指標，則至遲 2035 年兩岸軍事力量將完全失衡，如此成功防衛的嚇阻力將喪失。雖然台灣也可以大幅增加國防經費以增強軍事力，但將排擠經濟發展或社會福利等財政支出，且缺乏經濟成長率的支撐，在軍備競賽中終將落後。綜合評估，防衛性嚇阻作為台灣安全戰略手段的適用性高，但卻有嚇阻力逐漸降低的問題。

二、戰略武力

選項 2 雖也是運用軍事武力，卻是透過大規模毀滅性武器的戰略性嚇阻，以降低大陸動武意願。

雖然國軍曾擁有質量優勢，但面對大陸軍力龐大的數量優勢，發展核武早期確曾為國安選項。1955 年國防部成立中山科學院籌備處第一所（即核能研究

---

16 〈中共十九大政治報告〉中的論述是：「到 2020 年基本實現機械化，資訊化建設取得重大進展，戰略能力有大的提升」，以及「力爭到 2035 年基本實現國防和軍隊現代化，到本世紀中葉把人民軍隊全面建成世界一流軍隊」。請參閱新華社，〈習近平：決勝全面建成小康社會 奪取新時代中國特色社會主義偉大勝利─在中國共產黨第十九次全國代表大會上的報告〉，十、堅持走中國特色強軍之路，全面推進國防和軍隊現代化，中華人民共和國中央人民政府官網，2017 年 10 月 27 日，http://www.gov.cn/zhuanti/2017-10/27/content_5234876.htm。

所），就是為發展核武而規劃。但 1988 年核能研究所由中山科學院「歸建」行政院原子能委員會後，[17] 核武政策確定放棄。當然這並不意味發展核武不能再度成為國安選項，或研發中長程飛彈攻擊三峽大壩或核電廠等設施以達到類似的毀滅性效果。然而前者違反 1970 年生效的《禁止核武擴散條約》，中華民國雖然不是簽約國，在美、中監視下將很難獲得核燃料。後者依據 1949 年日內瓦公約的附加議定書（第一議定書），明確規定不得攻擊「含有危險力量的工程和裝置」條款，違反者將被歸為戰爭罪；且極可能遭到大陸的毀滅性報復，台灣將無法承受。而要發展或佈署大量能成功穿透解放軍反導系統的中長程飛彈，所需經費也將排擠傳統軍備的部署。綜合評估，風險太大，適用性低。

三、民主價值

選項 3 為軟權力概念，是以台灣民主、自由的生活方式，吸引大陸民眾認同，在降低大陸動武正當性的同時也降低其動武意願。

然而，因為行政缺乏效率，核心民主大國的經濟普遍缺乏成長，相對中國大陸卻以自身經驗走出了另一條路，扛起世界經濟復甦的責任，當前已出現「民主退潮」的質疑聲浪。[18] 習近平在〈中共十九大政治報告〉中強調「增強道路自信、理論自信、制度自信、文化自信」的論述，[19] 所謂「民主」的價值對大陸民眾的的吸引力有減弱趨勢。此一趨勢未來即便扭轉，從大陸網民對台獨高漲的「武統論」推估，也將無法抵銷怒潮澎湃的民族主義。推動「民主價值」的優點是不受環境限制也不排擠資源，在復合式安全概念下可列為選項之一，但不具決定性。

四、外部資源

選項 4 是引進美、日等國力量，尋求加入美、日同盟，以嚇阻大陸不敢輕

---

17 此為核能研究所對外的官方說法。請參閱核能研究所，〈組織沿革〉，核能研究所官網，https://www.iner.gov.tw/index.php/1800.html，2016 年 8 月 17 日更新。

18 湯錦台譯，Joshua Kurlantzick，《民主在退潮：民主還會讓我們的世界變得更好嗎？》（*Democracy in Retreat: The Revolt of the Middle Class and the Worldwide Decline of Representative Government*）（台北：如果，2015）。

19 新華社，〈習近平：決勝全面建成小康社會 奪取新時代中國特色社會主義偉大勝利—在中國共產黨第十九次全國代表大會上的報告〉，二、新時代中國共產黨的歷史使命，中華人民共和國中央人民政府官網，2017 年 10 月 27 日，http://www.gov.cn/zhuanti/2017-10/27/content_5234876.htm。

啟戰端。

美國與我中華民國一向有傳統友誼，1954 年訂定《中美共同防禦條約》以阻止共產主義擴散，至 1980 年因美國採取「聯中制蘇」戰略，改與大陸的「中華人民共和國」建交後一年始廢止。近年來則為因應大陸的強勢崛起，美國國家安全戰略大幅轉變，2017 年 12 月川普總統政公佈了其首份國家安全戰略報告，定位中國大陸為其「戰略競爭對手」（Strategic Competitor）；[20] 對台行動也愈趨友善，包括簽署《2018 年財政年度國防授權法》（The National Defense Authorization Act，NDAA），要求五角大廈評估重啟軍艦停靠台灣港口的可行性、加強資深官員交流，及邀請台灣軍隊參加空對空「紅旗」軍演；2018 年 3 月簽署生效的《台灣旅行法》，也解禁了 1979 年與我斷交後受限的美台官員互訪。

從 1954 年以來美、中、台三方關係演變的過程，顯示台灣的外部資源受國際戰略結構的制約。正如同 1954 年美國與我簽訂《中美共同防禦條約》是因為韓戰後共產主義的擴張、1979 的斷交是基於美蘇冷戰的結構，當前美國的友台措施同樣是美、中對峙結構下的產物。然而戰略強調的是能動性，結構性產物應視為「環境」而不能作為戰略工具操作。雖然當前美國再度重視台灣，但環境與簽訂《中美共同防禦條約》的 1954 年不同。當時無論國軍或解放軍實力都差美軍一截，美國介入台海可有效維持兩岸和平；現在則中國實力飛躍已成為美國戰略競爭對手，未來不能排除美國亞洲霸權地位被中國超越（overtake）的可能性，屆時「向台灣說再見」的棄台論將再現。[21] 僅回應美國基於其國家利益而出台的友台措施，並無法帶給台灣真正的安全保障，反而將帶來大陸反彈的風險。中國大陸駐美公使李克新即曾表示：「美國軍艦抵達高雄之日，就是我解放軍武力統一台灣之時」。[22]

---

20 財訊快報，〈美國家安全報告宣布中國為「戰略競爭對手」〉，鉅亨網，2017 年 12 月 19 日，https://news.cnyes.com/news/id/3993272。

21 美國攻勢現實主義學者米爾斯海默（John Mearsheimer）即曾在 2014 年的《國家利益》雙月刊上以「向台灣說再見」為題發表過論述。請參閱 John J. Mearsheimer, "Say Goodbye to Taiwan," *The National Interest*, March-April 2014, http://nationalinterest.org/article/say-goodbye-taiwan-9931,

22 聯合晚報，〈陸駐美公使：美軍艦抵高雄之日 武力統一台灣之時〉，聯合新聞網，2017 年 12 月 9 日，https://udn.com/news/plus/10172/2865468。

　　台灣的重要性來自於地緣戰略，位於第一島鏈的中間位置具關鍵性，若全面倒向大陸，則中國將無阻礙的進入太平洋；若全面倒向美國，則封鎖中國東海與南海的兵力轉用。在太平洋海權與亞洲陸權對峙的結構中，屬於「即便不屬於我，亦不能讓對方拿去」的地位，這意味著倒向任何一方，都有被另一方報復的風險。因此台灣以動態平衡，維持在中間地位最有利。即便美國改變其「一中政策」願提供台灣安全保障，甚至駐軍台灣，台灣也必須思考與美國軍事同盟後，中國因沿海精華區、台灣海峽、巴士海峽等生命線都遭美軍危脅，必須排除下而採取軍事反撲的風險。如此台灣將陷入「安全困境」：為保障自身安全的行動，因降低對方安全感而導致自身的更不安全。

　　外部資源有其不可依賴性，操作不慎也有相當大風險，作為安全戰略工具的適用性不如想像中那麼高。

五、兩岸交流

　　選項 5 是運用建構主義概念，透過互動與交流增進了解，以營造兩岸友善的康德無政府文化，降低大陸動武意願。

　　溫特（Alexander Wendt）的建構主義將無政府文化區分為霍布斯、洛克、康德三種無政府狀態文化，角色邏輯分別是敵意、競爭與友誼。[23] 在描述洛克文化如何往康德文化前進時，溫特也指出了相互依賴（Interdependence）、共同命運（Common fate）、同質性（Homogeneity）、自我約束（Self-restraint）等四種主要變數。[24] 溫特的理論提供了一個很好的國家安全戰略的目標概念：營造康德文化的安全環境；相互依賴等四個變數也是具有操作性的「手段」概念。

　　然而，實踐的經驗卻顯示，兩岸交流之初並沒有帶來同質性的認同，反而凸顯了彼此的差異性。馬英九時期兩岸互動增加，但台灣媒體對大陸觀光客不文明行徑的批評也快速增加，台人對大陸的好感度降低，甚至出現年輕人反

23　Alexander Wendt, Social Theory of International Politics (Cambridge: Cambridge University Press, 1999), pp. 346-349.

24　Alexander Wendt, Social Theory of International Politics, pp. 360-2.

中的太陽花運動。反而蔡政府執政後，兩岸官方互動雖然近乎停滯，台人對大陸的好感度卻增加，甚至超過負面觀感比例。[25] 對大陸人來說，「天然獨」的說法已表示台灣年輕人大多主張台獨，網路上民粹性的相互叫陣較以往增加很多，伴隨的即為「武統論」的快速增長。這是否意味著洛克文化往康德文化前進時並非直線而是有曲折反覆的過程，亦或執政者操作不慎反將轉向霍布斯文化前進。建構主義是否需理論性的修正，還有待學界進一步研究。

六、一中憲法

選項 6 是堅持中華民國憲政體制，以挑戰大陸動武合法性，降低其意願。

如果習近平在〈中共十九大政治報告〉中宣示達 54 次的「法治」概念確能落實，則就中國大陸的法律而言，對台動武的合法性是出現《反分裂國家法》第 8 條所描述：「台獨分裂勢力以任何名義、任何方式造成台灣從中國分裂出去的事實，或者發生將會導致台灣從中國分裂出去的重大事變，或者和平統一的可能性完全喪失，國家得採取非和平方式及其他必要措施，捍衛國家主權和領土完整」等現象時。因而從邏輯上反推，只要不出現這些現象則大陸沒有動武的合法性。雖然大陸對其法律條文擁有解釋權，例如「和平統一的可能性完全喪失」的條件如何界定？不排除隨著兩岸實力差距愈來愈大而愈益寬鬆。若具武統條件與實力時，甚至也可能直接修法。然而，台灣若堅持具「一個中國」內涵的中華民國憲政體制，對大陸動武的合法性就將形成挑戰。因為對崛起大國來說，「戰略可靠性」具有重要意義，若中國大陸拋棄和平統一的承諾，在台灣維持一中憲法下為統一而動武，將影響周邊國家的觀感，不利於其霸權地位的爭取。

中國大陸在國際權力如何發生轉移的問題上曾進行理論探討，閻學通即依據西方現實主義再參酌中國傳統的仁義觀，提出「道義現實主義」論述。他引入「政治領導」和「戰略信譽」這兩個重要變數，在肯定物質實力是崛起成功

---

25 依據聯合報民調中心 2017 年 11 月 20 日發布的民調數據，台灣民眾對於大陸人民的觀感首次逆轉成為好評居多，49% 對大陸民眾印象佳，比去年增加了 5 個百分點，持負面觀感的人則由 45% 降為 37%。世界日報，〈首度逆轉！49% 台灣人對大陸民眾印象好〉，奇摩新聞，2017 年 11 月 20 日，https://tw.news.yahoo.com/ 首度逆轉 -49- 台灣人對大陸民眾印象好 -111000474.html。

基礎的同時，進一步提出政治領導是國際實力對比變化的根本，即崛起國的道義或戰略信譽可提高其國際政治動員力，從而改變國際格局，以至建立新的國際規範和國際秩序。[26]「戰略信譽」既然在大國崛起中扮演重要角色，習近平也曾在接受國外媒體採訪時強調「堅持正確的義利觀，義利並舉，以義為先」的原則，[27]就必須遵守其既定承諾，以相應其崛起大國的地位。《反分裂國家法》既法有明文，台灣維持一中憲法，將挑戰其動武的合法性。

堅守一中憲法有低風險、不排擠資源的優勢，作為台灣安全戰略工具的適用性高。

七、貿易鏈結

選項7是運用新自由主義學者基歐漢（Robert O. Keohane）與奈伊（JosephS.Nye）所提出的「複合相互依賴」（complex interdependence）理論，他們認為在全球化下各國經濟已形成的相互依賴現象，而複合相互依賴的其中一個特徵就是武力不再是有效的政策工具，在發達的工業化國家之間，相互攻擊的恐懼已不復存在。[28]

雖然基歐漢與奈伊提出敏感性（sensitivity）和脆弱性（vulnerability）作為該國依賴另一國程度的指標，[29]但缺乏操作性的模型來衡量大陸對台灣的依賴到底有多深，以至不能武力犯台。同時，本文也不認為即便大陸對台灣的依賴極深，若《反分裂國家法》第8條所律定的動武條件明顯出現，大陸將不會以動武作為最後手段。因為對崛起的中國來說，「戰略信譽」的意義將超過經濟貿易的可能損失。但可理解的是，台灣在複雜的全球互賴體系中角色愈重要，則大陸以動武手段解決台灣問題的意願就愈低；因為除了本身的經貿損失之外，

26 閻學通，〈道義現實主義的國際關係理論〉，《國際問題研究》（北京），2014年，第5期，頁102-128。

27 〈習近平接受拉美四國媒體聯合採訪〉，人民日報，2014年7月15日，第2版。轉引自閻學通，〈道義現實主義的國際關係理論〉，頁128。

28 Robert O. Keohane and JosephS. Nye著，門洪華譯，《權力與相互依賴》（北京：北京大學，2002），頁12-14。

29 所謂「敏感性」是指一國政策變化導致另一國發生有代價變化的速度與所付出代價的多寡；「脆弱性」則是指在政策變化下行為體獲得替代選擇的相對能力以及所付出的代價。敏感性和脆弱性愈高，則該國對另一國依賴的程度愈大。請參閱門洪華譯，《權力與相互依賴》，頁24-256。

還必須顧忌作為全球供應鏈的台灣「斷鏈」後，所造成對其他國家科技產業損失的連鎖反應。

4月16日，美國商務部因大陸中興通訊違反限售伊朗、北韓等遭禁運國家的禁令且多次違反承諾，下令禁止美國企業販售其元件、軟體以及技術後；中興通訊即表示將面臨生存危機。[30] 這說明了供應鏈遭「斷鏈」對當前高科技產業的威脅性。因此，台灣在全球供應鏈上的獨佔性愈強，所牽動的產值愈大，大陸動武的「投鼠忌器」效果也就愈高。

對國家安全戰略而言，強化貿易鏈結能有效降低動武意願的效果，且有低風險、不排擠資源的優勢，作為戰略工具的適用性高。

參、台灣安全的複合式防盾

複合性安全戰略是運用所有包括軍事力量在內的戰略工具，以維護國家安全，因此除高風險選項外皆有適用性。同時，任何選項都有排擠作用，即便不排擠資源，也有政策性如立法優先、獎勵措施等排擠效應，故應選用較有利的戰略工具，賦予政策優先性，以避免無差別下的排擠。透過前一節戰略工具的適用性分析，本文主張台灣應建立複合式防盾的概念，以作為國家安全戰略的核心。

一、排除高風險選項

首先，台灣的安全戰略應排除高風險選項。正如本文前述：戰略研究立場是採取國家利益導向的現實主義思維，並要考慮環境的轉變與可能轉變。這意味著在國家的現實利益，而非道德、理想、傳統…或其他意識形態的考量下，必須依據當前與未來環境的可能轉變而調整安全戰略；以前適用性高的戰略工具，或許已不再適用或適用性降低。而在複合性戰略概念下，所謂不適用是因為其風險高。

就戰略邏輯來說，「否定其動武能力」與「降低其動武意願」間的優先性

---

30 Money DJ，〈中興通訊聲明：美國商務部禁令將嚴重危及生存〉，科技新報，2018年4月20日，technews.tw/2018/04/20/zte-said-they-cannot-survive/。

是依據雙方實力大小的差距而定。正如同在野外，遇見一條毒蛇、一隻野狗或一隻黑熊，選用的策略將不同：對蛇可以用登山棍驅趕、對野狗可再加上撿拾石頭嚇阻，但遇到難以抵擋的黑熊，較佳策略是避免激怒牠，降低姿態的設法脫離。對前兩者來說，因為有體力上的優勢，可選擇「否定其動武能力」；後者則明顯體力劣勢，故選擇「降低其動武意願」。若面對黑熊採取否定其動武能力策略而持棍棒揮舞，顯然風險太高。因此，當海峽兩岸實力天平愈向大陸傾斜時，「否定其動武能力」就成為高風險選項。

依據前一節分析，具高風險的選項為 2. 戰略武力與 4. 外部資源，兩者都具有否定大陸動武能力的嚇阻功能，但也都有提高大陸動武意願的風險，易引發難以承擔的毀滅性報復或較大規模的武力攻擊，應先予以排除。

然而，不選擇以爭取美國協防作為鞏固台灣安全的選項，並不表示要拒絕美國的「友台」行動，而是在美、中對峙的權力結構中要尋求動態平衡，例如在增加美國互動的同時，在其他方面能表現出對中國大陸的善意，以避免被誤判為完全選邊而出現報復行動。保持戰略的能動性，避免結構的限制，才是真正意義的「棋手」。

二、鞏固國安三支柱：武盾、法盾與矽盾

在其他選項中，1. 戰術武力仍應是主要防盾之一，簡稱「武盾」。因為「防衛性嚇阻」雖屬於嚇阻的概念，但強調的是非攻勢、非報復的防衛性，引發大陸反彈的風險低。

雖然兩岸實力天平已愈益向大陸傾斜，但在解放軍完成其現代化改革前，國軍仍具有局部的質量優勢，有機會依靠非對稱戰力否定解放軍攻勢而有嚇阻力。即便解放軍完成現代化改革，一支相對弱小但善戰的武力仍能造成入侵者的傷害。軍事武力的嚇阻力雖將愈益降低，但不會消失，仍有其重要功能，故可作為主要防盾之一。

選項 6 的一中憲法，則可以做為另一個主要防盾的「法盾」。雖然堅持具「一個中國」內涵的中華民國憲政體制，並不符合大陸堅持的最終統一目標，

但足以挑戰其動武的合法性。台灣亦可以重申中華民國體制下的相關事務，例如釣魚台群島或南海 11 段線主權聲索，以作為對大陸表達善意的籌碼。

雖然因目前我邦交國只剩 20 個國家，國內有部分人士以「中華民國」的國際地位遭受質疑，而有更改國名的看法。然而國際間不承認「中華民國」，並不意味著否定我政府擁有「獨立身分與健全功能」，否則不會有 119 個國家及地區給予中華民國護照免簽證、41 個國家可落地簽證方式入境。[31] 反而在當前中國大陸的實力擴展下，若修憲更改國名，獲得國際承認的機會更低，且在《反分裂國家法》制約下有立即遭大陸動武的風險。具有「一個中國」內涵的中華民國體制能挑戰大陸動武的合法性，是維持台灣安全另一個支柱。

相對於遵守一中憲法的消極性，選項 7. 的強化貿易鏈結，具有更積極的降低大陸動武意願的作用。「中國崛起」的首要條件就是經濟的持續高成長，這不僅需要本身製造業能夠向高科技產業發展，也需要全球經貿形勢保持穩定，任何供應鏈的「斷鏈」都能打亂其佈局。因此，台灣愈在全球供應鏈上有特定獨佔性，讓大陸動武時「投鼠忌器」的效果就愈高。台灣的確有半導體產業在全球佔重要地位的優勢，這形成台灣安全的第三根支柱：矽盾。

台灣半導體總產值為全球第 2。積體電路（Integrated Circuit, IC）上下游產業鏈完整，從上游的 IC 設計到後段的 IC 製造與 IC 封測，專業分工模式獨步全球，總 IC 產值全球排名第 2（市佔約二成），僅次於美國，超過韓國和日本。IC 設計產值全球排名第 2（市佔約二成），晶圓代工產值全球排名第 1（市佔約七成），IC 封測產值全球排名第 1（市佔約五成），記憶體產值全球排名第 4（市佔約一成）。[32]

中國大陸則是全球最大的半導體消費國，且嚴重依賴進口。2017 年 IC 產品需求達到 1.40 兆元人民幣，但國內自給率僅為 38.7%。[33] 2016 年 IC 進口金

31 中華民國外交部，〈免簽證資訊〉，外交部官網，2018 年 3 月 21 日，https://www.mofa.gov.tw/Upload/Visa/ 中華民國國民可以免簽證前往之國家 (地區)_1070321.pdf。

32 經濟部，〈晶片設計與半導體產業推動策略規劃〉，簡報，2017 年 10 月，file:///C:/Users/max12/AppData/Local/Packages/Microsoft.MicrosoftEdge_8wekyb3d8bbwe/TempState/Downloads/106 智 慧 srb 簡報 day1- 晶片設計與半導體產業推動策略規劃 - 經濟部 .pdf。

33 鉅亨網，〈中國半導體產業說不出的痛：技術水平上不去，美國又在背後捅刀〉，科技報橘，2018 年 4 月 3 日，https://buzzorange.com/techorange/2018/04/03/china-semiconductor-device/。

額為 2,277 億美元，出口金額為 675 億美元，貿易逆差為 1,602 億美元；相對同期大陸原油進口額為 6,078 億元人民幣，中國大陸 IC 逆差已經接近原油的兩倍。[34] 半導體是電子產業的核心技術，大陸對外依賴性高，將制約其發展。

中國大陸也知道此一致命弱點，2011 年 12 月發布《積體電路產業「十二五」發展規劃》、2014 年 6 月發布《國家積體電路產業發展推進綱要》，投入大量資金並從全球徵募人才以發展半導體產業，台灣曾任職台積電後被三星挖角的梁孟松即被聘為中芯國際的執行長、原南亞科技總經理高啟全跳槽到紫光集團。雖然要滿足進口替代需求還有段遙遠的路，但 2017 年半導體產值將達到 5,176 億元人民幣，年增率 19.39%，預估 2018 年可望挑戰 6,200 億元人民幣的新高紀錄，維持 20% 的年成長速度。[35]

反觀台灣則有成長趨緩的現象。2017 年全球半導體市場全年總銷售值達 4,122 億美元，成長 21.6%；台灣 IC 產業產值達新台幣 24,623 億元，僅年成長 0.5%，幾近持平，成長幅度遠低於全球。[36]

今年 3 月 12 日，美國總統川普以「保護國家安全」為由，禁止新加坡網通設備廠博通（Broadcom）以 1,170 億美元對美國行動晶片廠高通（Qualcomm）的併購，使半導體產業的發展正式進入國家安全領域的清單。台灣也應從國安角度，正視半導體產業的發展。

武盾、法盾、矽盾，是台灣國家安全的三個主要支柱，政府應賦予政策優先性，投入更多關注，以避免無差別下的排擠作用。

三、其他輔助性戰略工具

在複合式安全戰略下，只要風險低、資源排擠低的戰略工具都具適用性，可做為國家安全戰略的輔助。

---

34 中國產業發展研究網，〈2017 年中國積體電路行業發展現狀及市場前景預測〉，中國產業發展研究網，2017 年 12 月 26 日，www.chinaidr.com/tradenews/2017-12/117267.html。

35 TechNews，〈中國半導體產業成長力道逾全球，2018 年產值年成長將達 19.86%〉，科技新報，2017 年 11 月 9 日，technews.tw/2017/11/09/china-semi-2018-trend/。

36 〈去年台灣半導體產值僅年增 0.5% 遠低於全球成長 21.6%〉，自由時報，2018 年 2 月 13 日，http://news.ltn.com.tw/news/business/breakingnews/2341339。

項次 3. 的民主價值，雖然在當前民主退潮下無法抵銷大陸的民族主義情緒，但是台灣民主自由的生活方式，仍能吸引相當數量的大陸民眾的認同。項次 5. 兩岸交流也有助於增進理解，降低敵意。長期而言，爭取大陸人民同情能降低中共動武意願，有助於我國家安全。繼續推動民主價值與兩岸交流，風險低亦不排擠資源，是國家安全戰略的輔助工具。

肆、結論

本文運用戰略研究途徑探討台灣的國家安全戰略，採宏觀檢視，對戰略工具的選取，以及工具與目標連結的論述，都只擇其概要，因而就學術性而言將有不夠全面與周延的缺失。然而戰略研究的屬性一開始就被定位為應用科學而不是純科學，[37] 研究時強調政策取向與實用性，[38] 故本文的論述主旨是期望透過此途徑再審思當前台灣國家安全戰略的研究，而提出建設性的建議。

戰略是目的與手段的連結；本文指出，當前台灣的安全戰略之所以缺乏共識，是因為研究者受其統獨立場影響，在未明言的政治理想引導下未明確指出國家安全的戰略目標，或刻意模糊，以致所提出的戰略構想各說各話。本文認為，戰略研究者應採取國家利益導向的現實主義思維，本身雖可以有自己的政治理念，但研究時應排除政黨（選舉）利益及個人理念干擾，不受意識形態拘束而進行分析。

基於在此研究立場，本文提出「在確保我獨立身分與健全功能下維持兩岸和平」，作為台灣國家安全戰略的目標。此一目標的政治立場最接近統獨光譜中間的「維持現狀」概念，而「維持現狀」不僅是當前執政的蔡英文政府所揭示的兩岸政策，且符合台灣長期的主流民意。本文認為國家安全目標必須有此共同的政治基礎，安全戰略才有達到共識的可能。

---

37 Bernard Brodie, "Strategy as a Science." *World Politics*, Vol.1, No.4(1949), pp. 467-488.

38 為戰略學界共識，請參閱 Bernard Brodie, *War and Politics* (New York: Macmillan,1973), p. 452; Colin S. Gray, *Strategic Studies and Public Policy: The American Experience* (Lexington: University of Kentucky, 1982), p. 2; Edward N. Luttwak, *On the Meaning of Victory: Essays on Strategy* (New York: Simon and Schuster, 1986), p. 234.

　　因此，本文以複合式戰略觀提出台灣安全戰略的工具箱概念，以否定大陸動武「能力」與降低動武「意願」為功能性指標，分析目前較多被提及的戰略工具之風險、環境與資源等限制因素，而提出台灣的國家安全戰略。主張台灣應排除具有高風險的建立戰略武力，以及仰賴外部資源等選項，而以建立「武盾」、「法盾」、「矽盾」等三支柱，作為安全戰略的核心，並以繼續推動民主價值與兩岸交流作為輔助，形成中共在思考以武力統一時，不便、不願也不能的戰略情境，以達到維持兩岸和平的戰略目標。而在傳統上武力已受到高度重視時，本文要特別凸顯遵守一中憲法的「法盾」，以及鞏固我半導體產業在全球供應鏈上不可動搖地位的「矽盾」，在國家安全上所扮演的角色。尤其今年3月美國以「國家安全」為由，禁止半導體大廠博通與高通的併購案後，更使後者的國家安全角色受到重視。相對遵守一中憲法屬於消極性的防盾，發展半導體產業的「矽盾」更具有積極性意義，值得政府在連結國家安全概念後，投入更多的資源與關注。

Tactic Analysis of Taiwan's New Southbound Policy

Horng-Ren Wang *

I. Preface

In recent years, mainland China has gradually transformed itself from a factory of the world into a market of the world. Over the past two decades, Taiwan enterprises have been facing difficulties in their business managing in mainland China because of rising environmental protection costs and labor cost rising.

The DPP government promoted the "New Southbound Policy" immediately after President Tsai Ing-wen on board in May 2016 officially. The main targets of the southbound policy include 10 ASEAN countries, 6 South Asian countries, and New Zealand and Australia, where In 18 countries, Indonesia, India, Vietnam, Malaysia, the Philippines and Thailand are listed as priority or priority countries. The government openly declares that it upholds the core concept of "long-term deep plowing, pluralistic development and mutual benefits" and integrates the resources with various influence power which are from ministries, local governments and private enterprises and groups also. Keep all focal points such as "economic and trade cooperation", "personnel exchange", "resource sharing" and "regional linkage" in perspective. We hope that we will create a new cooperation model for mutual benefit and win-win results with ASEAN, South Asia and New Zealand and Australia and establish "awareness of the economic community".[1]

The Economic and Trade Offices of the Executive Yuan are responsible for policy coordination and promotion of implementation. They also make public

* Ph. D. Student, Graduate Institute of International Affairs and Strategic Studies, Tamkang University
1 〈「新南向政策推動計畫」正式啟動〉，行政院，105 年 9 月 5 日，http://www.ey.gov.tw/News_Content2.aspx?n=F8BAEBE9491FC830&s=82400B39366A678A。

statements on the four major spheres of work of the "New Southbound Policy" as follows:

1. Economic and Trade Cooperation

Change the patterns of using ASEAN and South Asia as export foundries in the past, expand cooperation with supply chain partners of countries, leverage domestic market with infrastructure projects and establish new economic and trade partnership.

(1) Industrial Value Chain Integration

In response to potential capacity of demand by local industries, it is necessary to strengthen the combination of competitive industries and supply chains in various countries. For example, starting from five innovation industries such as "ETC" (Electronic Toll Collection) Electrical charging system, smart healthcare and smart campus. Also "IOT" (Internet of Things) export. The establishment of a "Taiwan Desk" to link up local resources, assist Taiwanese businesses to organize agglomerations in the area, set up a single window for the new South to trade and play a platform for bilateral trade opportunities of "finding, creating, integrating and promoting".

(2) Leverage Domestic Demand Market

Make best usages of cross-border e-commerce with physical access, expand sales of high-quality consumer goods with affordable prices, promote the export of emerging service industries such as education, health, medical care and catering, and mold the image of Taiwan's industrial brand.

(3) Cooperation in Infrastructure Projects

Build an overseas output collaboration platform to focus on infrastructure construction and package plant export by flagship team to output what we build for infrastructure projects such as power plants, petrochemicals and environmental protection and seek strategic alliances with third-country manufacturers.[2]

2. Personnel Exchanges

Emphasize 「people」as the core, deepen the exchange and cultivation of bilateral young scholars, students and industrial personnel, and promote complementarities and sharing of human resources with partner countries.

(1) Cultivation of Education

Expand Taiwan Scholarship to Attract Asean and South Asian Students. Cooperate with domestic industries in establishing 「Learning and production special classes」and 「Foreign Youth Technical Training Courses」. Provide jobs hiring for trained pass students. Promote the teaching of small and medium-sized new residents of Chinese language. Also encourage universities to strengthen the Southeast Asian language and regional trade training on human resource point of view.

(2) Industrial Manpower

In view of the relocation of foreign workers who come to Taiwan to do specialized or technical work, the system of reviewing and setting up a reviewing system is considered. Eligible persons may extend their residence years and encourage their participation in technical training and application for licenses. Both sides ensure that staffs from overseas go back to Taiwan to provide social security coverage, and simplify procedures for going abroad to Taiwan to strengthen the media supply and demand of talents to help domestic enterprises find talent.

(3) New Residents Powers

Take advantage on the language and culture of the first generation of new residents, how to help them obtain relevant licenses and employment (such as mother tongue teaching and sightseeing). Encourage education institutions to

---

2 Ibid.

open specialized southern departments or programs to give students who have the advantage of a south-oriented language may take extra chances to cultivate the second generation of new residents as south-facing seeds.[3]

3. Resources Sharing

We should use bilateral soft power such as culture, tourism, medical care, science and technology, agriculture and small and medium-sized enterprises to get more opportunities for bilateral and multilateral cooperation so as to enhance the living quality of partnership countries and expand the room of depth for economic and trade development of our country.

(1) Medical

Promote bilateral cooperation in medical certification, new drugs and medical materials with the Association of Southeast Asian Nations, South Asia, Australia and New Zealand. Provide proper supporter for ASEAN and South Asian countries in nurturing health professionals.

(2) Culture

Promoting Taiwan's cultural brands through video, radio and online games.

And encouraging local governments to exchange and cooperate with ASEAN, South Asia and New-Omani cities.

(3) Sightseeing

Open ASEAN and South Asian countries people to get tourist visa with more flexing rules. Propaganda with diversify ways for tourism and improve the the quality of tour guides to establish a friendly environment for Muslim tourism.

(4) Science and Technology

Establish a platform for exchange of science and technology. Enhance the cross-border links between science parks and corporations. Also promoting

3 Ibid.

technological exchange of wisdom and disaster prevention and others.

(5) Agriculture

Establish "Taiwan International Agricultural Development Company" to expand overseas markets with national brands. Provide agricultural technology assistance, promote biological materials and farm machinery, enhance the operating capacity of partner countries.[4]

4. The Regional Chain

Expand multilateral and bilateral institutional cooperation with partner countries. Strengthen consultation and dialogue. Change the past single-mindedness model and make good use of the network of non-governmental organizations, expatriates and third-country powers to jointly promote regional stability and prosperity.

(1) Regional Integration

Active negotiate with ECA or individual economic cooperation projects with India and major trading partners of ASEAN. Renewing and strengthening signed bilateral investment and taxation agreements and establishing early warning and contingency mechanisms for major incidents to effectively control possible risks.

(2) Consultation Dialogue

Promote multi-level and comprehensive agreements and dialogues with ASEAN, South Asia, New Zealand and Australia. Have constructive dialogues and consultations on relevant issues and cooperation matters at the appropriate time with the other side.

(3) Strategic Alliances

To adjust the allocation of foreign aid resources, improve the mechanism for promoting foreign aid, and expand the participation of industry players in

---

4 Ibid.

the construction of a national economy plan; To enter into the markets of the Association of Southeast Asian Nations, South Asia and New Zealand and Australia, we should co-work and build alliance with third party (such as Japan and Singapore) and strengthen cooperation with private enterprises And NGO groups.

(4) Overseas Chinese Network

Establishing databases and exchange platform for expatriates (including graduates from Taiwan, local Taiwanese businessmen and overseas Chinese). Making the economic with trade network to be more functional between overseas Chinese and Taiwanese businessmen and strengthening links with Taiwanese enterprises.[5]

However, the DPP has been in power the bureau for past two years. The new southbound policy seems into dilemma. Taiwanese businessmen do not feel the impetus of the policy when people come to foreign countries and this policy do been criticized by the opposition parties for political saliva. Based on the Legislative Yuan Budget Center "2018 Central Government Overall Budget Assessment Report" referred to the work plan of the new southbound policy , there comes easy way to find out the economic trade cooperation, personnel exchanges, resource sharing and regional links four major start respectively and the policy has been romoting the operstion list NT $ 7190000000 (7.19 Billion NT dollars) since 2017, the budget comparing with next year increase NT$2740000000 (27.4 Billion NT dollars), and it means almost up to 61.6%.[6]

According to the report at the end of July 2017, New Southbound implemented a total of nearly 2.55 billion NT dollars in total budget of 4.5 billion NT dollars, the 2.55 billion NT dollars reflex to accounting for 56.3% of all budget. If we look at the implementation rates of the ministries, only the implementation rates of the original people's assembly, the Ministry of Education,

---

5 Ibid.
6 彭琬馨，〈立院預算中心：新南向預算執行率欠佳〉，自由時報，2017 年 10 月 2 日，http://news.ltn. com.tw/news/politics/paper/1140179。

the Ministry of Economic Affairs, the Ministry of Communications, the Ministry of Health, the Environmental Protection Department and the Ministry of Science and Technology will exceed 50%. And the remaining 14 departments performed very poor budget implementation rates.

Among them, the implementation rate of the Ministry of the Interior, the Presidential Office will be the lowest, only up to 0.51% and 3.94% respectively.

Even legislator Zhuang Rui-xiong is the member of the ruling party also pointed out that the new southbound has enough budget but no any sign of implementation will be seen.

"I am afraid it is difficult to have concrete results if things proceed the same way as used to be" said by DPP Legislator Tsai Si-in, Legislative Yuan need supervise and call for the relevant units to take actions and push harder.[7]

Therefore, the formulation and implementation of this new southbound policy need to be analyzed more rigorously. This new southbound policy is an important economic and trade policy at present. On the strategic point of view and will affect the overall development of Taiwan no matter now or in the future. If this policy can exert its effect. Then it will benefits Taiwan.

However, on December 13, 2017, Huang Chung-yan as a spokesman for the presidential office said that the new southbound Office will terminate the task force from next year. The reason is that the new southboud policy has entered a phase of full promotion from the task planning stage. The policy of the relevant ministries and commissions of the Executive Yuan has also gradually demonstrated its effectiveness. The new Southbound office of the Presidential Office has achieved its phase objectives earlier than expected.[8]

---

7 Ibid.
8 〈總統府：新南向辦公室階段目標達成 明年解編〉，Yahoo 奇摩新聞，2017 年 12 月 13 日，https://tw.news.yahoo.com/%E7%B8%BD%E7%B5%B1%E5%BA%9C-%E6%96%B0%E5%8D%97%E5%90%91%E8%BE%A6%E5%85%AC%E5%AE%A4%E9%9A%8E%E6%AE%B5%E7%9B%AE%E6%A8%99%E9%81%94%E6%88%90-%E6%98%8E%E5%B9%B4%E8%A7%A3%E7%B7%A8-101629657.html。

If you see the inquiry record of the Legislative Yuan March 6th, 2018,[9] Will find that all government departments have inconsistent saying on the new southbound ODA (Official Development Assistance)policy. It shows that government's preparation is still insufficient.

According to the current budget execution and presidential office terminated the new southbound office; we can suspicion reasonably the DPP government's determination and review this important policy very carefully. New southbound is not just a slogan and it is also an important national strategy.

II. Why Southbound?

As a nation's grand strategy and policy formation, the new southbound policy must have a strong strategic meaning. In his book 《Three Discussions of Sun Tzu》 Niu Shouzhong mentioned that strategy was the earliest idea of thinking and then comes to a complete plan, and the following step will be a concrete one for action taken, with the whole picture in short, we must have a complete context from idea to plan to concrete actions.[10]

Quoted the ancient Chinese strategic thinking, the first is "assured victory", that is we must have the conditions for successful victories in advance. Then we will be "complete victory". It will be able to use many political and economic or psychological methods other than armed forces then achieve success without bleeding. The last is the 「victorious warfare」. No shot has been, once shot, it must with full success.[11]

Reviewing the policy, discuss the gains and losses base on past experience and history, we should consider that history as a strategic and scientific action items, even artistic, and even attain the level of philosophy. To analyze the future direction of policy and the actual actions have many scientific principles of

---

9 高金素梅（吉娃斯·阿麗）個人臉書，取自 https://m.facebook.com/story.php?story_fbid=1664125916963976&id=276861822357066&_rdr 。

10 鈕先鍾，《孫子三論：從古兵法到新戰略》（台北：麥田，2007 年），頁 276。

11 軍事科學院戰略研究部，《戰略學》（北京：軍事科學，2001 年），頁 95。

judgment and many coordinated arts with each others in between.[12]

Many of the basic principles of "war preparedness" in ancient China are very suitable for assessment and reference here:

1. First, beforehand:

As to the future problems that may occur, a scientific analysis required. The formulation of strategic decisions and tactical decisions response in advance.

2. Second, widely cover:

Based on subjective and objective conditions, try to win the relative advantage.

3. Third, the proportionality:

Decision-making should be equal to its own strength, not be forced, beyond the scope of capacity.

4. Fourth, self-seeking:

Adhere to rely on our own ability and efforts. Can not rely on foreign aid.

5. Fifth, the hidden nature:

We usually combined with the wartime principle, as far as possible without exposure.

6. Sixth, whole picture:

Strategy must be systematic and comprehensive which has been covered with whole picture.[13]

These theories and principles installed at the beginning of the strategy establishment. Must be fully considered and judged. Also needs widely analysis.

---

12 鈕先鍾，《孫子三論：從古兵法到新戰略》，頁 24-27。
13 史美衍，《古典兵略》（台北：洪葉文化，1997 年），頁 52-55。

Unite the core values, goals and directions. Shape the strategy can only be more perfect, more pragmatic, so as to achieve the desired effect and achieve the goal.

Government's strategic thinking on the new southbound policy needs to review the current problems of Taiwanese businesses then set the future goals and face the future problems simulates.

Former Japanese ambassador to China and former president of Itochu Corporation, Mr. Niwa Uichiro analyzed three aspects: business, government and people. Forming a triangular relationship with a close link between the three parties.[14]

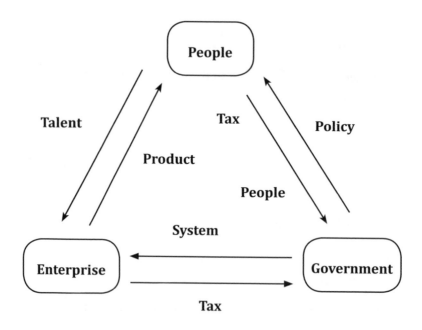

Figure 1: Government-Enterprise-People Cycle

Source: 丹羽宇一郎，《中国で考えた2050年の日本と中国》，頁135-138。

---

14 丹羽宇一郎，《中国で考えた2050年の日本と中国》（東京：文藝春秋，2016年），頁135-138。

This triangular relationship can be used as the basis for overseas investment analysis. First, the merits and weaknesses of the enterprise itself must be assessed. Then the local human culture can be integrated must also be considered. Finally, the local government's institutional policies and administrative efficiency are the entireness. The core point of foreign investment needs the overall analysis and evaluation then to make concrete actions.

China started its reform and opening up in early 1980. It attracted the participation of all countries on the world since early 1990. Until now China's economic keep growing high. However in recent years, mainland China has gradually transformed itself from a factory in the world into a market in the world. Most of the Taiwanese businessmen investing in the mainland are manufacturing industries. Facing the change in laws and environmental protecting regulations and rising wages, the cost has risen sharply and the operation has been difficult and now many production bases have to be shifted to other relatively low cost countries.

As we can review the trade amount record between China and Taiwan for last twelve years., it keep growing every year. Taiwan continues to maintain huge trade surplus. But it needs to return to the most basic strategic thinking. Political conflicts between Taiwan and China are still serious. Taiwan has always been subjected to China's military threat but foreign trade in recent years must rely on China.

Since 1990s, in the beginning of President Lee Teng-hui, government emphasized that trade should be decentralized and can not overly rely on China.

Since 1990 till now, all the last three presidents in Taiwan had emphasized the diversification of their dependence on China's economy and trade. The southbound policy has become a priority for Taiwan to get rid of its dependence on China's economy and trade.

Table 1: Trade Amount between Taiwan and China
(Hong Kong & Macao included)

Unit : US$100000000

Year	Taiwan export	Taiwan import	Trade surplus
2006	894.5	266.9	627.6
2007	1007.8	298.7	709.1
2008	999.8	329.1	670.7
2009	840.0	255.6	584.4
2010	1150.4	375.9	774.5
2011	1242.6	452.9	789.7
2012	1188.1	435.8	752.3
2013	1213.6	442.6	771.0
2014	1286.8	510.0	776.8
2015	1126.8	467.4	659.4
2016	1124.1	453.3	670.8
2017	1303.0	516.0	787.0

Source: Bureau of Foreign Trade, an Analysis of International Trade Situation in 2016, https://www.trade.gov.tw/App_Ashx/File.ashx?FilePath=../Files/Doc/d55ba47d-186d-45a9-a3d2-4d976b50ebb6.pdf; Ministry of Economic Affairs, https://www.dgbas.gov.tw/public/Data/8115162139198HDY1K.pdf.

The main targets of the southbound policy include ten ASEAN countries and six South Asian countries and New Zealand and Australia. Among these 18 countries, which Indonesia, India, Vietnam, Malaysia, the Philippines and Thailand are listed as priority and key points also.

III. The Merits of Taiwan Businessmen

In the past most foreign investment by Taiwan businessmen was in the manufacturing sector. Most of them went to set up factories in Mainland China since early 1990. In order to make good use of the cheap local land and labor and processing exports for OEM (original equipment manufacture) in major markets such as Europe, the United States and Japan. Other local industries are minority. Although Taiwan recognizes itself as a service industry, it does not have much service output so far.

With the rise of mainland China, there are already many competitors in processing and manufacturing. If there is only the searching for cheap land and cheap labor continue in the future, the competitiveness will continue to decline. Will the technology be continuously upgraded in the future? The core competitiveness continued to grasp then Taiwanese enterprises will not become the nomadic people.

The current new southbound policy will allow the government and Taiwanese businessmen to think again about how to transform their industries. It not only bases on export processing but also exports various industries and deepens their respective countries to establish a complete sales system and brand identity. To achieve effective market share. This is the real strategic goal of the new southbound policy.

All kinds of industries must be with more technology-oriented, information IT flow and international view points. At the same time, they must also be geographically integrated into local industries and prepare for long-term business operations, it comes to take root impact for meaning results.

Both the government and the industries must reconsider carefully about whether the industries we used to invest overseas have lagged behind in terms of technology and scale, and whether complete with new southbound policy or not. Should we take many industries for granted in investment list which are still highly polluting, dangerous and not be welcomed in everywhere?

Dustries are still highly polluting, dangerous and not welcomed in everywhere? Should we eliminate these industries and reevaluate the industries that suit our investment?

Does our government have well prepared the right policies for other industries?

Let us take a look at the current financial industry in Taiwan. Compared with other industries, the financial industry in Taiwan has lagged behind its counterpart in other industries in the past 10 years and lagged behind the development of the financial industry in the neighboring major competitive countries. As a result, in 2012, over the past 12 years, Taiwan The financial industry's share of GDP fell by 1.8 percentage points, far less than the sharply rising Hong Kong by 6%, Singapore by 2% or even by 1%.[15]

Taiwan's economy is highly open and its foreign trade is highly dependent. Its economic growth rate still stands at 3.72% over the same period. It shows that the demand for financial services derived from Taiwan's own economic growth and foreign trade and economic activities during this period failed to meet the needs of domestic industries and directly caused financial sector of Hong Kong has become the biggest beneficiary. Financial and insurance sectors in Asia including Hong Kong, Singapore and South Korea account for an upward trend in the service-sector GDP. Only Taiwan and the long-term economic slump in Japan have shown a decline, and Taiwan has even declined more clearly than Japan.[16]

In the recent decade or so, the employment growth in the financial industry in Taiwan has been slow. Compared with the backwardness of the neighboring major competitive countries, Taiwan occupies the last place with the four little dragons. The average annual growth rate from 2004 to 2012 was 8.02%, 3.06%, 1.78% and 0.03% respectively in Singapore, Hong Kong, South Korea and Taiwan. The financial industry in Taiwan is far behind with the modern economy in the post-industrialized era. Usually it takes a lot to boost economic growth through the development of the service industry. In this point of view the financial services industry in Taiwan has obviously failed to meet the needs of its own economic growth.[17]

At present, the status of Taiwan's financial industry do reflex what role will the government want the financial industry to play in the new southbound policy?

---

15 呂清郎，〈開放不足 台灣金融業難發展〉，中國時報，2013 年 6 月 10 日，http://www.chinatimes. com/newspapers/20130610000122-260210。

16 Ibid.

17 Ibid.

In strategic considerations, set up a foreign country is conservative service desk business is good or positive groundbreaking is better? Will South Asia into the hinterland or into our new market? At present, there seems to be no specific strategy. Relevant government departments and enterprises should be cautious in assessing them as the basis for their future implementation. Financial related industries must cooperate with government policies and relevant laws and regulations. Take this industry for example, The government must declare its policies and adjust the relevant laws and regulations so that the financial industry can cooperate and expand.

Under the trend of globalization, Taiwanese enterprises should not only go global and regard overseas as a market but also take global competitiveness as an important assessment indicator. Having globe layout, Taiwan businessmen the perspective of the Southeast Asia and South Asia, are proceed the process of globalization of production, service of globalization, globalization of marketing and the operation of capital globalization. However, the banking industry in Taiwan also has a global layout with the globalization of Taiwanese businesses. The layout of the two banks in Asia, the Americas, Europe and Africa shows a convergence of the two, the current focus is mainly on serving local Taiwanese customers. Consequently, the globalization of the Taiwan businessmen and the Taiwan banking industry carries out different operational logic. In the future, the overseas business expansion of the banking sector depends on the further opening up of the government's policies and relevant laws and regulations. Therefore, in the new southbound policy we must have more progressive and broader strategic thinking.

In the past Lee Teng-hui and Chen Shui-bian both presidents, promoted similar southbound economic policies and encouraged manufacturers to move forward in Southeast Asia and less in the mailand China. The results are not satisfactory.

The industry based on actual business demands. Foreign investment or mainland China-based, the proportion of China's exports to take off from 14% in

1990, the era of flat government has soared to 38% in 2011 closer to 41% . The same is true for investment. In 1991, investment in mainland China accounted for 15.6% of the total amount of overseas investment, while the era of Chen Shui-bian government was 66% and in 2012 it was as high as 73%. The meaning of the representative is that the pulling force of the economic trend basically comes from market signals such as price and cost rather than government policies. It is no surprise that the past economic failure of the southbound policy failed. And the new southbound policy promoted at present has to be evaluated more fully in order to achieve greater gains because of changes in the subjective and objective conditions.[18]

Taiwan still has advantages over the hard power and soft power of Southeast Asian countries. But if there is no concentration of power then set up the major principles and directions of projects, countries, cities, etc. The invested resources will disappear rapidly as if they were thrown into the sea. And these inputs of power and resources must ensure the maintenance of longterm competitiveness so that investment can be effective.[19]

IV. Southeast Asia and South Asia Differences between Countries

Southeast Asian countries vary greatly in their national conditions. All countries have different traits. In recent years the Vietnamese economy has taken off. However, accompanying bubbles and inflation have become more and more serious. The land planning and budget allocation in Vietnam has always been It is "The party has the final say" that means communists control it and the parliamentary supervision such as the rubber stamp. As a result, the ruling communist party has created a large number of vested interests in the huge

---

18 李淳，〈新南向政策，關鍵在於「心」而不是「新」〉，Yahoo 奇摩新聞，2015 年 9 月 30 日，
   https://tw.news.yahoo.com/%E6%9D%8E%E6%B7%B3%E5%B0%88%E6%AC%84-%E6%96%B0%E5
   %8D%97%E5%90%91%E6%94%BF%E7%AD%96-%E9%97%9C%E9%8D%B5%E5%9C%A8%E6%96
   %BC-%E5%BF%83-%E8%80%8C%E4%B8%8D%E6%98%AF-%E6%96%B0-224000850.html。
19 Ibid.

pocketball of pledging the national interest from the Party Central Committee to state-owned enterprises and banks. Vietnam's public construction lags far behind the actual demand. The performance of the government is extremely poor. The political structure of the issue is not resolved. And the economic structure of Vietnam will never be truly innovated.[20] To whom are facing the problems that southbound going of Taiwan businesses must pay attention.

Indonesia is the most populous Muslim country on the world. It is also the largest Muslim country in the current cooperation among Taiwan businessmen. The cultural differences among them are also greatest. Religion plays a very important role in the life of Muslims. Devout followers go to the mosque 5 times a day bow kneeling before kneeling to take off your shoes and then clean the body.

The general mosque are to maintain a quiet, solemn atmosphere. For more will be in the Islamic month of Ramadan. During the fasting month, all Muslims eat after sunset. They should not eat or smoke after sunrise until 12 hours before sunset. If anyone violates the canon then he will be punished.

Indonesia also traditionally has the issue of anti-Chinese rule. Indonesia's leader Islamist leader Bachtiar Nasir launched a protest in Jakarta last year with 100,000 people in Jakarta, triggering ethnic tensions and antagonisms on the ground.

In a recent interview, Nasir pointed at the rich Chinese people in their interview as 「The biggest problem」 in Indonesia and threatened the next goal is to solve the problem of "The Chinese are in control of economic power". Nasir refers to Indonesian Chinese, who make up less than 5% of the country's total population but possess many large enterprises and large amounts of assets. However Indonesian Chinese do not become "more generous and more equitable" and regard it as the most serious issue in Indonesia. He said Jakarta government is trying to attract investors with a capital construction project of 450 billion

---

20 張翔一，〈越南還能投資嗎？〉，《天下雜誌》，第 401 期，2011 年 4 月 13 日，http://www.cw.com. tw/article/article.action?id=5002548。

U.S. dollars (about 1.3 billion New Taiwan dollars) to reinvigorate Indonesia's economy. But it will not be able to substantially help Indonesia in its foreign investment, especially in China of the general population. "the government should ensure that the country is not sold to foreigners, especially Chinese".[21]

Not only Indonesia but also are various political and religious issues in Burma.

Internal chaos and bloody conflicts have taken place from time to time. To invest and cooperate would be better not only to avoid the occurrence of problems but also pay attention to our own security.

These intricate cultural and social issues are difficult to rule out in these countries at present but are deeply rooted in the longterm problems. As a result, enormous difficulties and risks in investment and cooperation are caused. This is an inevitable challenge in the implementation of the new southbound policy. Taiwan government how to evaluate and guide enterprises to make appropriate coordination and adjustment in partner countries so as to reduce the difficulties arising from similar problems is a major issue that both governments and enterprises must face together.

V. External Competition Analysis

Let's take Vietnam as an example. Since ten years ago Vietnam's economic structure consisted mainly of the textile, shoe-making and petrochemical industries. Taiwan was the most important source of foreign capital for Vietnam. At that time, it could be said that the majority of Vietnam's industries were supported by the Taiwan businessmen. These Taiwanese businessmen entered Vietnam around 2000 because of the old southbound policy at that time. And the

---

21　〈印尼最大的問題是什麼？印尼宗教領袖：華人太富有了〉，ETtoday 新聞雲，2017 年 5 月 14 日，https://www.ettoday.net/news/20170514/924183.htm?t=%E5%8D%B0%E5%B0%BC%E6%9C%80%E5%A4%A7%E7%9A%84%E5%95%8F%E9%A1%8C%E6%98%AF%E4%BB%80%E9%BA%BC%EF%BC%9F%E3%80%80%E5%8D%B0%E5%B0%BC%E5%AE%97%E6%95%99%E9%A0%98%E8%A2%96%EF%BC%9A%E8%8F%AF%E4%BA%BA%E5%A4%AA%E5%AF%8C%E6%9C%89%E4%BA%86。

Vietnamese government attached great importance to Taiwan's investment and even set up a Taiwan Affairs Committee to handle matters related to Taiwan. However even Japan, South Korea and many European and American countries also actively investing in Southeast Asia and South Asia in recent years. The economic structure of Vietnam has undergone great changes. The mobile phone industry has become Vietnam's largest export item. Only Samsung company accounted for 22.7% of Vietnam's total export value last year. South Korean enterprises plow deep into Southeast Asia. They use the cheap labor of South East Asia and South Asia. Also they use their power of group effort to occupy the local market. Korean brand appliances are everywhere in Vietnam.[22] So far we have seen the results and gains that South Korean enterprises have made considerable investments in Vietnam.

According to South Korea's "Korea Trade and Investment Promotion Commune", South Korea's total investment in Vietnam from 1988 to the end of 2016 was as high as 50.5 billion U.S. dollars. Accounting for 30.8% of the total foreign capital investment. As for Japan, it was 42.4 billion U.S. dollars and Singapore was 38.2 Billion U.S. dollars. Taiwan is 31.8 billion U.S. dollars.[23]

The Korean approach is to cultivate the local area and make a longterm plan for 30 years with a strong economic ambition. The basic reason why Taiwanese and Southeast Asians have the advantage of geography and blood relatives (new residents and overseas Chinese) but has continuously lost their leading position.

The fundamental reason is that most Taiwanese enterprises still only make use of local production. If you want to get greater benefits, we should refer to South Korea's in-depth local market and long-term planning strategy.[24]

---

22 何則文，〈只用五年就翻轉越南經濟結構：在越南，看見韓國「可怕的」南向經濟戰略〉，科技報橘，
   2017 年 2 月 23 日，https://buzzorange.com/techorange/2017/02/23/korea-taiwan-policy-in-vietnam/。
23 〈臺灣對越南投資額 318 億美元〉，中廣新聞網，2017 年 6 月 12 日，https://tw.news.yahoo.com/%E
   8%87%BA%E7%81%A3%E5%B0%8D%E8%B6%8A%E5%8D%97%E6%8A%95%E8%B3%87%E9%A
   1%8D318%E5%84%84%E7%BE%8E%E5%85%83-094340834.html。
24 何則文，〈只用五年就翻轉越南經濟結構：在越南，看見韓國「可怕的」南向經濟戰略〉。

South Koreans devote themselves to understanding local culture and adapting to local life. Starting points are far more than Taiwanese. In Vietnamese universities, the major amount of foreign students are Korean. Many Taiwanese staffs have been sent to Vietnam for more than 5 or 6 years. They can not speak a single Vietnamese because they usually use the translator or Chinese staffs to communicate with the Vietnamese at the grassroots level to get in touch with the real Vietnamese. The only chance Taiwanese staffs contact Vietnam society is to go to downtown shopping and dinning while Sunday. Taiwanese people who can integrate into the city is actually a minority. Most Taiwanese become a local foreigner forever. Compared with Koreans there is a clear difference.[25]

All the countries in Southeast Asia and South Asia are involved in the China's future longterm strategic plan "Belt and Road". Although the "Belt and Road" focuses on mostly infrastructures. Taiwan's new southbound policy emphasizes the integration of various talents, technologies and business opportunities, and there are somet differences in general direction. But in many ways Taiwan can also receive many benefits by means of the "Belt and Road" initiative.

However, in the hope of making use of the "Belt and Road" initiative.

We must still pay attention to the competitiveness of the belt and road initiative and our new Southbound policy. If there is any overlap, it will endanger our competitiveness. Commercial strategy and military strategy is actually the same, the market is the territory. And the market competition is extremely cruel zero-sum game. Then it become the commercial Red Sea, cruel competition in strength of more unfavorable to Taiwanese companies.

---

25 Ibid.

Table 2：Taiwan and Mainland China Foreign Trade with ASEAN 10

Unit : US$ 100 million

Period	Taiwan			Mainland China		
	Exports	Imports	Total	Exports	Imports	Total
	Growth Rate			Growth Rate		
2011	515.4	328	843.4	1,700.80	1,927.70	3,628.50
	23.2	13.6	19.3	23.1	24.7	23.9
2012	570.9	318.5	889.4	2,042.70	1,958.20	4,000.90
	9.5	-3.8	4.3	20.1	1.5	10.2
2013	592.2	330.6	922.8	2,440.70	1,995.40	4,436.10
	3.7	3.8	3.8	19.5	1.9	10.9
2014	601.7	347.3	949	2,720.70	2,083.20	4,803.90
	1.6	5	2.8	11.5	4.4	8.3
2015	516.4	290.4	806.8	2,774.90	1,946.80	4,721.60
	-14.2	-16.4	-15	2	-6.5	-1.7
2016	512.9	271.5	784.5	2,559.90	1,962.20	4,522.10
	-0.7	-6.5	-2.8	-7.7	0.9	-4.1
2017	585.8	310.6	896.4	2,791.20	2,357.00	5,148.20
	14.2	14.4	14.3	9	20.1	13.8

Note:ASEAN 10 Nations: Singapore, Malaysia, Philippines, Thailand, Indonesia, Vietnam, Brunei, Lao, Myanmar, Cambodia.

Source: ROC Customs Statistic; PRC Customs statistics.

We can compare trade amount between China and Taiwan with ASEAN 10 countries. In recently years, Taiwan keep almost the same and China grows up every year. And now about 10 times bigger amount than Taiwan. Taiwanese enterprises needs to be careful, do not direct competite with China.

Now business schools are emphasizing "Blue Ocean Strategy", that emphasize the real lasting victory is not only to win current competition, but to create a new space for undeveloped markets. This strategy is "Value Innovation". That is to create differences and more value. So that opponents can not catch up. It is strategically called "asymmetric warfare". In its basic connotation, it refers to facing the enemy, trying to evade the superiority of the enemy and exerting his own superiority to achieve military victory in attacking the enemy's weak side. Regardless of whether weak or strong are both practices of asymmetric warfare. The key point lies in how to utilize and utilize their own director and the Sun Tzu "Create advantages to win the final victory".[26] The new southbound policy must be based on the concept of blue ocean and asymmetry and create many "Invisible Champions" that Taiwan now possesses so competitive. Definition of "Invisible Champions" means enterprises have a strong lead in one area and difficult to go beyond. That is the reason Taiwan's manufacturing industry is still able to stay ahead of the world. Only in this The basic principle then the policy can be sustained for a long time.

With the help of China's "Belt and Road" initiative, it is still necessary to pay attention to the political effects caused by Taiwan's implementation of the southbound policy. The southbound policy will create an opportunity for Taiwan to intensify its interaction with the southbound country. It will also give rising to the political impact on the interaction between the two countries worries. We will be aware of political confrontation between Taiwan and China never stopped. In a written report sent by the "National Security Bureau" to the Legislative Yuan, the report pointed out:

Right now China strongly supports the "Belt and One Road" and partly overlaps. "Belt and One Road" may put pressure on the layout of Taiwan's new southbound policy. However, China focuses on infrastructure hardware construction and Taiwan is focused on e-commerce, personnel exchange and others. There are some differences between the two contents and modes.[27]

The Vietnamese national president Tran Dai Quang who went to the Belt and Road forum in Beijing 2017. Met with Chinese leader Xi Jinping. After the meeting Article 12 of the joint communique issued by the two. It said : "Vietnam reaffirms that it firmly pursues the one-China policy and supports the peaceful

---

26 馬煥棟，〈從中共軍事擴張論我海軍未來發展方向─以不對稱作戰思維探討〉，《海軍軍官》，第 35 卷第 2 期，民國 105 年 5 月，https://www.cna.edu.tw/tw/download.php?f=45758，頁 8-62。

27 鍾辰芳，〈台灣國安局："一帶一路"對台灣新南向政策形成壓力〉，美國之音，2017 年 5 月 17 日，https://www.voacantonese.com/a/taiwan-obor-20170516/3853993.html。

development of cross-Strait relations China firmly resists any major separatist activities aimed at Taiwan's independence". He said Vietnam will not develop any official relations with Taiwan. China appreciated this statement. China has consistently reiterated that it always upholds the one-China principle and opposes Taiwan's independence and consistently opposes any country establishing official diplomatic relations with Taiwan.[28]

In 2017, it even came out that the ruling party DDP in Taiwan and the Cambodian opposition party are cooperating in a political move that touches the sensitive nerves of the ruling party in Cambodia. It caused a negative reaction. Since then resistance or even a rebound has been inevitable. Similar issues should be avoided in the implementation of the new southbound policy.[29]

In other words, the operation of the new southbound policy must be careful not to fall into the political issue. Taiwan does not have any official state diplomatic relations in South Asia and Southeast Asia at present. In Cambodia and Laos, Taiwan does not even have any official economic offices. It is difficult for Taiwan government to help its enterprises officially. In such situation, Taiwan government must take care on how to implement the southbound policy and will avoid a coordinated solution to China's political issues. And it will offer actual commercial cooperation and dispute settlement and enough confidence and practical assistance to Taiwanese enterprises. This is a very urgent issue.

Not only China can offer Taiwanese enterprises external supports. It is a good direction if Taiwanese enterprises can proceed southbound together with Japanese and South Korean companies. Since 2012 Japan is initiating the "Indo Pacific Strategy". It will be good for Taiwanese enterprises cooperate with Japanese enterprises and contribute to each other's southbound policies. The United States has advocated their Asia-Pacific policy for many years. Recently in November 2017, President Trump travelled East Asia and declared the policy

---

28 Ibid.

29 〈冒進新南向：不能只想著快速收割〉，聯合新聞，2017 年 9 月 20 日，https://udn.com/news/story/7338/2711407。

of the Asia-Pacific region has extended to the "Indo Pacific". European countries have also taken an active part in economic development in South and Southeast Asia. How Taiwan can work with these countries and to transform competition into an aid? It is the direction of our efforts in the future.

VI. Conclusion

As early as the early 1990 when President Lee Teng-hui took power, he had already started to push for a southbound development policy. He hoped that Taiwan's foreign investment would not be completely dependent on the mainland China and that the southbound development policy should be used as a part of "Pragmatic Diplomacy".

He hoped that through its relations with Southeast Asian countries can develop economic and trade cooperation and expand our diplomatic space.

The government has promoted the implementation of the "Outline for Strengthening Economic and Trade Work in Southeast Asia" since 1994 and the various cooperation related to trade and investment with Southeast Asian countries. It has now entered the implementation phase of the seventh phase of the program of work.[30]

However up to now the external environment has greatly changed. China's political and economic strength has grown tremendously. By the new policy "Belt and Road", China has been gained many substantive influences in Southeast Asia and South Asia. If Taiwan's current new Southbound policy is to exert its functions, to let all industries in Taiwan take root in Southeast Asia and South Asia in order to facilitate longterm operation and expansion. We must first make longterm strategic analysis and judgment than evaluate their strengths and weaknesses in detail and make the appropriate adjustments. Enable Taiwanese enterprises all had the international competitiveness before they step out of Taiwan. They could further utilize the assistance from other countries outside the

---

30 〈第七期加強對東南亞地區經貿工作綱領〉，經濟部國際貿易局經貿資訊網，2015 年 3 月 26 日，
   https://www.trade.gov.tw/Pages/detail.aspx?nodeID=798&pid=319248。

region and then make detailed analysis of the most basic language, culture and economic environment in Southeast Asian countries. In the process of advancing globalization at the same time, more attention should be paid to the details of the localization, which can be more integrated into the local culture. The enterprise body is embedded in the host country. Both the manufacturing industry and the consumer and service industries can develop smoothly and have longterm local competition force.

We can say that from 1990 till now, after four Taiwanese presidents. All declared southbound policy. The basic content and purpose are the same.

But for Taiwan subjective conditions and the environment have many changes. Foreign investment requires more rigorous analysis and evaluation.

Based on strategic analysis, to look farther, wider vision, deeper thinking constitute the overall three degrees of space. While declaring the new southbound policy, the government must do more analysis and planning in the future to give Taiwanese enterprises more chances and risks in undertaking the southbound policy, therefore, the enterprises, the government and Taiwan can win the triple win-win situation.

中國「一帶一路」倡議對台灣新南向政策的影響：以印度為例

湛忠吉 *

壹、前言

2009 年美國總統歐巴馬就任，並將國家的戰略重心由反恐轉向亞太地區，同時提出「重返亞洲」的政策，而其主要目標不言可喻就是針對日漸強大的中國，藉由強化與亞洲盟國軍事與經濟的關係達到防範、遏止與圍堵中國的目的，這項政策於 2012 年修正為「亞太再平衡戰略」；美國為了實踐本項戰略，在軍事上除積極拉攏中國周邊的韓國、日本、菲律賓、泰國、印度與澳大利亞等國家組成軍事同盟，並強化其軍事與防衛力量，同時藉由雙邊與多邊演習，強化盟國實力希藉此達成在軍事上圍堵中國的目的；另在經濟上美國加速推動「跨太平洋夥伴關係協議」（簡稱 TPP），以高門檻的經貿合作模式阻止中國的加入，以此取得亞洲自由貿易區的主導權與控制權，達成遏止中國在亞洲繼續擴大的經貿實力與影響力。[1] 2017 年 1 月美國總統川普就任，隨即簽署退出 TPP 行政命令，並在許多場合強調美國優先的概念與不再容忍不平等的貿易，同時也要求歐盟、日本、韓國等國家，必須提高防衛經費的支出，這也讓外界質疑美國為維護自身國家利益，是否已開始實施戰略收縮，減少對外公共財的提供，也讓部分專家開始質疑，後美國時代世界是否因此而會陷入「金德柏格陷阱」。

反觀中國卻在此時舉起全球化的大旗號召世界各國不要築起貿易壁壘，並積極推進「一帶一路」倡議，同時邀請世界各國一起分享中國三十幾年來改革開放的經濟果實，這也讓世界各國逐漸向中國靠攏，而中國在世界的影響力也逐漸水漲船高，在美中勢力此消彼長的情勢下，台灣政府「新南向政策」推動是否會遭遇中國的制肘與相關國家所奉行的「一中」政策的挑戰或阻擾？中國現今積極推動「一帶一路」倡議，並強調合作與互助同時主動提供沿線參與

---

* 淡江大學國際事務與戰略研究所博士生
1 瞿少華，〈美國亞太再平衡理論根源及發展趨勢分析〉，《理論報導》（南昌九江），第四期，2014 年，頁 22。

國家公共財（建設沿線國家基礎設施與強化經貿往來），另由於世界貿易組織（WTO）目前因各國利益考量不一，造成無法繼續前進窘境，且近來美國總統川普上台後，更對 WTO 功能提出質疑，並多次違反 WTO 組織相關規定（如單方面徵收懲罰性關稅、發動 301 調查），雖經該組織判定美國敗訴，但美國仍然不理相關判決。因此中國未來能否透過「一帶一路」倡議整合沿線國家，成立取代現有以美國和西方國家主導的 WTO 組織，有效解決目前世界各國的貿易相關問題；而如果中國未來成功整合沿線各國，成立取代 WTO 之相關組織，而中國是否就如同「金德柏格陷阱」所談到的成為取代美國為新的世界霸主，成為穩定世界經濟運行的重要力量，而屆時台灣現行推動的「新南向政策」是否會受到影響與挑戰？

　　本文所要探討的議題有以下幾點：一、中國現在與印度的緊張關係是否對台灣政府推動的「新南向政策」有利？二、中國推動「一帶一路」倡議是否影響台灣「新南向政策」的推動？三、台灣「新南向政策」，有無可能搭上中國「一帶一路」倡議的便車順勢發展，並帶動我經濟的快速發展與轉型，藉此達到兩岸經濟雙贏的目的。四、如果未來中國成功整合「一帶一路」倡議沿線國家，成立取代現有世界貿易組織（WTO）的機構，台灣經濟與「新南向政策」是否會受到邊緣化與嚴重挑戰。中國「一帶一路」倡議與台灣「新南向政策」雙方在規模與投入上都並非在同一個檔次，另加上目前兩岸之間政府彼此關係日趨緊張由目前「冷對抗」有逐漸邁入「熱對抗」的趨勢，而在這種情勢下，中國「一帶一路」倡議未來的發展，實在值得我們深入研究與探索，藉由本篇研究，了解彼此政策的優點與缺點所在，供未來政府執行相關政策參考。

貳、「金德柏格陷阱」概念及其運用

　　「金德柏格陷阱」這一詞是源自於美國的一位著名的經濟學者金德柏格，他同時也是霸權穩定論的倡導者之一，而這個概念是在 1973 年金德柏格撰寫的《*The World in Depression 1929-1939*》書裡被提出來；「金德柏格陷阱」的論點主要強調「資本主義」市場經濟體系的穩定運行是需要成本的，而市場本身無法提供這種穩定市場運行的成本，更不可能指望企業來承擔這樣的成本，

這時市場相對應的國家就變成穩定市場運行的行為主體，但是在國際間，各國卻不會主動的為世界經濟體系的穩定，而花費所必須的成本，原因是因為這種付出往往也得不到應有的利益。因此金德柏格認為這種成本的付出，必須由一個超強的國家或霸主國不計收益而付出這種維護世界市場穩定的成本（簡稱公共財），也因為這樣的成本是維護世界市場的穩定而形成的「公共財」；而這個超強國或霸主國願意付出這樣的公共財絕對不可能是大公無私的，而其主要目的就是要維持自身在國際政治與世界市場的領導地位，同時藉此獲取更重要的利益。[2]

金德柏格認為 19 世紀英國就是世界的霸主國也是提供公共財的國家，但因為進入 20 世紀後，英國在經歷第一次世界大戰後國力迅速衰退，無法再有效提供公共財了，而當時世界上最強的就是美國，但美國卻不願意取代英國成為世界霸主，更不願意接替英國付出維護世界經濟體系所需的費用，也因如此而導致發生了 1929 至 1939 年世界經濟大危機與大蕭條，並間接導致國家主義、經濟民主主義及種族主義的法西斯主義的興起；因此金德柏格得出的結論認為當舊的霸主國衰退無法繼續對世界經濟提供公共財，此時必須有一個新的霸主國來替代原有的霸主國提供公共財，如此才能維持世界經濟體系的穩定，否則可能會重蹈 1929-1939 年代的世界經濟大蕭條情況。[3]

近年來隨著中國的快速崛起，與「亞洲基礎設施投資銀行」及「一帶一路」倡議的推進，並獲得世界多數國家認同與響應，這讓美國深感芒刺在背；另美國新任總統川普上台後強調「美國優先」政策，並對美國原有的外交與貿易政策實施大幅的調整與改變，同時也單方面的退出與廢除美國原先加入的組織與協議（如 TPP、巴黎協議等），這些改變除引起美國傳統盟國與盟友的抨擊，也讓世界各國對美國這樣的調整產生極大的疑慮與恐懼，而些改變除象徵美國國力與國家自信的衰退外，更因受到美國國債不斷的飆高，且對外貿易逆差不斷擴大，而國內各項設施與建設又呈現百廢待興，而美國為維護世界霸主地位，長期對世界提供公共財，這樣的做法也造成美國國內民眾極大的不滿。2017 年

---

2 Charles Kindleberger, *The World in Depression 1929-1939* (Berkeley and Los Angeles:University of California Press, 1973), pp. 5-18.

3 Ibid., pp. 108-117.

川普的上台，也說明了美國國內對美國傳統菁英領導的政府產生極大不滿與不信任，期望藉由川普的上台，改變美國未來；然而世界也注意到美國的這項改變，因此世界有許多學者開始討論後美國時代，然而環顧世界，目前最有能力接替美國這個角色的國家也唯有中國，而現在最主要的問題是崛起的中國有無能力與意願取代二次世界大戰後美國在世界上霸主的角色，並向世界提供維持世界經濟運行的「公共財」，另一方面現今的中國是否已經具備等同美國相同實力向世界輸出公共財；如果中國具備但沒有無意願向世界輸出公共財，那未來世界經濟是否會陷入這位美國學者所說的「金德柏格陷阱」，並再次導致世界經濟大蕭條的情況出現；另一方面，如果現今的中國如果有意願，但尚未具備美國目前經濟實力，那未來美國退出提供「公共財」的角色後，世界的經濟發展將會如何？

「一帶一路」倡議是 2013 年底，中國國家主席習近平所提出的中國對未來世界經濟發展的擘劃與藍圖，而這些沿線國家大都是未開發或是開發中國家，本身的經濟實力薄弱且部分國家還處於內亂狀態，也因如此才導致歐美大國對這些國家與地區極少著墨，然而中國要發展「一帶一路」倡議除必須與當地政府能密切合作外，同時也必須協助這些國家建設國內基礎設施與對外通聯道路，這也是攸關該項戰略能否成功的重要關鍵所在。「一帶一路」倡議的推動，也意味著未來中國必須對這些沿線國家提供更多的公共財，打通彼此的貿易的障礙與節點，才能吸引沿線國家對此倡議的配合與支持，而中國藉此達到實現偉大的「中國夢」與和平崛起的目的。兩岸自新政府上台後，雙方氛圍急轉直下，目前官方往來幾近中斷，民間交流尚未受到影響，且兩岸在外交的角力也不斷升級，因而中國的「一帶一路」倡議在執行的過程中，是否會對台灣政府大力推動的「新南向政策」產生排斥效果，還是「一帶一路」倡議與台灣「新南向」政策有互補效果，兩者的推進事實上不受兩岸關係的影響，在本篇中將以兩國對印度經貿現況加以說明並分析。

中國自 2012 年底提出「一帶一路」倡議後，並先後邀請相關國家成立絲路基金、亞投行及金磚國家新開發銀行等機構，藉此支撐此項倡議建設須所需的資金，並以此達到分散風險的目的；2017 年 5 月 14 日中國辦理第一屆「一

帶一路國際合作高峰論壇」，與會人員共有 29 國國家元首，130 餘國 1500 名
代表與會，也可看出「一帶一路」倡議受世界各國重視的程度。依據中國 2018
年 1 月 25 日商務部發布資料顯示，2017 年中國與「一帶一路」沿線國家貿易
額為 7.4 萬億元人民幣，同比 2016 年增長 17.8%，增速高於國外貿易額 3.6%，
而中國企業對外直接投資 144 億美元，與沿線國家新簽承包工程合同達 1443
億美元，同比 2016 年增長 14.5%。[4] 由以上數據可得知目前中國「一帶一路」
倡議與沿線國家經貿發展迅速，勢必加速「人民幣國際化」與「中國企業走出
去」，另也會大幅提升中國的國際地位；而將此概念運用在「金德柏格陷阱」
理論，隨著中國國際地位與實力日益提升，且世界貿易組織（WTO）對於世界
貿易問題逐漸無法或公平處理，那隨著「一帶一路」倡議規模又不斷加大，未
來針對類是問題勢必要有一個新的處理機構或機制，未來中國無論是被迫還是
主動，主導類似組織的產生與建立，中國目前都具備這樣的實力與資格，然而
考量現在兩岸的緊張關係，在可見的未來台灣想加入該組織的機會相對較小，
在此情況下，台灣的經貿或「新南向政策」是否會被邊緣化？

　　本文為說明上述情況如果發生，台灣的經貿情況可能會出現甚麼狀況，來
推論「新南向政策」在此狀況下未來的發展；本篇文章以當初台灣加入 WTO
組織前與加入後的差異來做比較，希在此情況下得到一個比較客觀的解答。

參、在中國「一帶一路」倡議下中印貿易

　　中國與印度是目前世界上增長最快的兩個經濟體，且兩國總人口總和約為
26 億之多，人口總數約佔世界的 40%，中印兩國在世界分工不同，因而兩國經
濟互補性大，另外目前中國也為印度第一大貿易國，印度也是中國第 18 大貿
易夥伴，與第 8 大出口市場。2000-2014 年中印雙邊貿易從 29.14 億美元，增
長到 705.79 億美元，而這 15 年間成長了 24.22 倍[5]。印度為南亞地區最大的國
家，且近幾年受惠於印度總理莫迪的經濟改革，使印度經濟呈現高速增長；而
中國自 2013 年底開始推動「一帶一路」倡議，而印度又地處於中國「一帶一路」

4　〈中國服務貿易指南網，「去年中國與“一帶一路”沿線國家貿易額增長 17.8%」〉，http://
　　tradeinservices.mofcom.gov.cn/article/tongji/guonei/buweitj/swbtj/201801/53469.html。
5　廖怡煒，〈一帶一路下中印合作的進展、挑戰與對策〉，《時代經貿》，第 24 期，2017 年，頁 36。

倡議的重要位置（為絲綢之路經濟帶與 21 世紀海上絲綢之路之交會點），因此其重要性也不言可喻，隨著中印兩國綜合國力快速的崛起，中、印兩國未來能否在此倡議上摒除舊有的恩怨，攜手合作將是影響 21 世紀亞洲能否成為世界經濟重心的重要因素。「一帶一路」倡議是中國擘劃中國未來數十年經濟發展的藍圖，也是試圖改變現有世界經濟發展的面貌，而此一項倡議的提出自然引起世界各國正、反兩種截然不同的聲音與評價。「一帶一路」倡議剛推出後，印度也表達自身參與意願，惟隨著中巴經濟走廊與瓜達爾港建設的推進，引起印度對中國此倡議的疑慮，進而推遲與暫停與中國「一帶一路」倡議的合作。然而對中國而言，「一帶一路」倡議，印度的參與與否將是影響該倡議能否在南亞順利推動的重要關鍵，也是中國在經濟與軍事上破除美國「亞太再平衡戰略」包圍的重要一環。

一、中、印兩國經貿現況

中國於 2002 年時對印度貿易呈小幅順差（3.97 億美元），2004 年後中國對印度貿易逆差已經達到 17.42 億美元，隨後兩國貿易逆差更是逐年加大，到了 2015 年時中國對印度逆差已達到 448.6 億美元，也因如此造成中、印兩國貿易摩擦的頻繁發生（如圖 1、圖 2）；而根據 WTO 資料統計 1995 年到 2013 年上半年印度對中國的企業提出反傾銷調查達到 157 起。[6]

中印兩國的在經貿結構上也呈現互補，中國對印度出口主要以技術密集產品為主，排名前幾大類產品，分別為：以電子、電氣產品為主的高技術製造業，比例高達 25.25% 以上，其次是以工程用產品為主的中技術製造業，占比 17.91%，第三高的則以加工產品為主的中技術製造業，占比 13.53%（如圖 3）；而印度對中國出口則以初級產品、資源型製造業、低技術製造業為等三類為主的比重分別為 28.23%、23.04% 和 20.94%，合併占到 72.21%（如圖 4）。

---

6 郭敏、陳潤，〈一帶一路倡議下中印貿易關係〉，《中國經貿導刊》，第 2 期，2018 年，頁 10-11。

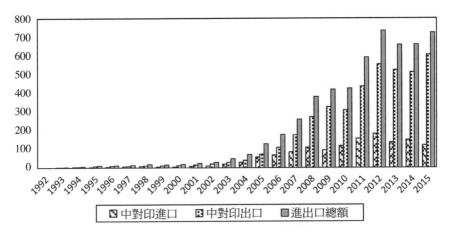

圖 1：1992-2015 年中印雙邊貿易額

單位：億美元

資料來源：UN Comtrade 貿易數據整理所得；郭敏、張小溪、劉霞輝，「中國社科研究
所 . 經濟研究所」，中國社會科學研究院經濟研究所，http://ie.cass.cn/academics/
economic_trends/201712/t20171204_3765987.html。

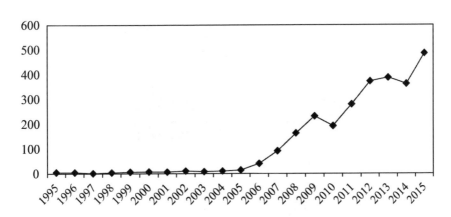

圖 2：1995-2015 年中國對印度貿易順差

單位：億美元

資料來源：UN Comtrade 貿易數據整理所得；郭敏、張小溪、劉霞輝，「中國社科研究
所 . 經濟研究所」，中國社會科學研究院經濟研究所，http://ie.cass.cn/academics/
economic_trends/201712/t20171204_3765987.html。

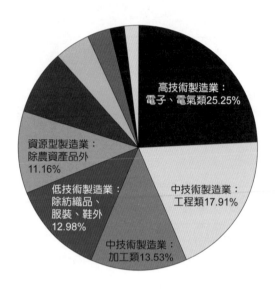

圖 3：中國出口印度產品分類及比重（2015 年）

資料來源：UN Comtrade 貿易數據整理所得；周晉竹，「中印雙邊貿易的競爭與合作」，
中國遠洋海運，第 9 期（2017 年），頁 31。

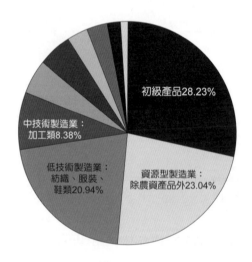

圖 4：印度出口中國產品分類及比重（2015 年）

資料來源：UN Comtrade 貿易數據整理所得；周晉竹，「中印雙邊貿易的競爭與合作」，
中國遠洋海運，第 9 期（2017 年），頁 31。

　　近年來中國對印度投資逐漸升溫，從 2010 年的 5.2 千萬美元增長到 2016 年的 16.12 億美元，六年間成長了三十倍，而結算到 2017 年 7 月底，中國對印度累積投資已達到 50.6 億美元，而印度對中國投資金額到達到 7.21 億美元；然而部分中國製造業甚至將產業鏈整體遷往印度（如手機業），而現階段中國對印度投資的項目，主要集中在汽車及汽車零附件、冶金、電氣設備、能源、機械及再生能源等領域。[7]

　　中、印兩國的經貿從 2000 年 20 餘億美元，發展到 2016 年的 700 餘億美元，且現在中國乃是印度最大貿易夥伴，同時也是 2013 年到 2016 年之間最大的貿易進口國（如表 1）：

表 1：**2013-2016** 年印度主要貿易夥伴（**%**）

年份 名次	印度貿易的主要出口夥伴				印度貿易的主要進口夥伴			
	2016	2015	2014	2013	2016	2015	2014	2013
1	美國 15.8	美國 15.1	美國 13.3	美國 12.4	中國 16.9	中國 15.6	中國 12.7	中國 11.1
2	阿聯酋 11.7	阿聯酋 11.5	阿聯酋 10.4	阿聯酋 10.2	美國 6.0	瑞士 5.4	沙特 7.1	沙特 7.8
3	香港 5.0	香港 4.6	香港 4.3	中國 4.7	阿聯酋 5.4	阿聯酋 5.9	阿聯酋 5.9	阿聯酋 7.1

資料來源：整理自中華人民共和國商務部網站資料，http://www.mofcom.gov.cn/

　　從上表可以看出 2013 年到 2016 年印度從中國進口的貨品比例每年均有大幅提升，這也表示中國貨品符合印度國內需要；從表 1 也可看出 2013 年到 2016 年美國均為印度最大的商品出口市場，而香港與中國在 2013 年到 2016 年也分別佔據印度出口的第三大市場，惟中國在 2013 年以後便退出了第三大出口的排名，改由香港取代；也因為印度出口市場的問題，造成印度對美國市場的依賴，也因此以印度角度來看，發展印、美兩國關係的急迫性與利益性遠大於發展印、中兩國關係。從本表也可以看出中國自 2013 年推出「一帶一路」倡議，中印兩國的進出口貿易並未受到此倡議有大幅的波動，反而中國於 2013

7 周晉竹，〈中印雙邊貿易的競爭與合作〉，《中國遠洋海運》，第 9 期，2017 年，頁 32。

年後退出了印度第三大貿易出口國的角色，這個數據也可以間接證明了印度對於中國「一帶一路」倡議的疑慮。[8]

二、中印兩國關係存在的障礙問題

（一）印度長期以來均視自己為全球性的大國，對自身的定位很高，並且視自己是中國的競爭對手，更期望自己是一個有聲有色的大國；而印度獨立初期與新中國成立之初處於同一水平，而中國未來的發展很自然地就被印度拿來當作參照。1962 年中印大戰，印度戰敗被視為奇恥大辱，而印度對此銘記在心，希望能一雪前恥超越中國，但因自身實力尚未具備，因此印度必須借助外力來平衡中國在區域的影響力。[9]

（二）外部因素對中印兩國之間關係的影響，由於美國為防制中國崛起挑戰美國在全球的領導地位，因此積極鼓動印度制衡中國，這主要因為中國的崛起對美國與西方國家造成威脅與挑戰，然而印度現階段尚無這方面顧忌；美國為平衡中國的快速崛起，起初是以日本作為平衡中國的主要角色，然而當日本因在經濟與軍事等方面被中國趕超後，逐漸力有未逮，這時美日兩國便看上經濟與軍事力量快速提升，且與中國是敵對國家的印度，透過各種手段拉攏印度與討好印度，並期其扮演對抗中國的角色。[10]

（三）由於印度長期將中國視為區域競爭對手，當中國提出「一帶一路」倡議以陸海並進的方式打開中國對外貿易通路，而這也對印度在南亞區域的大國地位造成挑戰，為因應中國挑戰印度也提出自己的整合發案：「次大陸國家經濟合作協定」、「西聯政策」、「季風計畫」等，另也積極推動與其他南亞國家的經濟走廊建設方案；另印度長期均視南亞為自己勢力範圍，並排斥國外勢力的介入，而近期中國為推進「一帶一路」倡議，加強與南亞各國的往來，尤其對「中巴經濟走廊」的推進與中巴對瓜達爾港的開發甚感不滿與警戒，認為中國是將自己的勢力範圍擴大到自己的領土與勢力範圍內。[11]

---

8 雷建鋒、範堯天，〈一帶一路倡議實施中的中印關係〉，《遼寧大學學報》，第 46 卷第 1 期，2018 年，頁 136-137。
9 同上註，頁 137。
10 同上註。
11 同上註。

肆、在台灣「新南向政策」下對台印貿易的互動

台灣由於地理面積狹小，國內內需也十分有限，而如需依靠內需來實現產業規模化發展是十分困難的，因此擴大對外市場，且發展對外經濟是台灣促進產業發展的必經之路[12]。

「南向政策」這一名詞，最早出現在 1994 年 1 月 10 日時任行政院長連戰，在「中國國民黨總理紀念月會」所提的行政報告中出現，而從 1994 年到新政府上台，「南向政策」共經歷三個階段。[13]

一、1994 年台灣開始推動南向政策，此一階段台灣在 1997 年時對東南亞的投資增長迅速，投資到達台灣對外投資的 27%，而當時雖然對中國投資速度放緩，但中國乃是台灣對外投資占四成的主要貿易國。因此當時此一政策雖然取得一定的成績，但成效依然十分有限；而 1997 年又適逢金融風暴席捲東南亞，造成東南亞國家嚴重的經濟衰退與市場萎縮，並引發了部分東南亞國家的排華運動，台商當下遭受到嚴重的經濟損失，使得台灣大量的資本開始撤出東南亞，也對未來台商到東南亞投資產生負面影響。[14]

二、1998 年 3 月第二輪「南向政策」開始實施，雖然當時政府給予很多投資優惠，但台商對於第二輪的「南向政策」始終興致缺缺，而此一階段台灣對東南亞投資不但投資金額沒有增加，反而大幅降低，而同一時期台商對中國的投資反而增加 78%。[15]

三、2002 年 7 月陳前總統開始推動民進黨版本的「南向政策」，要台灣走自己的路，中國不是台灣對外的唯一市場，但因產業未能配合，故造成了陳前總統執政期間，從原本台灣對中國的出口依存度由不到兩成升到四成，且赴中國投資佔台灣對外投資的比重，從近五成升到近六成。[16]

---

12 李紅梅、許振、黃蓉，〈四力博弈下的新南向政策問題研究〉，《台灣研究》，第 4 期，2017 年，頁 74。
13 趙凱，〈蔡英文的新南向政策評析〉，《九鼎》，118 期，2017 年，頁 52-53。
14 同上註，頁 53。
15 同上註。
16 同上註，頁 54。

　　2016 年 5 月 20 日新政府上台後，為能降低對中國的經濟依賴，積極推動「新南向」政策，擴大與南向國家交往，而這些新南向國家包括：菲律賓、泰國、緬甸、馬來西亞、新加坡、印尼、越南、柬埔寨、印度、孟加拉、澳大利亞、紐西蘭、寮國、不丹、巴基斯坦、尼泊爾、汶萊及斯里蘭卡等 18 國，其中蔡總統特別提到發展台印關係的重要；然而由於中印兩國在 20 世紀 60 年代兩國爆發戰爭，致使雙方關係惡化，有鑒於此當時台灣政府試圖改善與印度關係，後來卻因中印關係緩和而失敗。且印度官方始終奉行著「一個中國」原則，對台灣一直保持著冷淡與疏遠的關係[17]，到目前印度仍然禁止部長與省長訪台。台印 2009 至 2017 年 5 月雙邊貿易額如表 2。

表 2：2009-2017（1-5 月）台印雙邊經貿

年份 項目	2009	2010	2011	2012	2013	2014	2015	2016	2017 (1-5)
對印度出口	25.3142	36.2838	44.2737	33.8448	34.2291	34.2498	30.3617	28.2309	13.0843
對印度進口	16.2387	28.3745	31.3648	26.2364	27.5134	24.8532	19.1126	21.8381	15.1234
對印出口成長幅度	-15.82%	+43.37%	+22.02%	-23.56%	+1.05%	-0.06%	-14.37%	-3.75%	+10.37%
對印進口成長幅度	-30.40%	+74.94%	+10.54%	-16.35%	+4.86%	-9.6%	-24.42%	+16.33%	+115.09%

單位：億美元

資料來源：依據印度國家檔案 - 國貿局自製而成，https://www.trade.gov.tw/App_Ashx/File.
　　　　　ashx?FilePath=../Files/PageFile/...pdf

　　由上述數據可得知，台印兩國貿易於 2011 年達到高峰，台灣對印度出口達 44,2737 美元，對印度進口達到 31,3648 美元，到 2012 年時我對印度出口出現急遽下降，降幅達 23.56%；2015 年由於受到全球景氣及商品價格大幅下跌影響，雙邊的進出口均呈現衰退的現象。依據上表可看出，雙邊的貿易額度自 2009 至 2016 年大都維持在 40-70 億美元，這也表示台印兩邊的貿易額還有極

17　馮崢，〈台灣地區與印度關係新動向：回顧與前景分析〉，《現代台灣研究》，第 2 期，2017 年，
　　頁 59-60。

大的發展空間。另以貿易額排名來說，2016 年印度為台灣的 17 大貿易夥伴，較前一年進步一名，同時該國也是台灣第 15 大出口市場，及第 21 大進口來源國；以比例來說，2016 年台印的雙邊貿易占台灣全球貿易的 0.98%（2015 年為 0.95%），出口比重為 1.01%（2015 年為 1.05%），進口比重為台灣的 0.95%（2015 年為 0.82%）。[18]

由於印度擁有與中國近乎相等的人口，且經濟發展快速，故未來發展潛力無窮；且受到中印兩國間長期矛盾影響，致使印度想要藉機拉攏台灣對抗中國，因而對台灣會相對友善；因此台灣政府推動在「新南向政策」便主動將其列為首要國家。根據統計，目前台灣在印度投資的企業約 90 餘家，累積直接與間接投資金額約為 15 億美元，而主要集中於電子、航運、通訊科技、鋼鐵、製鞋、金融、營建及汽車零部件等產業，而投資地點集中於新德里、孟買、班加羅爾、清奈等地區，而台灣政府為了配合「新南向政策」，幫助台商打開印度市場，台灣貿易協會在孟買、清奈及加爾各答 3 大辦事處設立了「臺灣商品行銷中心」，協助台商擴大對印度的出口，但到目前為止成效有限，台灣對印出口僅占印度進口額的 0.9%。[19]

經統計截止 2016 年底台印雙邊貿易僅占台灣整體對外貿易的 0.98%，而台商在印度投資只佔外人直接投資金額 0.03%；台灣對印度出口的物品，主要是以工業製品，包括塑膠及其製品、電機電子產品與設備及其零附件、燃料、機器及零部件等。而台灣從印度主要進口為農工原材料及半成品及有機化學產品等；台灣目前在印度所有外國投資者中只排第 45 名，這也顯示台灣產品在印度的能見度並不高；另近年來台灣赴印度旅遊人數每年約 2 萬餘人，而這數字僅佔印度總入境人數的 0.2%，到 2017 年台灣入境印度人數到達 3.4 萬人次，另印度來台人數在 2014 年首次突破 3 萬人次，雙方旅遊人數自 2012 年到 2016 年間每年均有成長，但這也顯示雙方往來人數還有非常大的進步空間（雙邊旅遊人數如表 3）。[20]

---

18 貿易全球資訊網，〈印度與我國經貿關係〉，http://www.taitraesource.com/total01.asp。
19 馮崢，〈台灣地區與印度關係新動向：回顧與前景分析〉，頁 63。
20 陳牧民，〈新南向政策中的印度與南亞〉，《全球政治評論》，第 55 期，2016 年，頁 20。

表 3：台、印兩國 2012-2016 年雙邊出國往來人數

年份 項目	2017	2016	2015	2014	2013	2012
印度出國來 台灣人數	34,962	33,550	32,198	30,168	23,318	23,251
台灣出國到 印度人數	25,965	22,399	20,066	18,873	20,655	6,548

單位：人

資料來源：整理自中華民國交通部觀光局統計資料，http://admin.taiwan.net.tw/public/public.aspx?no=408

為增進雙邊合作機制，台灣目前與印度在經濟交流與溝通有以下幾個管道：

（一）台、印經貿對話會議

2005 年台灣與印度建立雙邊次長級經貿對話會議（原名稱為台印次長及經濟諮詢會議），這個會議主要由印度商工部產業政策與推廣部門的次長，與台灣經濟部次長一同主持，到 2016 年底已辦理 11 屆，而雙方討論議題涵蓋產業合作、資通訊專業人才交流、反傾銷與防衛措施的調查案、金融監控、中小企業、標準與符合性評鑑、太陽能、機工具、食品加工、投資與區域經濟整合等議題；另雙方為檢視合作議題進展，已建立每年舉辦兩次的檢討會議之機制，以掌握合作議題的進展與增進交流管道。[21]

（二）台、印工作貿易小組會議

台、印兩邊為維持雙方已洽簽的台、印經濟合作協定動能與排除關稅與非關稅障礙，雙方同意建立貿易工作小組會議之定期對話，第 1 次已於 2015 年 4 月在新德里辦理，而第 2 次會議於 2016 年 12 月 20 日與 27 日在台北召開，主要針對貿易、投資、產業合作及非關稅貿易障礙等議題實施意見交換。[22]

---

21 貿易全球資訊網，〈印度與我國經貿關係〉，http://www.taitraesource.com/total01.asp。
22 同上註。

（三）台、印民間經濟聯席會議

　　台、印民間經濟聯席會議由印度工商總會與中華民國經濟合作協會共同主辦，截至 2016 年底已舉辦 15 屆會議，雙方由產業與企業代表就兩邊投資與經商環境、政策、產業合作機會進行交流與合作。23然而 1995 年以來台印關係發展迄今，雙方經貿關係與人員觀光交流雖有進展，且經貿聯繫也日益緊密，隨著新政府大力推動「新南向政策」，且將印度列為重點發展國家，未來兩邊關係勢必會遭遇新的契機與挑戰。

一、台、印雙邊關係發展契機

　　（一）印度為世界上人口僅次於中國的第二大國，且市場潛力大，並擁有強大的經貿實力，而兩邊的經貿合作有極大的強化空間，且印度人口又具備勞動力、年齡等優勢，同時工資較低廉僅為中國的二分之一到三分之一。

　　（二）台、印目前沒有簽訂任何 FTA 協定，未來政府可洽詢對方意願，簽訂雙方 FTA 協議，除可強化台商在此投資保障，更有助於台灣有效開拓印度市場，降低彼此貿易壁壘，促進雙方合作互惠。

　　（三）目前由於中、印兩國交惡，印度官方與台灣官方合作，相對以往比較不受中國相關壓力影響，近期美、日兩國積極拉攏印度組成印太戰略聯盟，因此如何運用即將成形的美印日澳「印太戰略」與相關國家經貿接軌，將是台灣政府努力的重點。

二、台印雙邊發展制約因素：[24]

　　（一）區域市場區隔明顯。

　　（二）低階人力，文盲比例過高。

　　（三）基礎設施不足。

　　（四）政府辦事效率低落。

---

23 同上註。
24 馮崢，〈台灣地區與印度關係新動向：回顧與前景分析〉，頁 63-64。

（五）區域文化差異過大。

（六）官方語言過多。

（七）稅負複雜，且部分產業鏈尚未建立。

（八）工資以外的各種賦稅較高，生產成本高於中國。

伍、從「金德柏格陷阱」理論檢視中國「一帶一路」倡議對台灣「新南向」政策可能產生影響

2008年美國爆發次貸風暴，引發全球金融混亂，並且間接造成歐洲主權債務危機，讓世界各國對美國領導與建立的金融與世界經濟秩序開始感到疑慮；2017年1月美國新任總統川普上台，並在上台第一個工作日即簽署退出跨太平洋夥伴協定（TPP），另也片面宣布退出「巴黎氣候協議」，在經濟上更是多次強調「美國優先」，強迫別國接受美國不合理的稅賦與經貿協定；在軍事上威脅盟國增加負擔美國在外駐軍負擔，以上種種行為除讓美國盟邦與盟友感到訝異與不滿，更讓世界各國感受到美國這個世界超級大國是否已經開始在走下坡，還是美國不願意再為世界付出維繫世界經濟運行的「公共財」。然而就是因為在這種情況下，「金德柏格陷阱」理論再次受到大家的關注。

2018年1月23日至26日「世界經濟論壇」在瑞士的達沃斯舉行，會中德國工業巨頭西門子的執行長凱瑟（Joe Kaeser）曾說：「不管你喜歡與否，中國的一帶一路將成為新的WTO。」[25]；如果以這個概念來說明，如果中國「一帶一路」倡議未來成功整合「一帶一路」倡議沿線國家，成立一個新的國際貿易機構取代現在功能不彰的國際世界貿易組織（WTO），而加上中國先前成立的「亞洲基礎設施投資銀行」、「金磚開發銀行」等機構，將可作為支撐中國在未來籌組類似組織的能力與實力。考量目前兩岸關係的緊張狀態，如果此一狀態未能改善在可預想的狀況下，中國不可能會讓台灣加入該組織，那屆時台灣的經貿或「新南向政策」推動是否會受到此一組織的影響？本文以台灣2002年加入WTO組織前後經貿狀況，作為未來台灣在遇到類似情況下，可能會對

25　中央通訊社，〈達佛斯論壇 紐時：聚焦中國一帶一路〉，中央通訊社，http://www.cna.com.tw/news/aopl/201801290355-1.aspx。

台灣的經濟與「新南向政策」產生的可能影響作有關分析，希藉此得出較為客觀的數據與結果，做為未來政府擬定相關政策的參考。

　　台灣自 2002 年獲准加入 WTO 組織，而當時台灣加入該組織的考量原因有以下七項：一、考量 WTO 組織佔全球經濟總量97%。二、參與該項組織可以讓台灣業者在國際公平的環境下開拓海外市場。三、可促進台灣貿易體系並與國際接軌。四、可參與國際世貿組織相關規則的多邊談判。五、透過 WTO 架構解決貿易糾紛與爭端，確保台灣利益與競爭力。六、加強台灣與非邦交國的往來。七、擴大台灣出口商機，並強化台灣競爭力。對於上述議題為能客觀分析，本文以台灣以 2002-2010 年這一年間的成果作為檢視依據，藉此瞭解台灣加入該組織後在工業、服務業、農業、對外投資與外人投資情況作一分析。

　　由於台灣是一個海島型國家、面積小並且缺乏天然資源，所以要發展國家經濟必須依靠對外貿易，這也是為何台灣必須加入 WTO 這個組織的緣故，也因為加入這個組織也讓我們國際之間的連結更為緊密，並能為台灣廠商爭取更有利的貿易環境[26]。台灣自 2002 年底加入 WTO 至 2010 年，GDP 年平均經濟成長率達 4.66%，除 2009 年受到全球金融風暴的影響外變成負成長，其餘各年的經濟皆為正成長；另受到自由化、全球化的發展方向，台灣在 2010 年經濟快速復甦，成長率上修，並創 20 年來的新高。除經濟總量的成長外，台灣的平均每人 GDP 亦以相同趨勢逐步成長，至 2010 年，台灣平均每人 GDP 已近 2 萬美元（如表4）。[27]

　　加入 WTO 後，台灣在農業部門表現除 2003、2009 這兩年呈負成長外，大致呈現微幅正成長；而工業部門則逐年成長，惟 2008、2009 年受到全球金融風暴而呈現負成長，這也顯示台灣工業發展受國際局勢影響已十分明顯；三大部門中，以服務業為成長最快者，產值較 2001 年增加 2 兆 2,339 億元，顯示入會後所面對的國際競爭，已成為台灣服務業發展的驅動力。2010 年，台灣農、工、服務業生產毛額占 GDP 比重分別約為 1.57%、31.37%、67.05%（如圖5）。

---

26 徐遵慈、吳佳勳、吳玉瑩、葉長城、吳子涵、姜博瑄、陳建州、許裕佳（中華經濟研究院），〈台灣加入 WTO 十年成果檢視之研究〉，《經濟部國際貿易局》，頁9。
27 同上註，頁6。

表 4：台灣入會以來總體經濟指標之變化

總體指標	2002	2003	2004	2005	2006	2007	2008	2009	2010
GDP(百萬美元)	301,088	310,757	339,973	364,832	376,375	393,134	400,132	377,410	430,096[R]
年成長率 (%)*	2.51	3.21	9.4	7.31	3.16	4.45	1.78	-5.68	
平均每人 GDP (美元)	13,404	13,773	15,012	16,051	16,491	17,154	17,399	16,353	
年成長率 (%)	1.95	2.75	9.00	6.92	2.74	4.02	1.43	-6.01	
失業率 (%)	5.17	4.99	4.44	4.13	3.91	3.91	4.14	5.85	

說明：* 年成長率以美金為單位計算；R：主計處公布資料為修正值。

資料來源：行政院主計處及中華經濟研究院研究計算得出。

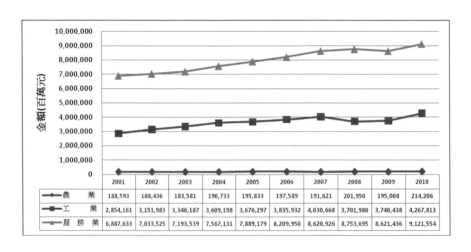

圖 5：台灣入會以來各產業 GDP 趨勢圖

資料來源：中華民國統計資訊網。

　　台灣加入 WTO 組織之後，三大產業就業人數也產生極大變化，台灣入會以後農業人口到從 2002 年的 70.6 萬到 2010 年降至 55 萬人，而工業就業人口則呈現逐年增加，每年平均成長率為 1.31%，惟在 2008、2009 等兩年減少，其主因乃是受當時金融風暴影響所致；另台灣服務業產值是成長最為快速的，其就業人數也為三大產業中成長速度最快，到 2010 年服務業就業人數為 617.4 萬人，每年平均成長率約為 1.79%（如圖 6）。[28]

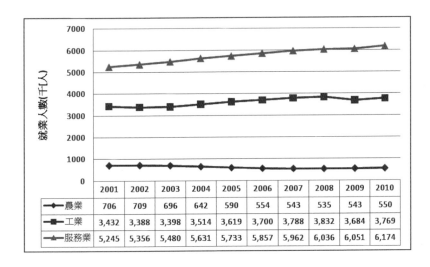

圖 6：台灣入會以來就業人數產業趨勢圖

資料來源：中華民國薪資與生產力統計月報。

　　台灣自 2002 年底加入 WTO 至 2007 年間，無論進出口皆有明顯的成長，在世界金融風暴發生之前，我國平均每年對外出口成長率約為 12.76%，進口則約為 14.13%。而在 2008 年金融風暴發生後，我進出口受到明顯下降，到 2010 年世界景氣又逐漸復甦，也讓台灣的進出口恢復成長，這也可看出台灣與世界各國的經貿連動性很高，因此世界經貿環境改變對台灣也會造成很大衝擊影響（如表 5）。[29]

28　同上註，頁 8。
29　同上註，頁 9。

表 5：台灣入會進出口貿易之變化

	2002	2003	2004	2005	2006	2007	2008	2009	2010
出口成長（%）	7.1	11.3	21.1	8.8	12.9	10.1	3.6	-20.3	34.8
進口成長（%）	4.9	13.0	31.8	8.2	11.0	8.2	9.7	-27.5	44.2

資料來源：財政部關稅總局及中華經濟研究院計算所得。

依據投審會資料台灣在 2002-2005 年外國投資金額漸漸升高，而到了 2005-2007 年間因對外國投資法規進一步鬆綁，同時實施金融改革，也造成外國人來台投資金額大幅提高；在台灣對外投資方面，除 2006 年外台灣自加入 WTO 後台灣對外投資金額都大於外國對我投資金額，而這其中以中國最為明顯（如圖 7）。

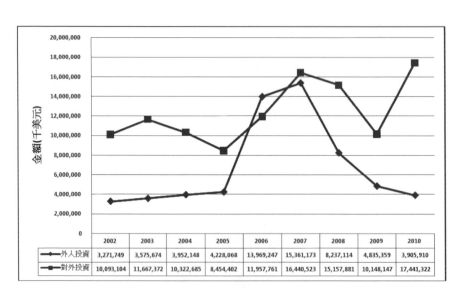

圖 7：台灣外人投資與對外投資趨勢圖

資料來源：經濟部投資審議委員會。

綜析以上資料可得知加入 WTO 雖然各有利弊，在利的方面有與各國貿易往來密切、加速台灣服務業的發展、出口貿易量加大、外資來台投資金額大幅

提升、經濟成長穩定等；在弊部分有農業就業人口降低、與世界經濟連動密切容易受世界經貿環境影響、對台灣對外投資金額大幅提升及對外進口物品也大幅提升等。台灣自 2002 年加入 WTO 後，讓台灣經濟快速成長，雖然也有一些副作用，但這也是現今經濟高度全球化的分工所無法避免的，也因為台灣加入 WTO 這個組織後，讓台灣外銷到世界各國的管道大幅提升，也讓我們在經貿上能與其他國家有對等與公平競爭的機會，同時加入 WTO 後讓我們國家在大國主導的國際經貿環境裡擁有發言權。台灣是一個地小人稠的國家，除缺乏天然資源外，更缺乏廣大的市場，因此對外的經濟發展，就成為台灣的重中之重，因此未來如果中國成功整合「一帶一路」倡議沿線國家，發展取代 WTO 的國際相關機構，如果兩岸情勢未能有效改善，那台灣想加入相關組織的難度也將可想而知；如果依「金德柏格陷阱」理論來看，且結合以上總總可能情況分析，未來如果台灣無法加入中國主導的類似 WTO 國際機構，台灣廠商可能會面臨貨品出口到外的競爭壓力（關稅）、國內市場缺乏吸引外商來投資意願、國內廠商向外發展缺乏投資保障、也可能導致企業、廠商及人才的加速外移，同時也無法享受到低關稅的優惠等，上述問題亦會導致台灣目前推行的「新南向政策」更加執行困難，而台灣突破方式也只能透過簽訂雙邊 FTA 或加入其他區域經濟組織來降低這項改變對台灣的經濟的衝擊。

陸、結論

由近期中美雙方貿易戰，我們可清楚看出台灣與中國並不處於一個同一量級的經濟體，且大國間的經貿大戰，許多時候受傷的都是周遭的小國，其主要原因就是台灣對中、美兩國的市場依賴度較高，且本身關鍵零組件技術也掌握在別人手中，另台灣產業大多以代工為主鮮少有自己品牌，因此雙方面任何一方面受到傷害，都會大大影響台灣經濟的發展。本文綜整以上資料，得到結論為中國與印度目前兩國因歷史與對彼此的不信任，致使中國「一帶一路」倡議在印度並未獲得青睞，也因如此兩國經濟近年來雖成長快速，且綜合國力也隨之水漲船高，但兩國經濟貿易額到 2017 年雙方貿易總額僅達 845.4 億美元，而同年中國與東盟國家的貿易總額 5,148 億美元，由此可知中印雙邊的領土主權爭議與中巴兩國的緊密合作，已造成印度對中國「一帶一路」倡議的疑慮，進

而影響中印兩國在經濟上的合作；另中國長期以來對印度享有巨大順差，也讓印度頗為介意，並多次發起對中國貨品的反傾銷調查，因而從以上資料來看，中國「一帶一路」倡議對台灣「新南向政策」似乎沒有直接影響。

中國「一帶一路」倡議雖未影響台灣對印度的經貿關係，然而台灣對印度的經貿也未因推動「新南向政策」有明顯的變化，而台灣與印度的貿易額大致維持在 40-70 億美元，這也顯示台灣政府在此地區還有非常大的努力空間，然而印度雖然人口眾多，經濟近幾年呈現快速成長，而本身內部問題也非常多如基礎設施不足、行政效率低落、官方語言較多、稅賦複雜等因素，也造成台灣廠商在此遭遇非常多的困難與障礙，進而影響兩國的雙邊貿易。以印度角度來看，中國與台灣兩邊與其貿易額相差懸殊，因此自然印度重視中國的程度絕對遠超台灣，而現今印度願意對我方較為友善，其主要目的是想利用台灣問題增加自己手上籌碼，以利未來能作為與中國斡旋的工具，因此可以預測未來如中印兩國關係獲得改善，勢必會影響台印雙邊的各項交流與往來，本文也在此建議政府必須維持與中國的良好關係，以避免遭遇因中國的快速崛起及在國際的影響力擴大的同時，而造成台灣在國際與經濟上的生存空間快速萎縮。

最後是以「金德柏格陷阱」理論來看，如果未來中國成功整合「一帶一路」倡議沿線國家，成立取代 WTO 的國際經貿組織，而台灣又遭排擠無法加入，台灣的經濟與「新南向政策」是否會受到影響；本項命題因無法準確預測未來發展，僅以台灣加入 WTO 後十年台灣的經貿發展來推論未來可能影響的層面可能會有哪些方面。以上述資料綜析可得知，台灣加入 WTO 雖然對部分產業有負面影響，但卻透過這個組織打開了台灣與 WTO 組織成員的的經貿關係，也大幅提升了台灣對外出口額，另也加速台灣企業赴外投資；由此可推論未來如果台灣無法加入 WTO 類似組織，台灣對外出口勢必降低，同時面臨別國徵收較高關稅的物品稅，也會降低台灣產品與外國同類產品的競爭，另部分企業為能享有該組織成員國的關稅優惠與出口貿易便捷，可能會加速在海外設廠或轉移，也會導致台灣產業進一步的萎縮；在參與國際經貿規則制定與意見表達方面，台灣將會因無法參與該組織喪失意見表達與規則修訂的權益；在貿易糾紛處理方面，如果台灣未來無法參與類似 WTO 國際經貿組織，當發生貿易糾

紛時，將無法提供一個國際協調平台實施協商，而對台灣相關糾紛處理將會產生不利影響；在市場開拓方面，因無法加入未來國際經貿組織機構，也會讓台灣無法透過組織相關法規與成員國正常交往，而如果要進入別國市場，台灣必須針對個別國家實施協商，藉此達到互惠互利，然而隨著中國的實力日益強大，未來別國要與台灣交往考量中國的態度，這也會增加別國與台灣交往的難度與障礙，更會大幅降低台灣企業向外拓展市場的競爭力。

綜悉以上資料，目前從中印與台印雙邊貿易額來看，中國「一帶一路」倡議在印度方面並未對台灣「新南向政策」造成影響與排擠，但從台灣加入WTO（2002-2010 年）期間的資料來看，能否加入國際經貿組織對台灣未來經貿影響顯著，且考量中國目前為世界第二大經濟體，且未來數年後即可能取代美國成為世界最大經濟體，因此對世界經貿組織影響力將也會水漲船高，且且不論中國未來是否會另組織有別於 WTO 的世界經貿組織，中國將是台灣未來不可得罪的重要國家，因而維持良好的兩岸關係，也是台灣在未來與世界各國保持順暢貿易往來的關鍵因素，並以此為基礎逐步與其他各國簽訂雙邊 FTA 才是能否成功的關鍵所在；以目前世界潮流來看，台灣與他國不論視發展政治或經貿關係都無法跳脫中國的影響，因此兩岸關係必定也是我政府首應謹慎面對的重要課題。

本文認為美國學者當初提出「金德柏格陷阱」這個理論，事實上有一個很大的理論盲點，這便是美國學者金德柏格當初並沒辦法證實 1929-1939 年經濟大蕭條的原因，就是因為英國衰退後，美國不肯接替英國的世界霸主地位為世界提供「公共財」而造成的，還是因為世界先經歷十年的大蕭條，世界各國為擺脫這種局面，在美國的號召下才而做出的妥協結果，相反的如果當時世界沒有經歷這十年的大蕭條，是否美國能否能如此輕易的號召世界各國成立有關的世界組織，進而主導世界經濟秩序與國際秩序，這可能就不得而知了。

綜合以上問題，可得知未來中國如成功整合「一帶一路」沿線國家成立取代 WTO 之組織，且兩岸又因關係不佳而被排擠，將會造成台灣經濟有被孤立的狀況，另應對國際金融風暴等相關能力亦會隨之降低，在發生貿易糾紛時，會因無中間協商平台，讓台灣經貿相關權益受損；在國際經貿規則制定與協商，

又因台灣無法與會參加，而讓台灣的經貿權益受到影響。因此，台灣必須正視自身經濟的缺陷與不足，在這全球化的時代，為了讓國家的未來更加光明，台灣政府時有責任與對岸保持良性互動與溝通，降低彼此衝突可能，並為兩岸未來創造雙贏的局面。

中國大陸政軍發展

解構中美國際互動下習近平的國家安全思維：
以「一帶一路」與「印太區域」的建構為例分析

翁明賢 *

壹、前言

從後冷戰時代以來，美國的國家安全戰略始終處於「威脅何在？」與「敵人為何？」的雙重戰略焦慮。首先，意識形態的對立，再者，正在崛起的強國與現存強國的對立。[1]美國學者布里辛斯基（Zbigniew Brzezinski）提出美國開國以來面臨不同時期的辯論，如何界定新的威脅？[2]尤其在 2001 年「911 事件」之後，面對國際恐怖主義興起，如何進行全球反恐需要各國支持，對於「威脅」的定義能否得到盟友的支持？其次，是否能夠尋找到有效聯合因應國際恐怖主義與大規模毀滅性武器的戰略基礎？[3]

2013 年以來，中國全力推動「一帶一路」倡議，到 2017 年轉型為每兩年一次的「一帶一路國際合作論壇」，從無組織形態，到建制化的常設機構，加上附屬的「亞洲基礎建設投資銀行」（AIIB），以及「絲綢之路基金」(Silk Road Foundation)，形成以北京為中心，有別於「歐洲聯盟」（European Union）的統合模式，橫跨亞歐兩洲、聯結中東與非洲的「政經複合體」（Political Economic Complex）。2017 年美國川普（Donald Trump）總統上台以來，在其首次「國家安全戰略報告」（National Security Strategy Report）載明，美國開始重新建構以華盛頓為中心，因應「中國崛起」的戰略威脅，聯合印度、澳洲、日本的「印太戰略」（Indo-Pacific Strategy），形成新型態的兩強區域集團的對抗。易言之，當前中美對抗則是在全球化相互依存下，又具有地緣戰略角度下的相互威脅的對抗態勢。

* 淡江大學國際事務與戰略研究所教授、台灣戰略研究學會理事長
1 倪樂雄，《尋找敵人：戰爭文化與國際軍事問題透視》（北京：經濟管理，2002），頁 3。
2 布里辛斯基（Zbigniew Brzezinski）原著，郭希誠譯，《美國的抉擇》（*The Choice: Global Domination or Global Leadership*）（台北：左岸文化，2004），頁 42-56。
3 同上註，頁 45-46。

英國學者 Burry Buzan 提出「區域安全複合體」（Regional Security Complex, RCS）的概念，來分析冷戰後各種區域安全形態，分析許多不同形態的「安全複合體」。本文嘗試結合社會建構主義「集體身份」的形成，建構一套比較當代中美建構的「一帶一路」與「印太區域」的異同與未來發展。

貳、建構區域安全複合體的解析理論與途徑

一、區域安全複合體的概念

Barry Buzan 與 Ole Waever 兩人提出「地區安全複合體理論」（Regional Security Complex Theory, RSCT），[4]有別於一般國際政治的解析層次：全球體系、國家與個人等層次有其局限性。Buzan 等人認為「地區」（region）層次的安全研究才是關鍵點，一方面可以銜接體系層次上的全球大國的互動關係。另外，可以緊扣地區層次上安全緊密相互依賴的群體。「地區安全複合體」被定義為：「一組單位，他們的主要安全化進程、去安全化進程，或是兩者緊密相互聯繫，以至於不能把它們的安全問題彼此分割開來，合理地進行分析或解決。」[5]

二、地區安全複合體的解析途徑

「地區安全複合體」的主要內核結構存在四個變項：第一、「邊界」：將「地區安全複合體」與近鄰區域區隔開來；一般地區或是區域的界定，一方面有傳統地理範圍的認定，另外也有政治外交上的安排。第二、「無政府結構」：一個「地區安全複合體」存在兩個以上的自治單位；所謂自治單位，類似一般民族國家的國際政治的「能動者」，而非一般更為廣泛的「行為體」，包括：國際組織、非政府組織等等。第三、「極性」：涉及單位之間的權力分配；在東北亞地區，中國與日本、南韓各有不同層次的影響力，中國崛起在亞太地區與美國戰略競逐，日本與韓國臣從美國，與北京互爭影響力量。第四、「社會性建構」：涉及單位之間的「友好」或是「敵對」模式。此外，從國際關係實

---

4 巴里布贊（Barry Buzan）、奧利維夫（Ole Waever）原著，潘忠歧、孫霞、胡勇、鄭力等譯，《地區安全複合體與國際安全結構》（The Regions and Powers: The Structure of International Security）（上海：上海人民，2009）。

5 同上註，頁 43。

證性的分析，從冷戰以來整體區域安全模式，[6] 包括：標準安全複合體、中心化安全複合體、大國安全複合體、超級安全複合體等四種型態，如表 1 所示。

表 1：地區安全複合體類型一覽表

類型	關鍵特徵	例證
標準化安全複合體	極性取決於地區大國	中東、南美、東南亞、非洲之角、南部非洲
中心化安全複合體 1. 超級大國安全複合體 2. 大國安全複合體 3. 地區大國安全複合體 4. 制度安全複合體	1. 以超級大國為中心的單極 2. 以大國為中心的單極 3. 以地區大國為中心的單極 4. 地區通過制度獲得行為體屬性	1. 北美 2. 獨聯體、可能的南亞 3. 無 4. 歐洲聯盟
大國安全複合體	以大國作為地區極的兩極或是多極	1945 年前的歐洲、東亞
超級安全複合體	強有力的地區層次安全態勢，源自大國對臨近地區的擴溢	東亞與南亞

資料來源：巴里布贊（Barry Buzan）、奧利維夫（Ole Waever）原著，潘忠歧、孫霞、胡勇、鄭力等譯，《地區安全複合體與國際安全結構》（*The Regions and Powers: The Structure of International Security*），頁 60。

基於以上對於「地區安全複合體」定義、與發展形態分析之後，本文提出以下主要命題與假設：命題：中美國際戰略互動態是牽動北京的國家安全戰略走向！假設：因應崛起的中國，美國建構戰略競爭對手，確立國家安全主要威脅來源與目標。是以，「一帶一路」與「印太區域」是一種「地區安全複合體」的特殊形態，其所屬國家之間相互往來，構成複合式相互依存關係。

參、影響中美國際互動的地緣戰略因素分析

首先，美國亞太戰略變動與調整。2017 年 1 月 20 日，川普（Donald Trump）正式就任美國總統後，推動各項國際與國內政策，影響整體世界戰略格局的演變，並聚焦於「美國優先」（America First）主軸，以復興美國經濟

---

6 同上註，頁 60。

為原則，否認先前美國歷任總統所建構的國際結構與秩序的安排，從而影響整體亞太戰略的結構與發展趨勢。

在亞太地區部分，在川普上台後，就不斷質疑以往美國處理北韓議題軟弱立場，公開要以武力制止平壤的後續行動。一方面透過聯合國安理會進行制裁決議案，要求各國進行對於平壤的經濟制裁，並希望北京發揮關鍵性效果，也是要中國表態制裁北韓的立場。不過，2018 年金正恩的元旦談話，開啓南北韓重新對話的開始，不僅共同組隊參加平昌冬季奧運，3 月 27 日，南北韓舉行第三次高峰會議，發表「板門店宣言」，也促成後續美國與北韓的「川金會」。

其次，美國不斷警告中國在南海地區的「填砂造陸」工程進展至島礁軍事化作業，加上新南海地圖測繪工作，根據 2018 年 4 月 22 日《南華早報》（South China Morning Post）報導，由中國官方資助的科學研究計畫，以衛星定位系統，將原本遞交到聯合國的南海九條斷續線，以實線的方式連起來。[7] 2018 年 4 月 12 日，習近平參與解放軍在海南島三亞海域進行了史上最大規模的海上閱兵，一方面是對 5 年來「建設強大海軍」工作的小結，也是對美國牽動南海情勢與台海局勢的反應。[8]

其次，中國創造性介入國際議題。2012 年習近平上台以來，中國領導人參與國際組織與活動的頻率日益升高，2014 年主辦了亞太經合會領導人非正式會議，2016 年主辦二十國集團杭州峰會後，2017 年舉辦第一屆「一帶一路國際合作高峰論壇」、2018 年 4 月舉辦博鰲亞洲論壇，被外媒評論為改變「韜光養晦」，走向「積極自主」的大國外交。[9]

王逸舟分析北京必須抱持「走向世界大國的自覺」，思考如何保障國內改革、發展與穩定，保證國家的領土完整與逐漸實現完全統一，在亞太地區乃至

7　〈聚焦南海 南海 U 形線從虛線變實線就可享有主權？〉，自由時報電子報，http://talk.ltn.com.tw/article/breakingnews/2406449。

8　〈多維觀點 解讀習近平南海閱兵的雙重政治符號〉，奇摩新聞，https://tw.news.yahoo.com/%E5%A4%9A%E7%B6%AD%E8%A7%80%E9%BB%9E-%E8%A7%A3%E8%AE%80%E7%BF%92%E8%BF%91%E5%B9%B3%E5%8D%97%E6%B5%B7%E9%96%B1%E5%85%B5%E7%9A%84%E9%9B%99%E9%87%8D%E6%94%BF%E6%B2%BB%E7%AC%A6%E8%99%9F-222001949.html。

9　〈習近平力推大國外交，全球野心引發憂慮與抵制〉，紐約時報中文網，https://cn.nytimes.com/china/20171023/china-xi-jinping-global-power/zh-hant/。

全球範圍，發揮建設性與負責任性的影響力。[10] 時殷弘 1995 年時提出中國應加入西方主導的政治經濟體制，採取「搭便車」戰略；2000 年前後，呼籲中國改變被動應付，塑造大戰略，在 21 世紀前期一心一意追求「基本安全與基本富裕」，並闡述了中國作為海陸大國面臨的戰略兩難。[11]

秦亞青認為中共自十八大以來，中國外交的總目標顯著特徵是基於：2013 年習近平提出的「人類命運共同體」與「新型國際關係」，並把它們結合在一起，並且放在這麼突出的位置。[12] 中共十九大報告中呈現「兩個重提」：重提「社會主義初級階段」，重提「和平與發展」的時代主題，顯示出中國在習近平領導下，針對國家安全態勢已經做出「戰略判斷」。[13] 換言之，中國對外關係已經發展至一個節點，應以進取精神實現對世界體系的「創造性介入」，一方面要求中國與世界體系對話並融入其中，要求中國參與完善和改造世界體系。[14]

肆、習近平主政下國家安全戰略思維與機制

2012 年習近平上台之後，在十九大工作報告中載明，堅持「總體國家安全觀」，以人民安全為宗旨，以政治安全為根本，強調堅決維護國家主權、安全、發展利益。[15] 2013 年 11 月，中共召開第 18 屆 3 中全會，決議成立中央國家安全委員會，由習近平擔任中央國家安全委員會主席，李克強、張德江任副

---

10 王逸舟，《全球政治和中國外交》（北京：世界知識，2003），頁 323。
11 時殷弘認為有以下結構因素：第一、中國經濟與軍事實力迅速增長；第二、中國對外部能源、礦產和市場依賴急劇加大；第三、美國的經濟衰退和在阿富汗、伊拉克的「過度延伸」都導致了它能力與威望的衰減，中國開始慢慢輕視美國的霸權；第四、國內民族主義情緒的作用。民族主義一方面是由於美國力量和意識形態的衰減；另一方面則是中國社會在轉型中變得多樣化，大眾媒體影響力越來越大，請參見〈時殷弘：習近平外交大戰略漸成型〉，紐約時報中文網，https://cn.nytimes.com/china/20150120/cc20shiyinhong/zh-hant/。
12 〈秦亞青：中國特色大國外交要著眼於『能讓世界變成什麼樣』〉，鳳凰網國際智庫，http://pit.ifeng.com/a/20171123/53513172_0.shtml。
13 同上註。
14 〈王逸舟：中國外交進入「創造性介入」時代〉，壹讀，https://read01.com/DRxQQ3.html#.WvuKUyN97oA。
15 其原文為：「『十』堅持總體國家安全觀。統籌發展和安全，增強憂患意識，做到居安思危，是我們黨治國理政的一個重大原則。必須堅持國家利益至上，以人民安全為宗旨，以政治安全為根本，統籌外部安全和內部安全、國土安全和國民安全、傳統安全和非傳統安全、自身安全和共同安全、完善國家安全制度體系，加強國家安全能力建設，堅決維護國家主權、安全、發展利益。」，〈習近平：決勝全面建成小康社會奪取新時代中國特色社會主義偉大勝利—在中國共產黨第十九次全國代表大會上的報告〉，新華網，http://www.xinhuanet.com/politics/19cpcnc/2017-10/27/c_1121867529.htm。

主席。[16] 2018 年 4 月 17 日，習近平召開中央國家安全委員會第一次會議強調，全面貫徹落實總體國家安全觀，為實現「兩個一百年」奮鬥目標，[17]並強調：「必須堅持統籌發展和安全兩件大事，既要善於運用發展成果夯實國家安全的實力基礎，又要善於塑造有利於經濟社會發展的安全環境。」[18]，對內「維穩」，對外「維權」成為國家安全的兩條主軸。

伍、「一帶一路」與「印太區域」緣起發起

一、「一帶一路」的緣起、構想與發展

2013 年習近平訪問哈薩克與印尼期間，先後提出共建「絲綢之路經濟帶」和「21 世紀海上絲綢之路」的重大倡議。2015 年 3 月 28 日，根據中國商務部發佈「願景文件」：「『一帶一路』建設是一項系統工程，要堅持共商、共建、共用原則，積極推進沿線國家發展戰略的相互對接。」[19]「一帶一路」共建倡議所涵蓋的地理範圍在於「亞歐非大陸及附近海洋的互聯互通，建立和加強沿線各國互聯互通夥伴關係，構建全方位、多層次、複合型的互聯互通網路，實現沿線各國多元、自主、平衡、可持續的發展。」[20]並涵蓋以下五大聯通：「以政策溝通、設施聯通、貿易暢通、資金融通、民心相通為主要內容。」

2017 年北京舉行「一帶一路國際合作論壇」，主辦單位提出成果清單涵蓋政策溝通、設施聯通、貿易暢通、資金融通、民心相通 5 大類，共 76 大項、270 多項具體成果。[21]與有關國家政府簽署政府間「一帶一路」合作諒解備忘錄，

---

16 〈習近平任中央國家安全委員會主席〉，風傳媒，http://www.storm.mg/article/26742。
17 〈習近平主持召開十九屆中央國家安全委員會第一次會議〉，軍報記者，http://jz.chinamil.com.cn/n2014/tp/content_8009685.htm。
18 同上註。
19 〈共建『一帶一路』願景與行動文件發佈（全文）〉，國際在線，http://big5.cri.cn/gate/big5/news.cri.cn/gb/42071/2015/03/28/6351s4916394.htm。
20 「一帶一路」點出的地理範圍：「絲綢之路經濟帶重點暢通中國經中亞、俄羅斯至歐洲（波羅的海）；中國經中亞、西亞至波斯灣、地中海；中國至東南亞、南亞、印度洋。21 世紀海上絲綢之路重點方向是從中國沿海港口過南海到印度洋，延伸至歐洲；從中國沿海港口過南海到南太平洋。」，參見〈共建『一帶一路』願景與行動文件發佈（全文）〉，國際在線，http://big5.cri.cn/gate/big5/news.cri.cn/gb/42071/2015/03/28/6351s4916394.htm。
21 〈『一帶一路』國際合作高峰論壇成果清單（全文）〉，新華網，http://big5.xinhuanet.com/gate/big5/www.xinhuanet.com/world/2017-05/16/c_1120976848.htm。

同時，「一帶一路」國際合作高峰論壇定期舉辦，成立論壇諮詢委員會、論壇聯絡辦公室等。

習近平要將「一帶一路」建設成為「和平之路、繁榮之路、開放之路、創新之路、文明之路。」[22]，並強調：「我們要構建以合作共贏為核心的新型國際關係，打造對話不對抗、結伴不結盟的伙伴關係。各國應該尊重彼此主權、尊嚴、領土完整，尊重彼此發展道路和社會制度，尊重彼此核心利益和重大關切。」[23]

2018 年 5 月 3 日，中國外交部強調：「已經有 80 多個國家和國際組織同中國簽署了合作協議。中國目前已在沿線國家建設了 75 個境外經貿合作區。從 2013 年至 2017 年，中國與『一帶一路』沿線國家的進出口總值達到了 33.2 萬億元人民幣，年均增長 4%，高於同期中國外貿的年均增速。今年一季度，中國與『一帶一路』沿線國家進出口增長 12.9%，其中出口增長 10.8%，進口增長 15.7%。這些事實都說明，共建「一帶一路」倡議源於中國，但機會和成果屬於世界。」[24]

二、「印太區域」的興起、概念與走勢

2006 年日本首相安倍第一次組織內閣時，就開始推行「自由與繁榮之弧」與「價值觀外交」，關注歐亞大陸外緣及印度洋周邊國家在日本外交戰略中的重要意義。2012 年 12 月 27 日，在印度報業辛迪加網站發表「亞洲民主安全菱形」（Asia's Democratic Security Diamond）[25] 理論，強調「兩洋交匯」的重要性和日本的角色。[26]

---

22 〈楊潔篪就『一帶一路』國際合作高峰論壇接受媒體採訪（全文）〉，新華網，http://www.xinhuanet.com/world/2017-05/17/c_1120990542.htm。

23 〈習近平對推動『一帶一路』建設提出五點意見〉，新華網，http://www.xinhuanet.com/world/2017-05/14/c_129604239.htm。

24 〈外交部：『一帶一路』機會和成果屬於世界〉，中國新聞評論網，http://hk.crntt.com/doc/1050/5/8/0/105058038.html?coluid=7&kindid=0&docid=105058038&mdate=0503200505。

25 According to Abe: "Peace, stability, and freedom of navigation in the Pacific Ocean are inseparable from peace, stability, and freedom of navigation in the Indian Ocean. Japan, as one of the oldest sea-faring democracies in Asia, should play a greater role – alongside Australia, India, and the US– in preserving the common good in both regions.", "Asia's Democratic Security Diamond," https://www.project-syndicate.org/commentary/a-strategic-alliance-for-japan-and-india-by-shinzo-abe?barrier=accesspaylog.

26 〈吳懷中：安倍政府印太戰略及中國的應對〉，海外網，http://theory.haiwainet.cn/n/2018/0326/

2013 年 3 月 23 日，安倍在美國「戰略與國際中心」（Center for Strategic and International Studies, CSIS）演講，正式提出「印太區域」（Indo-Pacific Region）的理念包括：「亞太地區、印太地區日漸富裕。而在這一區域，日本作為規則的推動者必須具有主導性地位。第二，日本今後也必須繼續做，任何人都能從中受益且充分開放的海洋公共財產等全球共同資源的守護者。正因為日本是一個有著如此願望的國家。第三，日本必須要和以美國為首的韓國、澳大利亞等志同道合的民主主義各國更加緊密地團結一致。」[27]

2016 年 8 月，安倍出席在非洲肯尼亞內羅畢舉行的第六屆「日本—非洲發展國際會議」上正式提出了「自由開放的印太」戰略構想，作為日本新的一項外交戰略。安倍強調：「穿越亞洲海域和印度洋，來到內羅畢就會知道連接亞洲與非洲的其實是一條海上之路。給予世界穩定與繁榮的正是這自由開放的兩大海域、兩大陸彼此結合產生的偉大動力。將太平洋與印度洋、亞洲與非洲的交流活動構建成與武力和威懾無緣的重視自由、法治和市場經濟之地，並使其富饒。日本有這個責任。」[28]

2014 年，日本積極展開與印度合作，將兩國關係提升至「特殊全球戰略夥伴關係」，2015 年 12 月，日印首長會談發表了面向「日印新時代」的聯合聲明，雙方同意推動印太戰略。安倍強調：「日印兩國蘊藏著無限可能的關係，我將與莫迪總理一起緊密合作，盡全力開創日印兩國新時代」，[29] 同時，日印兩國在東京會談中同意將安倍提倡的「印太戰略」（Indo-Pacific）與莫迪的印度「向東行動」（Act East）結合起來，構築起從亞太到印太的「鑽石型安保體制」，透過「海洋民主國家聯盟」，制衡中國在亞太地區戰略能量，並牽制中國戰略影響力外延拓展到印太地區。[30] 2017 年 12 月 4 日，在東京舉行的日中兩國經濟界會議上安倍表示：「在自由開放的印度太平洋戰略下，也可以與

---

c3542938-31286051.html。

27 〈日本歸來（在美國戰略與國際問題研究中心（CSIS）的政策演說〉，*日本國首相官邸*，2013 年 2 月 23 日，https://www.kantei.go.jp/cn/96_abe/statement/201302/23speech.html。

28 〈安倍總理在第六屆非洲開發會議（TICAD VI）開幕式上的演講〉，*日本國首相官邸*，2016 年 8 月 27 日，https://www.kantei.go.jp/cn/97_abe/statement/201608/1218963_11159.html。

29 〈亞洲週刊 2016 年 11 月 20 日‧【日本印度以印太戰略牽制中國】〉，*facebook*，https://www.facebook.com/yzzkgroup/posts/610249782519574:0.

30 同上註。

倡導一帶一路構想的中國大力合作。」[31] 安倍已將他提出的對外政策「自由開放的印度太平洋戰略」與中國大陸主導「一帶一路」經濟帶構想聯繫起來加以推動。

第二、印度與印太戰略發展關係。2014 年 11 月，印度總理莫迪參加當年度東協高峰會議時提出，新德里不再只是採取「東望政策」（Look East Policy），「向東採取以行動為導向的政策」（Act East Policy），[32] 此種戰略目標在於聯結東協國家與日本關係，強化印度在兩洋的戰略地位。不過，印度始終對中國的「一帶一路」倡議採取消極態度，2017 年的首度一帶一路國際合作論壇，新德里拒絕參加。印度對於「亞洲基礎建設投資銀行」卻相對積極，繳納的股本僅次於中國，成為第二大股東。在短期利益面前，印度覺得加入亞投行有利可圖。[33]

2018 年 4 月 18 日，在北京參加第五次中印戰略經濟對話的「印度國家轉型委員會」副主席庫馬爾（Rajiv Kumar）重申印度政府拒絕參與中國「一帶一路」倡議的立場，因為「一帶一路的旗艦項目—中巴經濟走廊—「穿過巴基斯坦佔領的喀什米爾地區，涉及到印度的主權問題。」[34] 另外，習近平大力提倡的「一帶一路」，雙方都持有不同觀點，涉及巴基斯坦控制的克什米爾地區，而印度聲索對該地區的主權。[35]

2018 年 4 月 27 日，中印兩國元首在武漢進行非正式會面，莫迪在推特上寫道：「習近平主席和我將就一系列雙邊問題以及全球要事交換意見。」[36] 如同印度常駐聯合國代表賽義德·艾克巴魯丁（Syed Akbaruddin）在華盛頓發表講話時說。「我們跟中國的關係就是這樣。我們與他們建立聯繫，也跟他們競爭。

31 〈印太新戰略 安倍盼加入一帶一路〉，中時電子報，http://www.chinatimes.com/newspapers/20171219000567-260119。
32 張棋炘，〈印度對外政策剖析：從『東望』邁向『東進』〉，《國際與公共事務》，第四期 (2016 年 7 月)，頁 67-102。
33 〈矯情 印度明確反對『一帶一路』〉，超越新聞網，http://beyondnewsnet.com/20171224/36804/。
34 〈印度再次拒絕中國「一帶一路」倡議〉，民報，http://www.peoplenews.tw/news/37ee0191-6c55-40e3-a43a-17be715e71a0。
35 〈這場峰會也很重要〉印度總理莫迪千里迢迢訪問中國 只為了和習近平『說悄悄話』〉，風傳媒，http://www.storm.mg/article/430942。
36 同上註。

在某些領域，我們一起協作；在另一些領域，我們求同存異，繼續向前。」「亦敵亦友」、「亦友亦敵」（frenemy）。[37]

第三、美國與印太戰略關係。川普提出任內第一份「國家安全戰略報告」勾勒出「印太戰略」倡議，完全否定「重返亞太」戰略佈局，以美國為主，結合印度、澳洲與日本，銜接以往既有的「四國安全對話」構想，從西向東，由海上來圍堵中國「一帶一路」國際發展戰略。根據上述報告所界定的印太區域的範圍：從印度西部海岸地區到美國東岸地區，屬於人口稠密與經濟爆發力的地區。美國希望此一自由、開放印太區域能夠延伸至早期美國開國時期的地理範圍。[38]

2018 年 5 月 7 日，美國國會眾議院軍事委員會主席索恩伯裡正式提出的「2019 財年美國國防授權法」（H.R.5515）議案，鼓吹美國政府要舉全力對抗中國日益擴大的軍事實力和政治經濟影響力，並通過加強美國、日本、澳大利亞、印度的四方軍演。[39] 關於印太事務部分，議案第 1242 條表述：中國將繼續追求軍事現代化項目，尋求近期的印太區域霸權，未來取代美國以取得全球主導地位。美方已經強調需要全政府的全方位能力來應對。美國國會要求川普總統應在 2019 年 3 月 1 日前向國會的適當委員會提交包含關於中國的全政府戰略評估和應對計劃。[40] 第 1245 條還要求美國繼續在印太地區開發和部署強大的導彈防禦系統，開展雙邊和多邊導彈防禦演習，以及增加導彈防禦系統與盟友

---

37 〈莫迪與習近平會晤，試圖緩解中印緊張關係〉，紐約時報中文網，https://cn.nytimes.com/world/20180427/narendra-modi-xi-jinping-china-india/zh-hant/。

38 其原文為：A geopolitical competition between free and repressive visions of world order is taking place in the Indo-Pacific region. The region, which stretches from the west coast of India to the western shores of the United States, represents the most populous and economically dynamic part of the world. The U.S. interest in a free and open Indo-Pacific extends back to the earliest days of our republic，參見 "National Security Strategy of the United States of America DECEMBER 2017", *The White House*, pp. 46-47, https://www.whitehouse.gov/wp-content/uploads/2017/12/NSS-Final-12-18-2017-0905.pdf.

39 具體決議如下：通過印度 - 太平洋地區穩定倡議，強化國防部的努力來籌劃和提供必要的軍事力量、軍事基礎設施和後勤能力。-- 通過印度洋 - 太平洋海上安全倡議來擴大和延伸海上安全合作，加強南海和印度洋的海上安全和海洋領域預警能力。-- 支持與日本、澳大利亞和印度的軍事演習。-- 提升安全合作，應對中國在非洲、東南亞和其他地區日益增長的影響力。請參見 "Reform and Rebuild: National Defense Authorization Act for FY2019", Microsoft Word - Chairman's Mark Summary FY19 NDAA v1.docx, https://armedservices.house.gov/sites/republicans.armedservices.house.gov/files/wysiwyg_uploaded/Chairman%27s%20Mark%20Summary%20FY19%20NDAA.pdf.

40 "Reform and Rebuild: National Defense Authorization Act for FY2019", Microsoft Word - Chairman's Mark Summary FY19 NDAA v1.docx.

能力一體化。最後，議案中提出，從 2020 年 1 月 1 日起，把「美軍太平洋司令部」的名稱改為「美國印度太平洋司令部」，並要求加強美軍具有中文專業技能人才的培養。[41]

三、「一帶一路」與「印太區域」國際評價

從 2013 年中國倡議「一帶一路」，到 2017 年舉辦國際合作論壇，累積不少相關協議、聲明與意向書，但也遭受不少批判。2018 年 2 月 17 日，慕尼黑國際安全研討會上，德國外交部長嘉布里爾，批評北京藉「一帶一路」打造有別於自由、民主與人權等西方價值的制度，自由世界的秩序正在崩解，「目前大陸是唯一擁有、而且堅定實現全球性地緣政治目標的國家」，他警告西方國家應該提出對策。[42]

陸、代結語：建構區域安全複合體的檢證與後續走向

根據前述論證，基本上可以解析本文命題：「社會建構主義」與「地區安全複合體理論」在地區安全層次的研究上具有「主體間互動」與「社會性建構」的關聯性。中美國際戰略互動態是牽動北京的國家安全戰略走向，以及「一帶一路」與「印太區域」是一種「地區安全複合體」的特殊形態，基於主導大國地緣戰略利益考量，影響內部的「友好」與「敵對」集體與角色身份關係。是以，因應崛起的中國，美國建構戰略競爭對手，確立國家安全主要威脅來源與目標。

其次，透過上述分析，理解「一帶一路」與「印太區域」未來走向：維持現狀、改變現狀？可得表 2「一代一路與印太戰略」地區安全複合體檢證表。

第三、確認中美兩國在亞太戰略競逐的「目標」各有不同，川普志在於時間競選諾言：「美國優先」，發展與振興美國經濟為原則，一切有助於「連任」為優先考量，「戰略抑制」中國在亞太地區的影響力。經濟上，削減美中貿易逆差，透過「智慧財產權」削弱中國製造 2025；型塑美中兩國在南海地區的遊戲規則，督促「南海行為準則」的推動，確認兩國擁有的海洋勢力範圍。

41 Ibid.
42 〈慕尼黑安全會議 德嗆聲中美〉，中時電子報，http://www.chinatimes.com/newspapers/20180219000187-260102。

表2：「一代一路與印太戰略」地區安全複合體檢證表

核心變項	一帶一路	印太區域
邊界	非大西洋區域均屬於發展範圍	非大西洋區域均包括在內
無政府結構	多國行為體	多國行為體
極性	中國為主	美國為主、日本、印度為輔
社會性建構	中國與巴基斯坦 中國與印度 中國與東南亞 中國與中亞國家 中國與非洲國家 中國與拉丁美洲國家	日本與印度 澳洲 美國？
演變趨勢	內核變革：從經濟利益共享到命運共同體內化	外在變革：美國的國家安全戰略發展動向

其次，中美在亞太戰略競逐也影響北京未來發展三階段目標：小康社會、社會主義現代化、社會主義強國。如何避免與華盛頓正面衝突，爭取更多戰略緩衝空間，透過「一帶一路」亞歐全球性鏈結，以「世界島」包圍美國，建構國際關係民主化下的「人類命運共同體」。

同時，中美戰略互動自然牽動其他兩角關係：「美台關係」與「兩岸關係」，主要在於：兩岸關係受制於是否接受「九二共識」？在以往各層次管道不通情勢下，美國通過「台灣旅行法」以及「國防授權法」有利於兩方軍事合作關係與高階人員的來往，以及，台灣如何透過「新南向政策」使得「一帶一路」與「亞太區域」有機結合。

換言之，透過名義下的「一帶一路」與「印太區域」「地區安全複合體」，台灣具有關鍵戰略地位。因為，北京推動的「一帶一路」並非僅止於中國本土向外擴張至亞歐大陸，經由台灣向東到南太平洋，飛越到整個中南美洲；而印

太區域戰略則是由西往東南，從印度聯結澳洲、日本與美國，超越第一與第二島鏈區域，涵蓋第三島鏈到達美國本土，台灣恰好在上述兩大「地區安全複合體」中心關鍵位置。是以，外交部亞太司從澳洲、紐西蘭到太平洋邦交國，到東南亞國家、巴基斯坦等範圍，成立「印太科」，有助於提昇與印太地區連結。[43]

43 〈吳釗燮：外交部增設印太科　周五揭牌〉，中國新聞評論網，http://hk.crntt.com/doc/1050/6/2/2/105062266.html?coluid=46&kindid=0&docid=105062266&mdate=0508171633。

中共十九大後兩岸互動的變化

謝志傳 * 唐欣偉 **

壹、中共十八大以降的兩岸互動

在 2012 年中共十八大結束後，兩岸關係仍持續增溫。可是在 2014 年與 2016 年，主張多與中國大陸交流的國民黨在台灣的選舉中連續遭到重挫。強烈反對國民黨政策的民進黨取得台灣多數民意支持後，台海兩岸間的交流能否像馬政府時期那樣熱絡，就成為一個問題。

經常在媒體上看到，同時也是一般百姓較容易直接感受到的，就是大陸地區人士來台人數的變化，從 2008-2015 年間的高速增長，轉變為連續兩年的下跌（圖 1）。

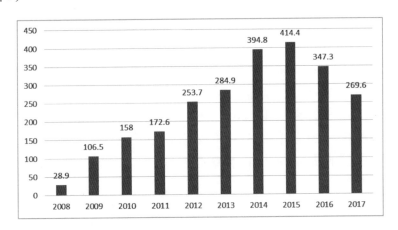

圖 1：大陸地區人士來台數量 (2008-2017)

圖明：Y 軸座標單位為萬人。

資料來源：依中華民國行政院大陸事務委員會兩岸經濟交流速報統計數據繪製。

---

* 國家政策基金會外交及國防組高級助理研究員
** 國立臺灣大學政治學系副教授

大陸地區人民來台的目的多為觀光。自 2010 年時起，中國大陸一直是台灣的最大觀光客源國，在 2013 年時人數相當於排名第二之日本觀光客的兩倍。但是在 2015 年突破四百萬的高峰後就連續兩年下降，至 2017 年時雖仍居第一位，卻已不及日、韓觀光客的總和。而自 2016 年起，台灣觀光外匯收入也在持續增長多年後出現下滑現象，從 2015 年巔峰的 4589 億新台幣減為 2016 年時的 4322 億，比 2014 年時更低。[1]

這樣的情形並不讓人感到意外。在 2016 年總統選舉前，現在執政黨的支持者就曾公開表示樂見此事。[2] 進入 2016 年後，我方批准入台的大陸考察團數量也比 2015 年時少。[3] 除了來台陸客人數下降外，在台灣大專院校的大陸研修生人數，也在民國 104 學年度增長到 34114 人的頂峰後，於次年度開始下降。[4]

台灣的國際活動空間也自 2016 年起開始受限。例如世界衛生組織（World Health Organization, WHO）在兩岸關係改善的 2009 年起，邀請我方以「中華台北」名義，作為觀察員出席世界衛生大會（World Health Assembly, WHA）。[5] 但在 2017 年，我方便未獲邀請。此外，國際民航組織（International Civil Aviation Organization, ICAO）曾於 2013 年邀請我方參加第 38 屆大會，可是卻沒有在 2016 年繼續邀請參加第 39 屆大會。[6] 聖多美及普林西比、巴拿馬、多明尼加這三國，分別於 2016、2017 與 2018 年和我國斷交，同時與中華人民共和國建交。由此便知北京不再放寬台灣的國際空間。甚至奈及利亞要求我關閉設在首都阿布加的代表處，遷往拉哥斯；[7] 台灣參與 2017 年五月在澳洲

1 交通部觀光局，〈觀光統計圖表〉，觀光局行政資訊系統，2018 年 4 月 25 日，http://admin.taiwan.net.tw/public/public.aspx?no=315。

2 即時新聞綜合報導，〈宋憂 200 萬中客不來阿里山 網友：超開心的啊〉，自由時報，2015 年 12 月 30 日，http://news.ltn.com.tw/news/politics/breakingnews/1556921。

3 焦點新聞，〈陸團被拒 傳我嚴審退件暴增〉，中時電子報，2017 年 1 月 24 日，http://www.chinatimes.com/newspapers/20170124000769-260301。

4 教育部，〈近年來大專校院境外學生在臺留學 / 研習人數（96~106 年度）〉，教育部統計處，2018 年 5 月 8 日，https://goo.gl/e4Bgnr。

5 裘兆琳，〈我國參與世界衛生組織之策略演變與美國角色分析：一九九七～二〇〇九〉，《歐美研究》，第 40 卷 2 期，2010 年 6 月，頁 431。

6 外交部，〈臺灣爭取有意義參與國際民航組織 International Civil Aviation Organization 並成為其觀察員〉，外交部參與國際組織，2018 年 5 月 8 日，https://www.mofa.gov.tw/igo/cp.aspx?n=3de11e74cbd6c44e。

7 即時新聞綜合報導，〈我奈及利亞外館將改名遷館 代表已返台〉，自由時報，2017 年 6 月 14 日，http://news.ltn.com.tw/news/world/breakingnews/2100127。

舉行的鑽石國際認證制度「金伯利流程」年度大會，被中國大陸代表團趕出會場，[8]。這與 2008 年到 2016 年初的情勢截然不同。

根據北京方面的說法，這是因為台灣從 2016 年起不再接受「九二共識」的緣故。北京自兩岸分治以來，幾十年來對兩岸關係的堅持就是一個中國原則。在聯合國的 2758 號決議案，與其他國家的建交公報皆載明一個中國原則。若中共從該原則往後退，很可能會遭遇強大的內部反彈。過去北京重申九二共識多次，如今漸漸轉變成一個中國原則，等於是快把九二共識這個覆蓋一個中國原則的詞語拋棄了，屆時想要回到九二共識恐怕也非易事。不過行政院長賴清德在立法院中表示「中國封鎖台灣的國際空間、挖我邦交國也不是一天、兩天的事⋯我們不一定要接受中國的理由。」[9] 顯然我國政府對於如何因應情勢，自有一套主張。

貳、民進黨的中國政策

近兩年來，台灣的觀光業或是國際空間，都受到不同程度的衝擊。然而，當前政府並沒有因此回復馬政府時期的大陸政策。這可能是基於以下三個原因：

首先是自信心。蔡英文主席於 2014 年接受媒體採訪時表示，民進黨最大的挑戰，就是九合一大選，若打好，連大陸都會朝民進黨方向來調整。[10] 2016 年民進黨不僅贏得執政權，連立法院也贏得過半席次。民進黨首次達到行政與立法一把拿的完全執政，況且民進黨在總統大選時宣稱要維持兩岸現狀，但絕口不提九二共識，因此在贏得選舉後，便認為沒有九二共識的維持現狀已經是台灣主流民意。蔡政府向大陸喊話，要尊重台灣的新民意，也就是沒有九二共識的維持現狀。

8　呂伊萱，〈中國代表團在澳洲國際會議咆哮 趕台灣代表出場〉，自由時報，2017 年 5 月 2 日，http://news.ltn.com.tw/news/politics/breakingnews/2054913。

9　即時新聞綜合報導，〈台多斷交原因 賴揆：不一定要接受中國的理由〉，自由時報，2018 年 5 月 5 日，http://news.ltn.com.tw/news/politics/paper/1198011。

10　謝雅柔，〈自信破表！小英曾說打贏選戰 陸會配合民進黨〉，中國時報，2017 年 6 月 17 日，http://www.chinatimes.com/realtimenews/20170617001248-260407。

其次是過去成功的操作經驗。民進黨籍市長陳菊與賴清德登陸時，在大陸官員面前表示支持台獨。[11] 回台後，兩岸交流依舊熱絡，陸客來台旅遊人數屢創新高，大陸還是持續採購台灣農產品。所以民進黨認為大陸對於民進黨是有期待的，料定大陸不敢隨意得罪台灣民眾的情感。但是，當時國民黨為了勝選，也希望大陸持續採購產品與鼓勵更多陸客赴台旅遊，而大陸也希望國民黨能持續執政，維繫住兩岸和平發展的局面，沒有限縮陸客與購買農產品。2014 年與 2016 年的選舉結果，大陸的善意沒有獲得台灣民眾的認同，而民進黨認為大陸不敢限縮陸客赴台旅遊，也不敢中斷購買台灣農產品。台灣的農民也保持同樣的看法。可是事實的發展與這些預期並不一致。[12]

第三，是去中國化的成功。蔡政府與民進黨認為只要培養出更多支持他們的年輕選民，再加上黨產會，就能使國民黨無法翻身。畢竟大陸無法介入台灣內部事務。時間久了，大陸必然會妥協與蔡政府恢復協商，找出民共的互動新模式。蔡政府採取親美日的政策，把台灣的安全寄託於美日的奧援，也認為大陸對於美日毫無辦法，所以不擔心切斷兩岸的連結。換言之，就是把大陸當成紙老虎，而兩岸政策也無須變動。《亞洲週刊》曾報導過日本記者發現台灣人對於美國與日本的信賴超乎想像，尤其是當兩岸發生軍事對抗時，咸信美國與日本必會出兵協助台灣。[13]

對民進黨而言，淪為在野黨的國民黨在過去兩年來，無論是面對黨產會的追討，或國民黨立委在立法院的表現，甚至換了新的黨主席之後，聲勢依舊低迷，無法對民進黨構成威脅。所以民進黨在兩岸上只要持續目前所謂「維持現狀」的作法，不主動挑釁大陸，北京對民進黨目前也是莫可奈何，這也是造成民進黨兩岸政策不變的主因之一。

第四，靜待中共十九大塵埃落定。中共於 2018 年召開十九大，蔡政府與民進黨認為中共內部勢必進行一場權力鬥爭，等到十九大之後再靜觀其變。

11 張永安，〈脫稿演出 賴清德在中國大談台獨共識〉，自由時報，2014 年 6 月 7 日，https://newtalk.tw/news/view/2014-06-07/48062。

12 陳家祥，〈觀光變「關光」！北市陸客上半年大減 41 萬人、倒 12 間飯店〉，Ettoday 新聞雲，2017 年 10 月 20 日，https://goo.gl/f9xHTW。

13 林永富，〈美日救台灣？日人：匪夷所思〉，中國時報，2017 年 5 月 8 日，http://www.chinatimes.com/newspapers/20170508000624-260309。

參、蔡總統的三新主張

蔡英文總統就職將屆周年之際，拋出「新情勢、新問卷、新模式」的兩岸關係互動新主張，[14] 呼籲對岸正視新局勢的客觀現實，共同維持和平穩定的狀態，兩岸也需要有一些結構性的合作關係。

不過，兩岸關係必須要海峽兩岸在共同的政治基礎上努力，而非把兩岸關係的惡化歸咎於中國大陸一方。蔡總統認為她在美國對於兩岸關係的表述，以及就職以來的發言都表達了極大的善意。然而，北京方面尚未明白地認同這樣的主張。

就內容而言，蔡的三新主張不脫 2016 年選舉結果論。蔡政府認為民進黨提出沒有九二共識的維持現狀而贏得選舉，北京單方面要蔡政府就九二共識表態，實則違背台灣的民意。因此，蔡英文才會提出答卷必須兩岸來共同面對，而不能要求蔡政府單方面回應。其實北京自從兩岸分治以來，幾十年堅持一個中國政策與反對台獨，似乎沒有討價還價的空間。習近平在會見國民黨主席洪秀柱時說，共產黨不處理台獨是會被人民推翻的。

習近平也曾多次重申：「兩岸同胞是血脈相連的骨肉兄弟，是割捨不斷的命運共同體；兩岸關係和平發展是維護兩岸和平、促進共同發展、造福兩岸同胞的正確道路；台灣任何黨派、團體、個人，無論過去主張過什麼，只要承認「九二共識」，認同大陸和台灣同屬一個中國，都願意同其交往；兩岸同胞以及海內外全體中華兒女要攜起手來，共同反對「台獨」分裂勢力，維護國家主權和領土完整，絕不容忍國家分裂的歷史悲劇重演；絕不允許任何人、任何組織、任何政黨、在任何時候、以任何形式、把任何一塊中國領土從中國分裂出去！」上述這段話表明了中共的底線。

原國台辦主任張志軍於 2017 年 5 月 8 日接受台灣媒體訪問時表示，「蔡所說的三新，其中有一個是肯定的，那就是自從去年 520 以來，兩岸關係出現新的變化，兩岸關係不是向好的方面，而是向壞的方面、令人擔憂的方面發展，

---

14 林敬殷、丘采薇，〈蔡英文拋兩岸「三新」主張 需有結構性合作關係〉，聯合報，2017 年 5 月 3 日，https://udn.com/news/story/9829/2439457。

這些兩岸民眾都不願意看到。」[15]

肆、國共兩黨關係發展

另一方面,共產黨與國民黨間的關係,也變得比過去幾年冷淡。在 2017 年,中國大陸對於國民黨幾位黨主席候選人的兩岸論述十分關注。北京透過學者與媒體不斷重申其基本立場,說明了大陸對於台灣各政黨的兩岸路線極其關注。北京的擔憂並非空穴來風,而是兩岸歷經八年的和平發展之後,台灣民眾用選票將北京對台灣的讓利予以否定,兩岸關係未來發展充滿變數。

媒體報導,前中國社科院台灣研究所長余克禮在罕見的首度對前總統馬英九開炮,痛批馬英九執政 8 年死抱「不統、不獨、不武」和「只經不政」政策路線,迴避兩岸政治談判,既加劇台灣的台獨分離傾向,更讓兩岸政治議題成為台灣政治禁忌;「對兩岸政治關係傷害很大,也讓兩岸的政治關係實際倒退。」[16]自此之後,許多大陸涉台學者不僅加大力度批判國民黨的兩岸路線,也對所謂「九二共識,一中各表」中的各表頗有微詞。國共兩黨的關係,可說進入重新建立互信的階段,而造成此一發展的原因,有幾點原因;

首先,國民黨與共產黨的領導人在 2005 年達成五項願景,兩黨的共同政治基礎是堅持九二共識與反對台獨。在國民黨未重新執政之前,國共兩黨在前述基礎上,透過不斷的交流與舉辦兩岸經貿文化論壇,累積互信。這也就是為何兩岸在國民黨重新執政後,兩岸關係快速向前邁進。在馬英九政府時期,兩岸的海基會與海協會簽署了二十三項協議,兩岸不僅實現大三通、大陸觀光客赴台旅遊、兩岸間的經貿與文化交流達到兩岸分治幾十年所罕見的盛況,正是因為過去的互信基礎。而馬政府的兩岸路線,被指為傾中賣台,國民黨卻無法提出適當的論述讓民眾信服,導致太陽花運動,使兩岸關係受到巨大傷害。

再者,國民黨的兩岸路線在 2016 年被選民所否決,證明了「不統、不獨、不武」和「只經不政」政策路線,不僅使得兩岸關係無法大幅度邁進,也無法

---

15 杜宗熹,〈蔡英文提兩岸三新論。張志軍:只有一個新是肯定的〉,聯合報,2017 年 5 月 8 日,https://udn.com/news/story/7331/2449225。

16 藍孝威,〈大陸學者余克禮:馬執政路線 傷害兩岸關係〉,中國時報,2016 年 11 月 26 日,http://www.chinatimes.com/newspapers/20161126000370-260108。

獲得民眾支持。當洪秀柱擔任黨主席時，提出要深化九二共識與簽署兩岸和平協議，遭到黨內不少人質疑。北京對於國民黨內有如此之發展，似乎也深感詫異，因此對於國民黨的路線也有諸多揣測。前一陣子，中國評論網不少評論文章，就是針對國民黨是否向民進黨路線靠攏，或是走向獨台，提出疑問。而此次國民黨提出的政綱，完全回歸至馬政府時期的路線，無異加大北京對國民黨路線的不信任，也才有學者提出，國共兩黨需重建互信一說。[17]

此外國民黨所謂的「一中各表」，中就是中華民國。北京方面認為這個論述已經被綠營拿來當作台獨的保護膜，藉著中華民國的外殼搞文化台獨，意圖切斷兩岸的連結，使兩岸漸行漸遠。大陸國台辦副主任龍明彪 2017 年 7 月 6 日在南京「中華民族抗日戰爭史學術研討會」開幕致詞時指出，分裂勢力在台灣推動各種形式的台獨活動，限制、阻擾兩岸民間交流合作，將給台海和平穩定構成重大威脅。北京認為各種形式的台獨，也包括獨台。

北京對於國民黨的兩岸論述如此關注，其原因在於國民黨一旦棄守其基本立場，放棄追求國家統一的理想，想要走中間路線吸引選票，勢必會使得大陸內部民眾加大對北京政府的批判。而國民黨對北京某方面來說，是代表兩岸同屬一個國家的意義，如果當國民黨也不再堅守此一路線時，武力統一台灣的聲勢可能高漲，畢竟和平統一是大陸的既定政策，如果和平統一無望時，中共為了政權的存續，必得做出相對回應。大陸國台辦主任張志軍 2017 年 3 月 6 日稱，今年兩岸關係最大挑戰，就是台獨勢力蠢蠢欲動，如果得不到有效遏制，「台獨之路走到盡頭就是統一」，而且這種統一方式會給台灣社會和民眾帶來巨大傷害。[18] 張志軍的言論已反映出北京的思維，如果國民黨的兩岸路線還是受到北京的質疑，那國共兩黨的關係發展，可能受到嚴重傷害。

簡言之，國共兩黨面對的政治情況不同。國民黨需要靠贏得選舉才能執政，而共產黨是制定遊戲規則之政黨，況且國民黨主張須獲選民認同，在選戰中才有勝利可能。兩黨思考模式與方向不同，才導致國共兩黨間目前的情況。

---

17 陳君碩，〈吳選擇性原汁原味 國共互信減〉，中國時報，2017 年 8 月 22 日，http://www.chinatimes.com/newspapers/20170822000727-260301。

18 陳君碩，〈張志軍：台獨走到盡頭就是統一〉，旺報，2017 年 3 月 6 日，http://www.chinatimes.com/realtimenews/20170306005251-260409。

伍、中共十九大報告傳達何種訊息

中共於 2017 年十月十八日召開十九大，中共總書記習近平做出十九大報告，其中關於兩岸的部分雖僅數百字，但卻透露北京想傳達給台灣的些許訊息。

習近平之報告，首先回顧過去五年來的對台工作成果。習稱在堅持一個中國原則和九二共識的基礎上，大陸推動兩岸關係和平發展，加強兩岸經濟文化交流合作，也實現了兩岸領導人歷史性會晤。在應對台灣局勢變化之際，仍堅決反對和遏制台獨分裂勢力，達到有力維護台海和平穩定的目的。這是過去北京對台工作的描述，而這些成果來自兩岸共同的政治基礎。

然而 2016 年後，兩岸關係不像過去那麼熱絡。大陸黨媒中國新聞社也在十九大之前發出專文，希望有心於兩岸關係的各方仔細琢磨十九大報告中涉台部分。

習近平在報告中明確指出未來中國大陸對台工作的重點，首先指出解決台灣問題是中華民族根本利益之所在。在國民黨執政時期，以和平發展取代「和平統一、一國兩制」，畢竟兩岸關係自 2008 年來如此緩和，沒必要重申一國兩制。而民進黨上台後，情況有所改變。因此，此次的報告直接點名，大陸對台政策就是朝「和平統一、一國兩制」，似乎沒有其他選項了。所謂尊重台灣現有的社會制度和台灣同胞生活方式，也就是一國兩制。

此外，習近平也呼籲民進黨政府只要承認九二共識的歷史事實，認同兩岸同屬一個中國，兩岸雙方就能開展對話，協商解決兩岸同胞關心的問題，台灣任何政黨和團體同大陸交往也不會存在障礙。這不僅是對民進黨政府喊話，也是對台灣所有政黨與團體或是個人表達北京的立場與態度。

習近平以兩岸命運與共與兩岸一家親表達大陸對於台灣的情感，同時也釋出未來大陸將跟台灣同胞分享大陸發展的機遇。大陸將擴大兩岸經濟文化交流合作，實現互利互惠，逐步為台灣同胞在大陸學習、創業、就業、生活提供與大陸同胞同等的待遇，增進台灣同胞福祉。這表示未來台灣民眾在大陸可以享受國民待遇，而大陸也將出台更多惠台政策，吸引更多台灣年輕人赴大陸發

展。[19]換言之，只要願意赴大陸發展的任何人，大陸都將歡迎。這可能使得已經遭遇困境的台灣產業界與教育界，面對更嚴峻的局面。

報告中所謂「我們堅決維護國家主權和領土完整，絕不容忍國家分裂的歷史悲劇重演。一切分裂祖國的活動都必將遭到全體中國人堅決反對。」就是展現對於台獨的基本立場。

至於眾所注目的六個任何：「我們絕不允許任何人、任何組織、任何政黨、在任何時候、以任何形式、把任何一塊中國領土從中國分裂出去！」則是大陸不放棄武統的承諾，也意味著 2017 年時兩岸情勢的險峻。

而推進祖國和平統一進程出現在報告中，則是昭告所有人，爾後大陸對台政策必須是有益於推動兩岸和平統一，且兩岸不能繼續原地踏步下去，也就是兩岸問題不能一代拖過一代了。

十九大的報告有軟有硬，軟的部分就是惠台政策，硬的部分則是比過去強硬，展現出北京處理兩岸問題的信心與決心。換言之，十九大報告傳達出北京從此要主導兩岸問題了。

陸、十九大之後中共兩岸政策之轉變

十九大之後，國台辦副主任劉結一於 2017 年 11 月 8 日的武漢台灣周開幕致詞於表示，「一個中國原則是兩岸關係的政治基礎。體現一個中國原則的九二共識明確界定了兩岸關係的根本性質，是確保兩岸關係和平發展的關鍵。」[20]這代表九二共識只是體現一個中國原則，兩岸的交流必須奠基在一個中國原則之上。

第三屆兩岸媒體峰會於 2017 年 11 月 24 日在北京飯店舉行，大陸國台辦主任張志軍在開幕致詞時表示，十九大報告確認了未來大陸對台工作是「六個一」，承認「九二共識」是破解當前兩岸政治關係僵局的不二法門。「六個一」

---

19 經濟日報數位部內容中心，〈中共十九大政治報告全文〉，經濟日報，2017 年 10 月 18 日，https://money.udn.com/money/story/5641/2764202。

20 中共中央台辦、國務院台辦，〈劉結一在第十四屆湖北·武漢台灣周開幕式上的致辭〉，中共國務院台灣事務辦公室，2017 年 11 月 18 日，https://goo.gl/2F86b4。

再次重申解決台灣問題，實現和平統一與一國兩制的目標；堅持體現一個中國的九二共識，明確界定了兩岸關係的根本性質；對台獨分裂活動畫下紅線；再次強調「兩岸一家親」理念，尊重台灣現有的社會制度和生活方式，願意率先和台灣同胞分享在大陸發展的機遇。[21] 張志軍的講話已經充分表明大陸未來對台政策的輪廓，更清楚地對民進黨政府表達兩岸交流的基礎。中國大陸社科院台灣研究所所長楊明杰 11 月 27 日在兩岸智庫學術論壇上表示，智庫學者研究兩岸關係時，要始終將「以人民為中心」作為推動兩岸關係發展，謀求和平統一的理論指南。[22] 楊明杰的論點更點名了大陸未來對台工作重點是台灣民眾，而非民進黨政府。假定民進黨政府未改變其政策，以後兩岸關係依舊是民熱官冷的局面。

然而不少大陸學者表示，大陸在十九大後，將會從台灣的國際空間著手。中國國際問題研究院副院長郭憲綱直言，這樣的說法沒錯。他強調，蔡英文政府使馬英九執政時期和大陸達成的「外交休兵」默契終止，在這種情況下，大陸和台灣邦交國建交的速度加快，巴拿馬就是一個很明顯的例子，中國大陸與教廷的交流也是一個信號。[23] 上海復旦大學台灣研究中心主任信強也透露，巴拿馬與大陸建交後，現在至少有 7、8 個國家，甚至 10 幾個台灣的邦交國想要與大陸建交。他強調，大陸方面必定會有動作。[24]

曾有台灣學生到聯合國人權理事會旁聽，但在換證入場時，對方指中華民國護照非有效文件，表明台胞證才是「有效身分證明文件」。有媒體指出，大陸開始陸續試點，在國際上壓縮中華民國護照使用空間。[25] 可見情況惡化之程度。

---

21 藍孝威，〈張志軍：十九大後陸對台工作有「六個一」〉，中國時報，2017 年 11 月 24 日，http://www.chinatimes.com/realtimenews/20171124003105-260409。

22 束沐，〈楊明杰：以人民為中心是謀統一的理論指南〉，中國評論通訊社，2017 年 11 月 27 日，https://goo.gl/YXJaoT。

23 蔡浩祥，〈學者直言 陸將加速與台友邦建交〉，旺報，11 月 23 日，http://www.chinatimes.com/newspapers/20171123000794-260301。

24 同上註。

25 陳君碩，〈陸打壓台護照 強化台胞證功能〉，商業週刊，2017 年 8 月 8 日，http://www.businessweekly.com.tw/article.aspx?id=20443&type=Blog。

柒、結語

隨著中國大陸實力的發展,對於掌控兩岸關係走向的信心也日益提升。當大陸祭出一波波政策,向台灣青年招手,而台灣的學校、企業吸引力又呈現下降趨勢時,磁吸效應就會開始出現。國際社會已經注意到台灣將是人才流失最大的地區。不過,北京對台動武的可能性,也會因而下降。

此外,共產黨與民進黨的領導人可能有一個共通點,就是都將鞏固自身的執政地位當作優先考量的目標。十九大之後的習近平,地位已比剛上台時穩固,有較大空間對外採取更具彈性的作法。而不一定要堅持強硬路線。例如北京對日本的態度,就有緩和的跡象。我們若以月而非年為單位來觀察,就會發現,大陸人士來台的數量,已經從 2017 年上半的谷底開始回升。

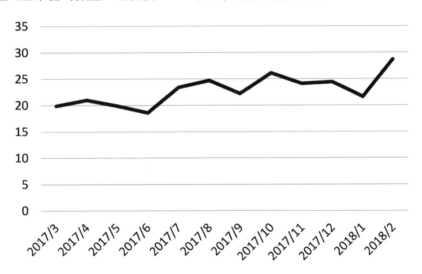

圖 2:大陸地區人士來台數量變化 (2017/3-2018/2)

圖說:Y 軸座標單位為萬人。
資料來源:依中華民國行政院大陸事務委員會兩岸經濟交流速報統計數據繪製。

這個回升的趨勢能否持續,有待進一步觀察。此外,本文僅探討中國大陸與台灣自身因素,並未著墨於具有關鍵影響力的美國。這將是值得在未來進一步觀察的重點。

軍改對共軍「大型島嶼聯合進攻作戰」之影響

揭　仲 *

壹、前言

　　共軍對武力奪取台灣的作戰並無統一的用語，包括早期的「聯合登島戰役」[1]、「島嶼進攻戰役」[2]，到「大型島嶼一體化聯合進攻作戰」[3]、「島嶼登陸聯合作戰」[4]和「大型島嶼聯合進攻作戰」[5]等；但對涉及的作戰形態和關鍵影響因素等，卻有頗為一致的看法，只是隨時間推移、兩岸軍力消長，尤其是共軍在現代資訊化作戰方面的進步，導致個別作戰型態的重要性、可運用戰術，與我方防禦態勢而有所變化。本文選擇用「大型島嶼聯合進攻作戰」，除該名詞出現較晚，以「大型島嶼」代表台灣也容易與其他小型島嶼作區隔。

　　共軍目前正推動的軍事改革（以下簡稱軍改），外界習慣以習近平在 2015 年 9 月 3 日宣佈將裁軍三十萬為起點；[6]但相關的論證與準備，早在宣布裁軍前就已陸續展開。除 2013 年 11 月 12 日中共第十八屆三中全會審議通過的《全面深化改革若干重大問題的決定》等指導文件外；[7]部分內容，包括七大軍區減併、[8]組織扁平化等，[9]還可前推至胡錦濤時期。事實上，這次軍改，是共軍在冷戰結束後，提升現代化資訊作戰能力過程中，體認到雖然在技術和裝備上已

---

1 黃政權、翁海濱、于治國、陳光，〈聯合登島戰役第二梯隊集團軍作戰準備階段衛勤保障難點與對策〉，《瀋陽部隊醫藥》，第 17 卷第 1 期，2004 年 1 月，頁 28；武祥、唐志虎，〈論聯合登島作戰資金供應保障〉，《軍事經濟研究》，2012 年第 10 期，2012 年 10 月，頁 73-75。

2 陳新文、楊潤華、羅明正，〈島嶼進攻戰役應急保障旅的運用〉，《軍事經濟研究》，2006 年第 1 期，2006 年 1 月，頁 76-78。

3 王克海、王兵、曹正榮，《一體化聯合作戰研究》（北京：解放軍，2005 年 4 月），頁 101。

4 劉廣華，〈信息化條件下島嶼登陸聯合作戰物資保障訓練淺析〉，《科技視界》，2015 年第 18 期，2015 年 6 月下旬，頁 162。

5 曹正榮、孫龍海、楊穎等主編，《信息化陸軍作戰》（北京：國防大學，2015 年 6 月），頁 109。

6 申孟哲、劉少華，〈裁軍 30 萬：中國的和平信心〉，人民網，2015 年 9 月 11 日，http://military.people.com.cn/BIG5/n/2015/0911/c1011-27570128.html。

7 第十五條即為「深化國防和軍隊改革」，參閱〈中共中央關於全面深化改革若干重大問題的決定〉，人民網，2013 年 11 月 15 日，http://cpc.people.com.cn/n/2013/1115/c64094-23559163-15.html。

8 徐尚禮，〈濟南軍區傳裁撤直屬中央軍委大陸軍事網站另傳因應俄及外蒙的北京、蘭州與瀋陽三軍區將合併〉，中國時報，2003 年 2 月 23 日，11 版。

9 王克海、王兵、曹正榮，《一體化聯合作戰研究》，頁 25。

經有了較快的發展，但「體制編制則成為制約戰鬥力生成的主要短板」；[10] 若不調整體制編制，則硬體的改良效果無法充分發揮。

鑒於我軍在資訊化聯合防衛作戰的成果，未來共軍若要以武力奪取台灣，勢必要打一場高度資訊化和高強度的「大型島嶼聯合進攻作戰」；目前共軍也正藉由軍改，提升遂行此一作戰的能力。因此，本文擬透過對中共、特別是共軍學者的著作，探討：（一）共軍「大型島嶼聯合進攻作戰」主要內容；（二）現階段共軍在遂行此一作戰上所面臨的問題；（三）共軍藉軍改，想在「大型島嶼聯合進攻作戰」上獲致那些能力、達成哪些目標，最後則概略說明對中華民國國防可能造成的影響。期望能獲得一個初步的框架，作為後續研究的起點。

由於此次軍改側重組織與軟體調整，本文不擬對硬體改良細節進行過多的探討，以聚焦在軍改所造成的影響上。

貳、大型島嶼聯合進攻作戰

根據中共、特別是共軍學者的著作，「大型島嶼聯合進攻作戰」是「諸軍兵種奪占大型島嶼的一體化聯合進攻作戰」；[11] 是中共在當今世上所面臨「信息化局部戰爭」的主要樣式之一，[12] 其內容可歸納為下列七點：

一、是「在國家統一和領土完整受到嚴重威脅，而其他手段不能解除這種威脅時組織實施的，具有戰略決戰性質」。[13]

二、將以台灣本島為核心，在周邊水下、海上、空中與陸地範圍內，運用諸軍兵種的力量進行政治、軍事、經濟等綜合打擊。[14]

三、主要包括「聯合封鎖作戰」、「聯合火力突擊作戰」與「聯合登島作戰」等三種作戰形態；[15] 但單獨運用任何一種，都很難以較小的代價取得勝利，

---

10 侯建偉，〈加快實現戰區主戰與軍種主建的功能耦合〉，解放軍報，2016 年 6 月 7 日，第 6 版。

11 王克海、王兵、曹正榮，《一體化聯合作戰研究》，頁 103。

12 同上註，頁 101。

13 曹正榮、孫龍海、楊穎等主編，《信息化陸軍作戰》，頁 112。

14 王克海、王兵、曹正榮，《一體化聯合作戰研究》，頁 103。

15 曹正榮、孫龍海、楊穎等主編，《信息化陸軍作戰》，頁 109、116；王克海、王兵、曹正榮，《一體化聯合作戰研究》，頁 101、107。此外，在多篇期刊文章中亦分別針對某個作戰形態的問題進行探討，例如莊益夫、李向陽，〈精確制導武器在兩棲火力支援作戰中的運用〉，《飛航導彈》，

必須以某作戰型態為主，再與其他的作戰形態相結合。[16]

四、不僅是遏制分裂活動的軍事行動，更是「以軍事打擊為手段」，結合政治、經濟與外交等方面之鬥爭，「迫使分裂勢力屈服的決戰」。[17] 因此，以政治攻心為上、軍事打擊為支撐，「軍事打擊要服從服務於政治鬥爭的需要」。[18]

五、作戰目的是要徹底屈服、摧毀台灣本島的分裂勢力、不讓分裂勢力死灰復燃，因此應該以奪占和控制全台灣為目標。[19]

六、將不可避免地受到「國際政治與外交鬥爭的制約」，特別是「強敵軍事干預」、周邊某些與中共有領土和海洋權益糾紛的國家「可能藉機提出無理要求或挑起事端」、邊境地區民族分裂勢力也可能「與境外反動勢力相互勾結」。[20] 因此，共軍將力求速戰速決，透過一場快速的決定性作戰，避免為各種干涉、干預與影響提供更多的機會，是「降低連鎖反應的有效舉措」。[21]

七、為「速戰速決」，共軍就必須提高作戰效益，力爭於最短時間內達成癱瘓、分化、瓦解我軍的效果。[22]

分析相關文獻，可發現：

一、將美國介入視為「必然」，介入型態從提供戰略情報預警與情報交流、[23]

2014 第 5 期，2014 年 5 月，頁 29-32。

16　曹正榮、孫龍海、楊穎等主編，《信息化陸軍作戰》，頁 109；王克海、王兵、曹正榮，《一體化聯合作戰研究》，頁 103；在本書中將「聯合火力突擊作戰」稱為「聯合火力打擊作戰」。

17　王克海、王兵、曹正榮，《一體化聯合作戰研究》，頁 104。

18　曹正榮、孫龍海、楊穎等主編，《信息化陸軍作戰》，頁 119；王克海、王兵、曹正榮，《一體化聯合作戰研究》，頁 104。

19　曹正榮等，《信息化陸軍作戰》，頁 119；王克海等，《一體化聯合作戰研究》，頁 107。

20　王克海、王兵、曹正榮，《一體化聯合作戰研究》，頁 105-106；魏岳江，〈架起“海上長城” 談談如何進行島嶼封鎖戰〉，《艦載武器》，2003 第 3 期，2003 年 3 月，頁 39；鄧臺宏、黃亮，〈對敵占大型島嶼的閃擊戰〉，《艦載武器》，2007 第 10 期，2007 年 10 月，頁 74。

21　曹正榮、孫龍海、楊穎等主編，《信息化陸軍作戰》，頁 115。

22　王克海、王兵、曹正榮，《一體化聯合作戰研究》，頁 101。

23　陳炫宇、任聰、王鳳忠，〈渡海登島運輸勤務保障面臨的問題和對策〉，《物流技術》，2016 第 10 期，2016 年 10 月，頁 167；趙峰、姚科、李加，《台軍基於信息系統聯合防衛作戰新型態：網狀化作戰》（北京：九州，2016 年 10 月），頁 98。

對共軍「聯合封鎖作戰」與「聯合登島作戰」進行干預、[24]甚至經濟制裁。[25]

二、雖然是「聯合封鎖作戰」、「聯合火力突擊作戰」與「聯合登島作戰」等三種作戰形態的結合運用，但隨著共軍軍力成長，「聯合火力突擊作戰」的重要性日益突出。

三、一旦戰端開啟，北京會以徹底占領台灣為作戰目的；除非軍事行動受挫，否則不會與台北尋求政治上的妥協。

此外，文獻完成時間越晚者，雖仍把切斷台灣對外交通當作「聯合封鎖作戰」的主要目的之一；但越來越強調是為了拘束我海軍艦隊活動，替緊接著、甚至同時進行的其他作戰形態創造有利條件；很少提及「以長期海空封鎖絞殺台灣」此一措施。雖然未說明原因，但可能理由包括：

一、除非台灣民心士氣極為脆弱，否則封鎖要發揮絞殺的效果需要一段時間，與速戰速決的作戰指導不符；也容易給外力介入提供所需的時間與理由。

二、當共軍意圖以長期海空封鎖絞殺台灣時，很容易因為我軍的對抗作為，使狀況螺旋上升為爭奪制空權與制海權的作戰。共軍反而會因為事先無聯合登島作戰的準備，喪失速戰速決的機會。

三、若選擇離台灣本島較遠的海域實施封鎖，容易讓美軍獲得在有利區域中，挑戰共軍封鎖兵力的機會。

因此，除非國際形勢對中共極端有利，或美軍軍力深陷其他地區無力應付，長期封鎖台灣應非當前及未來共軍對台作戰的優先選項。

參、現階段共軍所面臨的問題

雖然共軍前南京軍區副司令員王洪光曾撰文指出，「部署在第一、第二甚至加上第三島鏈的美軍力量，都不足以阻止大陸解放台灣」、「100 小時內可拿下台灣」；[26]但鎂光燈外，共軍學者所撰寫的專書與期刊文章卻指出，現階

---

24 竇超，〈艱巨的作戰談談我軍未來登陸作戰面臨的問題〉，《艦載武器》，2008 第 9 期，2008 年 9 月，頁 37、40。

25 鄧臺宏、黃亮，〈對敵占大型島嶼的閃擊戰〉，頁 74。

26 楊家鑫，〈解放軍中將撂狠話 100 小時攻下台灣〉，中國時報，2018 年 4 月 10 日，A8 版。

段共軍要發動「大型島嶼聯合進攻作戰」，將遭遇很多難題。這些難題又可區分為「我軍的防衛力量」與「共軍自身能力的不足」兩大類。

一、我軍的防衛力量

在中共、特別是共軍學者的著述中，對我軍防禦能力的評價頗高，尤其是在「戰略預警」、「聯合制壓」、「一體化聯合防空作戰」、「抗登陸防衛體系」等方面。

（一）戰略預警

中共學者專家指出，在共軍遂行「大型島嶼聯合進攻作戰」時：[27]

敵可通過多種手段和途徑或借助他國勢力的戰略偵查能力，結合自建的高、中、高空戰役偵察網及低空、地面戰術偵察網，以綜合性的戰場偵察監視與空中預警系統，對登陸方的戰略戰役後方、沿海地區及縱深實施全方位、全天候的偵察監視與跟蹤搜索。

換言之，在美國的協助下，我軍已初步形成：[28]

涵蓋陸海空天電網的大縱深、多維域、全天候聯合偵察預警體系，偵察範圍涵蓋大陸全境，具備對我東南沿海地區重大戰備活動早期預警、對台島周邊數百海里區域內艦機動態空時感知的能力。

因此，在我軍擁有強大的戰略預警與偵察監視能力、且能得到美國情報支援的情況下，共軍想隱蔽戰役企圖，造成戰略奇襲的可能性微乎其微。[29]

（二）聯合制壓

共軍學者對我軍以具遠程打擊能力的「雄風二E巡弋飛彈」、「萬劍彈」、研發中的超音速巡弋飛彈為核心，結合資電作戰與特種作戰之「聯合制壓作戰」的關注與日俱增。認為隨我軍彈藥數量及打擊距離不斷增加，對共軍防空類目

---

27 陳炫宇、任聰、王鳳忠，〈渡海登島運輸勤務保障面臨的問題和對策〉，頁167；趙峰、姚科、李加，《台軍基於信息系統聯合防衛作戰新型態：網狀化作戰》，頁98。
28 趙峰、姚科、李加，《台軍基於信息系統聯合防衛作戰新型態：網狀化作戰》，頁234-235。
29 寶超，〈艱巨的作戰談談我軍未來登陸作戰面臨的問題〉，頁35。

標、岸置反艦飛彈陣地、彈道飛彈陣地與雷達站等「構成的威脅越來越大」。[30]

共軍近年也越發關注我軍「聯合制壓作戰」，對其兵力在集結與裝載階段所造成的威脅，[31] 對後勤保障系統造成極大的挑戰：[32]

> 未來條件下渡海登島作戰中，敵將集中使用各種先進的作戰手段，特別是火力殺傷範圍更大的遠程制導武器，採取多種形式，實施空襲、破壞和阻止我方實施登陸作戰，機場、港口、後勤基地、交通樞紐以及醫院等各種後勤固定目標和後勤力量，將遭敵全程全方位突襲。

共軍學者甚至指出，在我軍的「聯合制壓作戰」下，「後勤保障系統甚至有癱瘓的危險」。[33]

（三）一體化聯合防空作戰能力

共軍學者對我軍的聯網作戰能力給予高度肯定，特別是在防空作戰上。例如共軍學者指出，我軍衡山指揮所不僅能收集海空軍作戰單位戰術行動的情形，還能透過 Link-16 數據鏈實現對重要海空作戰平台的即時指揮；[34] 並可全方位、即時顯示本島周邊海、空動態，[35] 構成以網路化資訊系統為依託，將防空預警探測、指揮控制及火力交戰等系統進行組合並形成有機整體，「為防空作戰的一方帶來顯著的信息優勢」、「為實施空襲作戰的一方帶來新的威脅與挑戰」。[36]

（四）抗登陸防衛體系

中共學者專家指出，我軍經過「數十年戰場建設」，形成了堅固的防禦體系；[37] 主島海岸地區構築「以永備工事為主、與野戰工事相結合的島嶼築壘防

---

30 趙峰、姚科、李加，《台軍基於信息系統聯合防衛作戰新型態：網狀化作戰》，頁 242。
31 曹正榮、孫龍海、楊穎等主編，《信息化陸軍作戰》，頁 113、122。
32 陳炫宇、任聰、王鳳忠，〈渡海登島運輸勤務保障面臨的問題和對策〉，頁 167-168。
33 同上註，頁 168。
34 趙峰、姚科、李加，《台軍基於信息系統聯合防衛作戰新型態：網狀化作戰》，頁 3。
35 曹正榮、孫龍海、楊穎等主編，《信息化陸軍作戰》，頁 30。
36 羅金亮、王雷、謝文杰、李鵬佳，〈對敵網絡一體化防空體系的聯合壓制策略〉，《系統工程與電子技術》，第 39 卷第 11 期，2017 年 11 月，頁 2191。
37 寶超，〈艱巨的作戰談談我軍未來登陸作戰面臨的問題〉，頁 36。

衛體系」；在便於登陸的方向和地段上，形成「幾乎與自然環境融為一體的海岸防禦陣地」。[38] 使共軍登陸部隊在航渡與兩棲作戰過程中，將面臨「敵方抗登陸體系的猛烈火力阻擊」。[39] 共軍學者還特別指出「敵潛艦」可能對登陸船團實施「遠程導彈攻擊」和「近距離魚雷攻擊」，對大型登陸艦船構成不小的威脅；其中又以魚雷攻擊的「隱蔽性更強、威脅更大，且更難以防禦」。[40]

二、共軍自身能力的不足

共軍學者也指出，自身在「運輸能力」、「精確打擊火力」、「一體化聯合作戰能力」等方面存在明顯缺失。

（一）運輸能力

區分為部隊與物資在裝載前的集結與前運，和跨越台灣海峽階段的海空投送等兩階段。共軍學者認為在這兩階段中，共軍的運輸能力都有不少改進空間。

在裝載前的集結與前運階段，由於部隊與物資原本都是分散部署，導致交通線非常複雜；然而「現有交通線路網結構普遍傾向民用多、貫徹國防要求少」，尤其在「重點輸送機場、港口、公路、鐵路基礎設施相對薄弱」、「與當前擔負的形勢任務還不適應」。[41]

在海空投送階段，則是制式海空運輸載具嚴重不足。共軍「軍用運輸機部隊比例很低」、「人員比例更遠低於世界水平」；[42] 空中投送能力更僅為美軍的 2%、俄軍的 3%。[43] 此外，共軍「專業裝卸在保障隊伍能力不足，致使在完成空中投送任務中軍用物資投送的裝卸載效率低下」；即便動員民航機，仍面臨「能保障大型民航運輸機起降的軍用機場數量則更少」、「軍方地面保障裝

---

38 曹正榮、孫龍海、楊穎等主編，《信息化陸軍作戰》，頁 112。
39 莊益夫、李向陽，〈精確制導武器在兩棲火力支援作戰中的運用〉，頁 29。
40 吳福初、徐寅，〈登陸海域水區警戒兵力對潛警戒能力及配置方法研究〉，《軍事運籌與系統工程》，2017 第 2 期，2017 年 4-6 月，頁 8。
41 朱光、王輝、劉永華、趙旺，〈海上方向軍事行動聯合投送保障研究〉，《軍事交通學院學報》，第 19 卷第 2 期，2017 年 2 月，頁 7。
42 周章文、粟銀、張登成、王壯壯、楊紀明，〈軍民融合式空中戰略投送體系建設的思考〉，《國防交通工程與技術》，2017 年第 6 期，2017 年 11-12 月，頁 6；侯小平、胡堅明、陳興德，〈部隊陸空聯合投送路徑優化〉，《軍事交通學院學報》，第 19 卷第 5 期，2017 年 5 月，頁 8。
43 周章文等，〈軍民融合式空中戰略投送體系建設的思考〉，頁 6。

備與民航運輸機的相容性較差，難以滿足大中型民航運輸機地面保障需要」等問題。[44]

在海上運輸方面，則「大型遠海保障制式艦船數量嚴重不足，能用於渡海登陸的船舶更是數量有限、噸位偏小，只能輸送少量部隊」，「無法完成大規模作戰的各類保障」。[45]

此外，當共軍實施「大型島嶼聯合進攻作戰」時，不但運輸保障對象多、力量構成異常複雜、運輸環節多、運輸轉換頻繁，卻欠缺「權威性很強的聯合運輸指揮機構」，[46] 也缺乏可管控複雜作業的「戰略投送全維信息化管理體系」。[47] 更嚴重的是，因應此一作戰所需的海空聯合投送「尚停留在理論研究層面」，關於「如何組織大規模、整建制、多軍兵種聯合協同海上實戰化訓練」，「很多方面都處於空白」。[48]

（二）精確打擊火力

近年共軍雖不斷提升地對地戰術彈道飛彈的精確度，和增加巡弋飛彈的數量；但從若干共軍學者的著述中，可看出目前共軍的精確打擊火力，還不足以在「聯合火力突擊作戰」中，直接摧毀和破壞台灣本島的要害目標。[49] 部分原因是目前共軍部隊中，先進作戰平台的比例偏少，難以為「信息化武器系統、精確制導彈藥、器材」提供作戰平台。[50]

（三）一體化聯合作戰能力

在軍改啟動前，共軍雖已建立了全軍自動化指揮網、全軍作戰指揮綜合數據庫系統，並改造了高級指揮機構的指揮設施，在軍兵種內建成了多個專用指

---

44 同上註，頁 6-7。
45 張健、吳娟，〈大規模作戰海上民用運輸船舶動員與運用〉，《軍事交通學院學報》，第 19 卷第 11 期，2017 年 11 月，頁 3。
46 陳炫宇、任聰、王鳳忠，〈渡海登陸運輸勤務保障面臨的問題和對策〉，頁 166。
47 周章文等，〈軍民融合式空中戰略投送體系建設的思考〉，頁 7。
48 朱光等，〈海上方向軍事行動聯合投送保障研究〉，頁 7。
49 曹正榮、孫龍海、楊穎等主編，《信息化陸軍作戰》，頁 116。該段文字原文如下：「島嶼聯合火力突擊作戰固然可以直接摧毀和破壞島上要害目標，甚至迫敵接受己方條件，但需要大規模的聯合火力突擊力量和信息化武器裝備為物質基礎」。
50 曹正榮、孫龍海、楊穎等主編，《信息化陸軍作戰》，頁 33。

揮控制系統，[51] 但在同軍種不同兵種間，和不同軍種間，還無法完全融合，使高級指揮機構間還無法全面達成訊息即時交換和建立戰場共同圖像。

例如第七十六集團軍副軍長曹均章少將就指出，當前在集團軍「一體化聯合作戰能力」方面，仍存在下列缺失：[52]

1. 基礎設施不強大。「脖子以下」改革後，資訊網路基礎建設存在明顯短板⋯現有衛星裝備，雖然在不同條件下能實現傳真、短消息、話音、資料、視頻等各種業務通信，但呼通率不高，加之有的衛星通信頻率資源有限，申請頻率困難，不能作為常用手段；戰術互聯網裝備新舊並存，型號多樣⋯集團軍所屬各兵種分系統接入困難，也不能滿足其他軍種隨偶接入。

2. 指控系統不統一。目前，集團軍部隊指揮系統多版共存⋯主要兵種都有自己獨立的指控系統，但還不能與一體化平台真正融入；氣象水文、測繪導航、後勤和裝備保障有各自的資訊系統⋯無法互聯，資料無法融合共用；偵察、工程、防化等兵種目前還沒有完備的指控系統，尤其是北斗指控系統，自成體系，與一體化平台存在斷路，無法實現真正定義上的「一張圖」。

3. 無論內聯或外聯，資訊系統和各類武器裝備平台參差不齊。

除陸軍集團軍外，在海軍航空兵的「一體化聯合作戰能力」方面，也存在下列缺失：[53]

1. 指揮資訊系統仍存在瓶頸，空空、空地與空艦間數據鏈路系統組網建設仍待加強。

2. 武器裝備信息化含量不斷提升，但性能參差不齊。

3. 偵察預警能力與大規模聯合作戰不相適應。

4. 作戰數據庫缺少主要作戰對手目標資料，輔助決策能力弱。

---

51 徐國興，《我軍信息作戰鬥力量建設研究》（北京：軍事科學，2013 年 7 月），頁 43。

52 曹均章，〈新體制下集團軍資訊化建設的現狀分析與對策建議〉，《國防科技》，第 38 卷第 6 期，2017 年 12 月，頁 5。

53 馬培蓓、紀軍、單嶽，〈提高海軍航空兵體系作戰能力問題研究〉，《國防科技》，第 38 卷第 4 期，2017 年 8 月，頁 117-119。

5. 現有機種指揮系統內部要素鏈接和系統間之互接，以及與綜合電子信息系統、初級戰術互聯網對接等問題，尚待研究解決。

6. 現有電子對抗能力不足。

綜上所述，在我方「戰略預警」與「一體化聯合防空作戰」具備優勢，「聯合制壓」具備一定能量，「抗登陸防衛體系」堪稱完善；而共軍在「運輸」與「一體化聯合作戰」不足，僅「精確打擊火力」具有數量優勢的情況下，現階段共軍只能以「聯合登島作戰」為核心，並以「聯合封鎖作戰」、特別是「聯合火力突擊作戰」，支援登陸部隊突破我軍抵抗登陸的防禦體系，確保建立登陸基地，為轉入本島作戰創造有利的條件。

但由於「聯合火力突擊作戰」能量不足，未必能在開戰初期達到長時間癱瘓、甚至瓦解我軍防禦體系的效果。同時，由於海空運輸能力有限、資訊化程度不足，使共軍無法立即集中各種力量，對台灣本島防衛作戰體系的重心進行重點攻擊；必需循傳統兩棲作戰模式，先確保灘頭陣地、建立登陸基地、待後方梯隊上岸後再轉往攻擊各要害目標，不僅成本較高，也難速戰速決。

肆、軍改的影響

共軍鑒於現階段還未具備「速戰速決」的條件；急欲透過本次預訂於2020 年完成之軍改，「著手」提升相關能力，並置重點於「軍民融合」、「提升資訊化程度」與「革新編制與訓練」。

一、軍民融合提升戰略投送能力

共軍學者認為，本作戰涉及將各軍兵種部隊與大量物資向作戰方向投送，因為分佈點多、交通線長、涉及區域面積廣大，故應充分利用軍隊、政府、企業的運力資源，發揮「全社會參與」的優勢，透過「資訊網路的融合與連結，努力使各軍種力量、地方力量和軍地資源形成合力、融為一體」；[54] 實現「遠距離迅速集結部隊力量，多方向同步組織聚焦行動勢能」。[55]

---

54 梁啟明、懷前進，〈深入做好應對海上方向局部戰爭人民戰爭動員準備〉，《國防動員》，2017 年第 9 期，2017 年 9 月，頁 24。
55 朱光等，〈海上方向軍事行動聯合投送保障研究〉，頁 7。

　　換言之，共軍希望透過「軍民融合」，不僅能提升集結與裝載階段的效率，大幅縮短部隊與物資在動員、運輸、分配和裝載所需的時間；然後再結合軍事運輸載具與民間運輸載具，實施「快速、高效的聯合投送，有力地搶佔行動先機，以投送達到提升作戰能力、形成有利態勢、達成戰略企圖的目的」。[56]

　　為達成目標，共軍已在 2017 年 1 月 1 日施行的《中華人民共和國國防交通法》中規定，「以大中型運輸企業為主要依託，組織建設戰略投送支援力量，為快速組織遠距離、大規模國防運輸提供有效支援」，[57]明確「建設骨幹在軍、主體在民、軍民融合、一體保障的戰略投送力量體系」。[58]

　　未來也不排除根據「戰區主戰」體制，在中央軍委和戰區建立「一體化聯合投送指揮體系」，對陸、海、空運輸實施集中統一的指揮管制。以中央軍委聯合作戰指揮中心作為聯合投送最高指揮機構，指揮具戰略意義的聯合投送行動，和下級的聯合投送指揮機構；戰區聯合作戰指揮中心則作為整個體系的籌畫中心，負責聯合投送計畫、決策等組織籌畫任務；重點地區設聯合投送指揮所，根據戰區授權，負責指揮區內裝（卸）載和中轉（銜接）運輸。[59]其中，戰區聯合投送指揮機構還要負起整合「各類交通運輸管理、指揮、保障信息資源」，以建立「交通信息資源數據庫」，[60]並規劃平時的「軍地聯訓」與海空聯合投送訓練。[61]

二、提升資訊化程度

　　共軍希望藉由資訊化程度的提升，在「部隊動員與機動的速度」、「複雜作戰行動的掌控」，和「提升精確打擊能量」等方面能取得明顯的進步，增加「速戰速決」的機率。

---

56 同上註。
57 李遠星、王丙，〈新時代戰略投送支援力量建設運用研究〉，《國防》，2017 年第 12 期，2017 年 12 月，頁 20。
58 屈百春、廖鵬飛、高志文，〈軍民融合加快推進戰略投送能力建設〉，中華人民共和國國防部，2016 年 9 月 5 日，www.mod.gov.cn/big5/regulatory/2016-09/05/content_4724631_3.htm。
59 李鵬、李聯邦、王磊、劉思陽，〈任務式指揮對我軍聯合投送的啟示〉，《軍事交通學院學報》，第 19 卷第 6 期，2017 年 6 月，頁 7。
60 朱光等，〈海上方向軍事行動聯合投送保障研究〉，頁 8。
61 張健、吳娟，〈大規模作戰海上民用運輸船舶動員與運用〉，頁 5。

（一）部隊動員與機動

共軍學者認為，藉由建立「戰略投送全維資訊化管理體系」，可打造一條無縫隙的戰略投送運輸鏈，達成下列效果：[62]

1. 利用軍隊指揮資訊網路、交通運輸企業資訊網路並加以柵格技術改造，連結由鐵、水、公、空運輸和管內部隊五大模組和軍委、戰區地區、聯保中心三級構成的軍交運輸信息平台。

2. 聯通互不隸屬的各運輸系統，形成一體化資訊平台，達到各個分系統同步感知、同步行動、自主協同、主動回應、精確配置，實現對軍事運輸的需求受理、任務下達、動態跟蹤、態勢分析、輔助決策和即時指揮，提高軍事運輸的資訊掌握、快速反應和指揮控制能力。

（二）提升對複雜作戰行動的掌控能力

當前共軍於實施「聯合登島作戰」時，雖已能採用結合平面登陸與垂直登陸的「立體登陸」；但受限於各種因素，還是只能循傳統兩棲作戰模式，先確保灘頭陣地、建立登陸基地、待後方梯隊上岸後再轉往攻擊各要害目標。而傳統兩棲登陸作戰模式所需登陸場的正面與縱深較大；例如師級單位正面約25到30公里、縱深則有15到20公里：旅級單位所需正面也有8到10公里、縱深則有4到8公里；對登陸場的海岸地形、水域與水文條件、登陸場後方地形等，都有頗嚴格的條件。[63]

共軍研究也指出，若循傳統兩棲攻擊，則台灣比較合適的登陸區是通霄和三灣間的海岸，因為從這區域到台北間的海岸並非不適合登陸的海濱泥地，且地勢開闊適合建立登陸場，有利於展開後續行動；[64] 若從台北、或台南和高雄間的地區進攻，都會很快地轉入城鎮作戰，不利於建立登陸基地。[65] 換言之，共軍若要循傳統兩棲攻擊模式，即便能解決登陸載具的問題，並以若干兵力實

---

62 周章文等，〈軍民融合式空中戰略投送體系建設的思考〉，頁7。
63 吳航海，《兩棲登陸作戰方案的規劃與評估分析》（碩士論文：南京理工大學導航、制導與控制研究所，2017年），頁24。
64 吳豔傑、陳重陽、吳晶，〈我國未來海上作戰環境對兩栖艦艇發展的新要求〉，《艦船電子工程》，第263期，2016年5月，頁3。
65 同上註。

施垂直登陸，台灣本島海岸地形也會對首波上岸兵力數量與登陸場選擇，造成嚴重限制。

　　為突破前述限制，共軍已開始研究「從不同方向、不同地域、採用不同登陸方式，實施全方位、全時域、全空域登陸」；[66]認為這方式「具有較大的隱蔽性，可以更快的速度超越敵抗登陸障礙，在對方預料不到的時間、地點和天氣條件下，在敵防禦地域的全縱深多個方向上登陸，極大地提高了登陸作戰的有效性」。[67]

　　共軍研究進一步指出，若採傳統方式，則我軍「僅需七至八萬人防守 15% 的重點海灘即可」；[68]若大量採用氣墊船和地效飛行器，依前述新登陸作戰的方式實施，我軍就「必須部署三十萬人守備 80% 的大部分海灘」，在共軍可登陸灘頭的「火力與裝甲兵規模也會因此減少四分之三」。[69]若共軍能投入大量的直升機，並使陸航直升機部隊透過數據鏈路和高度資訊化的指揮系統，與海、空軍各兵種「無縫鏈結」；[70]則我軍還要防守共軍可能機降登陸的地點，且陣地防禦方向與火力指向都要重新調整。[71]換言之，若共軍能運用高度資訊化的指揮控制系統，有效管制複雜的作戰行動，就能發動多元化的「聯合登島」作戰，大幅增加我軍防衛作戰的困難度。

　　但此種「多路並舉、立體多點登陸、分進合擊」的登陸攻擊，將使作戰管制，包括船隊編組與海上機動、船隊掩護、即時火力支援、登岸後的部隊掌握與全作戰過程後勤保障等變得極端複雜。為解決此一問題，共軍在此次軍改中，積極提升資訊化，以建立和運用指揮自動化系統；除持續調整相關缺失，也置重點於「指揮自動化向下延伸」、「建設與充實各類型數據資料庫」和「大數據運用與決策自動化」等，[72]最終是要達成從中央到各重點作戰單位的指揮決

---

66 陳炫宇、任聰、王鳳忠，〈渡海登島運輸勤務保障面臨的問題和對策〉，頁167。
67 同上註。
68 趙峰、姚科、李加，《台軍基於信息系統聯合防衛作戰新型態：網狀化作戰》，頁210。
69 同上註。
70 孫峰剛，〈探究數據鏈下的陸航作戰指揮控制新模式〉，《中國高新區》，2018年第6期，2018年3月下，頁274。
71 趙峰、姚科、李加，《台軍基於信息系統聯合防衛作戰新型態：網狀化作戰》，頁210。
72 揭仲，〈從資訊化作戰角度看中共改革：組織調整、指揮自動化、戰略支援部隊〉，（論文發表於2017淡江戰略學派年會第十三屆紀念鈕先鍾老師國際學術研討會—新形勢下的台灣國家安全戰略研

策自動化，提升不同軍 (兵) 種、不同武器載台之整合與對複雜作戰過程的掌控能力。

（三）提升精確打擊能量

共軍學者認為，要遂行「大型島嶼聯合進攻作戰」，必須在作戰的全過程、與作戰的各種型態中，持續實施猛烈的火力。[73] 火力不僅在「聯合火力突擊作戰」中具關鍵的地位，在「聯合封鎖作戰」與「聯合登島作戰」也是如此。而未來共軍若要大量使用如氣墊船、地效飛行器及直升機等快速登陸工具，將會大幅加快登陸作戰的節奏，對火力支援的即時性需求更加緊迫；同時，未來在第一波登岸時極可能採小群、多路、分進合擊的方式，使火力支援的精確性遠比密度重要。

要全過程提供精確的火力支援，除了擴充各類型精準彈藥的數量，還需克服「打擊目標信息獲取及傳遞」、「確保投射載具的行動安全」與「火力打擊效果評估」等三大問題。[74] 而透過軍改提升資訊化程度後，將有助於克服前述三大問題。

三、革新編制與訓練

此次軍改，共軍陸軍作戰部隊主體朝「軍—旅—營」三級制落實；且戰鬥部隊旅、營級單位，絕大多數改編為合成旅與合成營，使指揮體制朝「資訊化」作戰的扁平式體制轉變；[75] 並使重新設計後的合成旅與合成營，能夠作為「聯合部隊的一個功能模塊遂行多樣化軍事任務，特別是能夠適應全頻譜作戰」。[76]

改編後的合成旅與合成營，理論上在 C4ISR 方面的資訊化程度高、且能得到先進通訊設施支援，能實現營、連、排終端間、及與網內其他終端互聯互

討會，台北，2017 年 5 月 6-7 日），頁 214。
73 王洪光，〈《島嶼戰爭論·術論》：制勝有術〉，《鐵軍》，2015 年第 10 期，2015 年 10 月，頁 32。
74 莊益夫、李向陽，〈精確制導武器在兩棲火力支援作戰中的運用〉，頁 30。
75 揭仲，〈軍改後的中共陸軍：以集團軍組織調整、合成營與合成旅為例〉，(論文發表於變遷與動力：現階段中國大陸的挑戰學術研討會，台北，民國 106 年 12 月)，頁 175。
76 張健、趙世宜、劉月，〈陸軍合成旅跨區實戰化演練公路投送保障〉，《軍事交通學院學報》，第 19 卷第 6 期，2017 年 6 月，頁 2。

通和態勢共用，快速處理情報與傳輸指令。[77]合成營還增設參謀長，建立四個參謀組的營指揮機構，具備較強的情報搜集處理、決策計劃、協調控制、綜合保障等能力，能獨立組織指揮戰鬥行動。[78]

由於合成營具備較強的機動能力與資訊化作戰能力，所需登陸場的正面與縱深又分別僅約三公里和二公里；[79]因此，只要共軍指揮體系能負荷，未來共軍進行「聯合登島作戰」時，首波就可用合成旅甚至合成營為單位，實施「多路並舉、立體多點登陸、分進合擊」。同時，合成營未來將裝備包含數據指揮鏈路在內的「一體化指揮平台」，[80]能接受戰場共同圖像，並透過指揮數據鏈路直接呼叫精準打擊火力支援，[81]彌補重型裝備不足的缺點。

當共軍陸軍的精銳單位全面實現「模組化」，成為彼此可互換、擴大、拼組，以編組適應各種不同任務的特遣部隊時。日後共軍就可依任務需要，以合成旅或合成營為基幹，抽調陸軍各集團軍特戰、陸航、兩棲、空軍空降兵、海軍陸戰隊之精銳，編組模組化的對台攻擊軍。若再加上共軍已開始調整「應急部隊」與「拳頭部隊」的訓練模式，避免因訓練周期固定，使作戰能力出現「波浪式反復」的情形，[82]使共軍日後在全年各階段都能保有許多戰力完整、可隨時出動的作戰部隊，大幅縮短對台攻擊軍的編組與動員時間。

此外，為應付我軍「聯合制壓作戰」對共軍集結與裝載階段的威脅，共軍已將「濱海地區防衛作戰」視為穩定大型島嶼聯合進攻作戰後方的重要舉措，以確保整個作戰體系的穩定。[83]

---

77 朱勝湘、甄海濤，〈數字化裝甲合成營作戰指揮面臨的問題和對策〉，《學園》，第4期，2012年2月，頁200。
78 同上註。
79 吳航海，《兩棲登陸作戰方案的規劃與評估分析》，頁24。
80 任興志、李勇，〈最小參謀長坐鎮中軍帳〉，解放軍報，2016年4月27日，第4版。
81 孫學佔、牛濤、陶連鵬，〈演兵場上謀新思變，第二十六集團軍某旅緊貼部隊轉型發展練兵備戰〉，解放軍報，2016年11月22日，第5版。
82 揭仲，〈中共軍事改革：現況與未來可能的發展〉，（論文發表於發展與挑戰—面對中共十九大學術學術研討會，台北，民國105年12月），頁108。
83 曹正榮、孫龍海、楊穎等主編，《信息化陸軍作戰》，頁113。

因此，共軍也調整海防部隊指揮體制，由軍改前的「師─旅─團─營」，調整為「旅─營」制，並由戰區陸軍機關的「邊海防處」領導。[84] 新海防旅採多兵種合成方式，由原本以海防炮兵為主的部隊，調整為由炮兵、防空、資訊、甚至特戰等組成的合成戰鬥部隊。[85] 例如東部戰區的某海防旅，所轄支援保障營包含偵察、通信、防化、工程、衛勤等 9 個專業兵種，還轄防空飛彈連與遠端火箭炮分隊。[86] 未來海防旅不排除納入共軍早期預警和區域聯合防空體系，以便有效履行「岸島防衛」、「前沿防空」、「近岸警戒管控」與「海上聯合作戰」等任務。[87]

伍、結語：對我防衛之影響

共軍此波軍改的總體目標，是希望於 2020 年前「在領導管理體制、聯合作戰指揮體制改革上取得突破性進展，在優化規模結構、完善政策制度、推動軍民融合深度發展等方面改革上取得重要成果」，以建構「能夠打贏資訊化戰爭」的現代軍事力量體系。[88]

不過，此次軍改牽動層面既多且廣，除領導指揮體制和組織編裝在 2020 年有可能完成外，其餘與「大型島嶼聯合進攻作戰」密切相關的項目，和與新組織編制配套的戰術戰法與訓練等，到 2020 年時都才起步未久。例如共軍直到 2017 年 9 月，才首度實施集團軍調整組建後首次的合成旅基地化、實戰化對抗演習；[89] 至於新《軍事訓練大綱》，則是到 2017 年底才完成各軍兵種大綱的編修。[90] 因此，共軍要將前述與「大型島嶼聯合進攻作戰」密切相關的內容全部吸收，還需相當時間。

---

84 〈東部南部北部戰區陸軍海防旅亮相，南陸某海防旅不久前剛成立〉，澎湃新聞網，2017 年 6 月 1 日，www.thepaper.cn/newsDetail_forward_1698406。

85 劉謙、李駿、王超，〈立體用兵，今日海防脫胎換骨〉，解放軍報，2017 年 7 月 3 日，第 1 版。

86 陳志民、劉清雲、範小敏，〈海防尖兵換羽新生〉，解放軍報，2017 年 8 月 20 日，第 2 版。

87 周宏君，〈新型陸軍海防部隊建設發展的戰略思考〉，《國防》，2017 年第 2 期，2017 年 2 月，頁 82-83。

88 〈中央軍委關於深化國防和軍隊改革的意見〉，新華網，2016 年 1 月 1 日，http://news.xinhuanet.com/mil/2016-01/01/c_1117646695.htm。

89 〈跨越 2017·朱日和陸軍合成旅基地化訓練落幕〉，中華網，2017 年 9 月 11 日，http://military.china.com/news/568/20170911/31358276_all.html。

90 〈全軍加緊推進新軍事訓練大綱編修 2018 年將按新大綱施訓〉，中華人民共和國國防部，2017 年 6 月 12 日，http://www.mod.gov.cn/topnews/2017-06/12/content_4789605.htm。

　　但隨共軍裝備的提升，加上軍改在組織、觀念、聯合作戰與戰術戰法等方面的精進，預計將替日後遂行「大型島嶼聯合進攻作戰」造成下列影響：

一、由於指揮自動化與軍民融合的廣泛運用，使啟戰前的運輸和集結階段，與部隊從平時轉換進入戰時體制等所需時間大幅縮短，以創造戰役行動的突然性，形成「戰術奇襲」，彌補無法造成「戰略奇襲」的缺失。

二、在「聯合登島作戰」方面，隨海空運送能量增加、各式專用與特種運具大量服役、指揮管制能力的精進，和部隊「模組化」；使共軍可以從空中和「岸對岸」的模式，直接將兵力投送在台灣本島要害位置，並立即發起攻擊。

三、在「聯合火力突擊作戰」方面，不僅兵器數量、質量均大幅提升，也因為「一體化聯合作戰」能量提升，讓「聯合火力突擊作戰」的打擊力、破壞力與效率均大幅增長。

　　因此，當時間越接近 2035 年，共軍對台作戰型態將逐漸轉變，改以「聯合火力突擊作戰」為核心，優先打擊能影響戰爭進程與結局的目標，直接謀取戰略效果。換言之，即嘗試「追求」下列目標：

　　啟戰時，除以海軍、空軍與火箭軍實施聯合封鎖，控制台灣周邊重要戰略水域；也會立刻動用各式遠距、精準或非精準火力，在戰略支援部隊聯合行動下，對台灣本島選定目標發動攻擊。此時的「聯合火力突擊作戰」雖未必能在交戰初期將台灣本島關鍵目標破壞殆盡，卻有可能使台灣防禦體系出現較長時間的癱瘓現象。

一、一旦我軍防禦體系出現較長時間的癱瘓，共軍海空軍會利用這段震盪期，確保局部海空優勢；讓特編的「模組化」攻擊軍，搭乘各式常規與特種輸具，從空中與海上多點進入，直接投送到台北附近，並從各進入點立即向台北地區各關鍵目標發起攻擊。攻擊過程中，第一線部隊透過戰場共同圖像與即時數據傳輸等資訊化設備，直接呼叫遠距精準火力與海空軍支援，彌補重型火力的不足。

二、在地面部隊朝台北各重要目標奔襲時，海軍、空軍、火箭軍與戰略支援部

隊持續掩護後續兵力投入，並將台北與台灣本島其他區域分割孤立。同時伴隨心理戰的運用，和特種人員對「分裂勢力」骨幹進行直接軍事打擊等措施，希望能迅速瓦解台灣的抵抗意志，以便速戰速決，創造有利的既成事實。

就算在共軍成功制壓台北並擊破我軍主要指揮體系後，其他地區我軍部隊仍 各自為戰；此時的戰果也可能讓美軍進一步的干預卻步，並有利於後續登陸梯隊迅速上岸，轉往台灣本島各地清剿。換言之，共軍仍可望在相對短的時間內，完成對台灣主要地區的控制。

中國人民解放軍戰略支援部隊之太空部隊研究

楊國智 *

壹、前言

　　習近平於 2012 年 11 月當選中共中央總書記和中共中央軍委主席，開始推行軍隊改革計劃。2013 年 11 月中共十八屆三中全會提出了「要深化軍隊體制編制調整改革，推進軍隊政策制度調整改革，推動軍民融合深度發展」，並將軍隊改革納入了全面深化改革的總體布局。[1]

　　2015 年 9 月 3 日，在紀念中國人民抗日戰爭暨世界反法西斯戰爭勝利 70 周年大會上，習近平宣布，中國將裁軍 30 萬人；次月，中央軍委常務會議審議通過了《領導指揮體制改革實施方案》；同年 11 月 24 日，中央軍委改革工作會議召開，軍隊改革由此正式啟動。[2]

　　此次軍改新成立的 3 個軍、兵種中，相比陸軍領導機構和火箭軍 2 大軍種的明確定義，各界對戰略支援部隊此一單獨成立的新兵種認識卻比較模糊，大陸官方對此兵種的解釋也相對稀少。但央視引述中共中央軍委主席習近平的話說，戰略支援部隊是集「天電網」於一體，在太空、天空、電磁等空間支援其他傳統作戰力量的兵種，為戰略支援部隊作了明確定義。[3]

　　神舟 11 號航天員（太空人）景海鵬 2017 年 10 月 18 日身著軍裝出現在中共 19 大會現場。被發現其佩戴少將軍銜臂章，臂章上有「戰略支援部隊」字樣。這顯示，經過新一輪的軍改，號稱「天兵天將」的航天員大隊已正式隸屬於戰略支援部隊。[4] 航天員是中國太空探索的中堅力量，堪稱中國「天兵」的先鋒。

---

* 淡江大學國際事務與戰略研究所博士候選人

1. 新京報，〈全會公報十大看點：明確全面深化改革總目標〉，Sina 新聞中心，2013 年 11 月 13 日，http://news.sina.com.cn/c/2013-11-13/025928692758.shtml。
2. 人民日報，〈習主席和中央軍委運籌設計深化國防和軍隊改革紀實〉，中國共產黨新聞網，2015 年 12 月 31 日，http://cpc.people.com.cn/n1/2015/1231/c64094-27997805.html。
3. 張國威，〈陸戰略支援部隊太空、網路爭雄〉，中時電子報，2016 年 1 月 5 日，http://www.chinatimes.com/newspapers/20160105000893-260301。
4. 賴錦宏，〈景海鵬臂章透玄機航太員改隸戰略支援部隊〉，聯合新聞網，2017 年 10 月 19 日，https://udn.com/news/story/11556/2766096。

戰略支援部隊則是在本輪軍改中誕生的最神秘、最年輕的部隊，從航天員大隊的隸屬關係來看，太空必定是戰略支援部隊與強敵對手過招的一個重要戰場和領域。[5]

本文以分析其太空部隊的發展研究，在太空戰略運用，影響其他戰略的成敗與否，若無太空部隊是否就無法發揮其所謂軍事戰略的完成，是值得研究探討之路。而戰略支援部隊所屬太空部隊之的建置，而目前所知為航天系統部，負責領導航天系統工作，屬副戰區級，它並非獨立軍種，能發揮多少戰力，也是值得論述的議題。

貳、中共太空部隊起源

一、中共太空制天權內涵

20 世紀中期以前，人類從未想到，有一天戰爭會在地球大氣層之上 100km（公里）以外的太空中進行。由於 V-2 飛彈的發明，科學家們就開始研究要如何防禦來自太空的威脅，和如何能把飛彈打得更遠更精準且威力更大；政治人物也開始體認到，太空是一個未經開發的處女地，誰能搶得先機，誰就掌握了政治及軍事上的優勢，到了 1960 年代，距人類發射人造衛星不過幾年間，當時的美國總統甘迺迪（John F. Kennedy）曾斷言：「誰能控制太空，誰就能控制地球」，從那時起，大國之間就開展了歷時數十年的太空競爭。[6]

中共解放軍出版社《爭奪制天權：太空的探研、開發與爭奪》，認為「軍事戰略」和「太空戰略」劃為同一戰略層層級。儘管太空戰略的影響領域十分寬廣，但在現階段，太空戰略研究的側重重點仍然是增強國防實力和確保國家安全方面的問題。因此，它們同屬於比國防戰略更為狹窄的戰略層級（如圖 1）。[7]

---

5 同上註。
6 李貴發，〈淺談太空作戰及台灣因應之道〉，《尖端科技》，第 319 期，2011 年 3 月，頁 66。
7 張健志、何玉彬，《爭奪制天權：太空的探研、開發與爭奪》，（北京：解放軍，2008 年 1 月），頁 184。

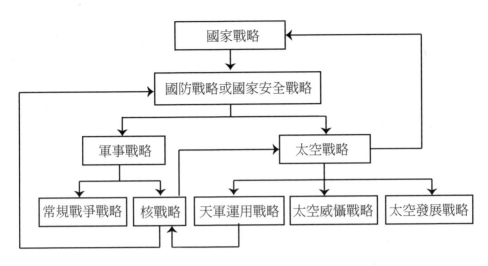

圖 1：戰略層次傳遞關係圖

資料來源：張健志、何玉彬著，《爭奪制天權：太空的探研、開發與爭奪》（北京：解放軍，
2008 年 1 月），頁 182。

　　中共在習近平 2015 年掌政期間在 5 月 26 日發佈《中國的軍事戰略》白皮
書說明「太空是國際戰略競爭制高點。有關國家發展太空力量和手段，太空武
器化初顯端倪。中國一貫主張和平利用太空，反對太空武器化和太空軍備競賽，
積極參與國際太空合作。密切跟蹤掌握太空態勢，應對太空安全威脅與挑戰，
保衛太空資產安全，服務國家經濟建設和社會發展，維護太空安全」。[8]「實行
新形勢下積極防禦軍事戰略方針，調整軍事鬥爭準備基點。根據戰爭形態演變
和國家安全形勢，將軍事鬥爭準備基點放在打贏資訊化局部戰爭上」[9]。中共也
認清在 21 世紀，世界航天大國的太空戰略將依然是根據本國國家安全戰略和
軍事戰略的需要，以一切可資利用的手段爭奪制天（太空）權。太空軍事力量
的任務是：在為地面軍事行動提供各種航太支援服務的同時，將為爭奪制天權
而進行天戰準備。奪取制天權將成為太空戰略的最高軍事目標。[10]

8　中華人民共和國國務院新聞辦公室，〈中國的軍事戰略〉，中華人民共和國國防部，2015 年 5 月，
　　http://www.mod.gov.cn/auth/2015-05/26/content_4586723.htm。
9　同上註。
10　楊忠成、賈俊明〈太空戰略〉，中國空軍百科全書編審委員會，《中國空軍百科全書》，（北京：
　　航空工業，2005 年 11 月），頁 88。

太空戰略，是關於發展和運用太空力量、奪取未來太空優勢的方略。從廣義上講，太空戰略是主權國家或國家集團為開發和利用太空，建設和運用國家太空力量所採取的方針、原則和政策的總和。太空戰略直接反映國家戰略的需求，並接受國家戰略的指導和制約。從狹義上講，太空戰略就是太空軍事戰略。按照現代軍事學術研究的沿革習慣，通常，軍事戰略包含三大戰略要素，即威懾戰略，發展戰略和運用戰略。同樣，太空戰略也包含著太空威懾戰略，太空發展戰略和天軍運用戰略三大組成部分。[11]

二、中共太空部隊定義內涵

中共解放軍官方並未公開提到「天軍」這個名稱，但是中共軍事學者著書常有「天軍」（或稱太空作戰部隊[12]、航天部隊[13]、外太空部隊[14]）該名詞。另外中共「中國航太科技集團有限公司」（前身源于 1956 年成立的國防部第五研究院，歷經第七機械工業部、航天工業部、航空航天工業部、中國航天工業總公司和中國航天科技集團公司的歷史沿革）對外網頁解釋「天軍」是「是獨立于陸、海、空之外的新軍種，是在宇宙空間作戰的軍隊，其主要任務包括太空作戰，支援空中、地面、海上作戰和開發宇宙空間等。『天軍』是一個多兵種的合成軍隊，除指揮機關和航天院校，還有許多兵種部隊，大體包括太空艦隊、地基部隊、航天和空天飛機部隊、火箭部隊、C4I 部隊等」。[15]

在中共之《中國空軍百科全書》中解釋未來的太空軍事力量「天軍」，將是人類高智慧、高技術的集合體，在未來武裝力量中佔據首要地位。另一方面，太空戰場極其廣闊深遠，它全面包容覆蓋傳統的陸海空戰場，具有「居高臨下」的空間優勢。「天軍」一旦控制了太空戰場，就能憑藉其高智慧、高技術和高空間優勢，全面瞰制陸海空戰場。制天權將主導制空制海權和制電磁，直接影響戰爭的進程與結局。爭奪制天權的主要手段是各種部署在太空以及地面、空

---

11 張健志、何玉彬，《爭奪制天權：太空的探研、開發與爭奪》，頁 200。

12 姜連舉主編，《未來作戰將在太空打響－拉直太空作戰那 N 多個問號》，（北京：軍事科學，2015年 12 月），頁 107。

13 周碧松，《浩渺太空的競相角逐》，（北京：軍事科學，2015 年 5 月），頁 231。

14 〈天軍「外太空部隊」〉，百度百科，https://baike.baidu.com/item/%E5%A4%A9%E5%86%9B/28190。

15 〈什麼是「天軍」〉，中國航天科技集團公司，http://www.spacechina.com/n25/n148/n272/n4785/c316588/content.html。

中、海上的太空進攻性與防禦性武器系統。爭奪制天權的主要作戰樣式有：太空信息戰、太空封鎖戰、太空軌道破擊戰、太空防衛戰和太空對地突擊戰等。在軍事航天技術的推動下，爭奪制天權作戰行動，將是以「天軍」為主體，其他軍種參加的聯合作戰。[16]

日本「讀賣新聞」2014 年 8 月 25 日報導，中國也準備在近期內成立「太空部隊」，以應對美國及其他國家太空軍事化。交惡的日中兩國幾乎同時宣布加入太空領域競賽，且積極強化與美俄兩國在太空領域的合作，太空領域的競賽正朝著日美對抗中俄的架構發展。中國成立太空部隊是習近平政權軍事機構改革計畫的一環。讀賣報導指出，中共中央軍事委員會將成立「中央軍委聯合作戰指揮中心」和「航天（太空）辦公室」兩個新單位，其中「航天辦公室」即是用來推動成立太空部隊。2013 年 3 月習近平訪問俄羅斯時，有意參考俄軍作法，為人民解放軍建構行使衛星系統的作戰體制。習近平六月出席一項空軍將領會議時，強調應將天空與宇宙形成一體，建設攻擊與防衛兼備的強大空軍，這對捍衛國家主權、安全及發展利益不可或缺。習近平的積極態度，等於為中國成立太空部隊定調。[17]

大陸學者認為必須建立中國特色的太空部隊，擁有外層太空打擊能力和作戰能力。[18] 大陸軍方已經把組建太空作戰能力的天軍，視為牽制美國在軍事科技上大幅領先優勢的「奇兵」，因此，其勢必會以更積極的作為，繼續推動天軍的發展。現階段，共軍運用太空科技發展的軍事能力項目包括：（一）太空資訊戰；（二）太空反衛星戰，以雷射攻擊衛星或運用電磁波干擾衛星通訊；（三）太空反飛彈武器；（四）運用太空武器攻擊地面、空中或海面的重要軍事目標。現階段，共軍正積極規畫在 2015 年完成天軍的組建部署，把太空科技與軍事力量結合起來，打造陸、海、空、天、電磁頻譜，以及網路的六維兵力結構。[19]

---

16 胡思遠，〈制天權〉，中國空軍百科全書編審委員會，《中國空軍百科全書》，（北京：航空工業，2005 年 11 月），頁 88。

17 林翠儀，〈「第四戰場」日中紛設太空部隊〉，自由時報，2014 年 8 月 26 日，http://news.ltn.com.tw/news/world/paper/807737。

18 周碧松，《浩渺太空的競相競逐》，頁 313。

19 曾復生，〈共軍核衛星對抗美軍閃擊全球〉，財團法人國家政策研究基金會，2013 年 1 月 23 日，https://www.npf.org.tw/printfriendly/11919。

參、中共太空部隊發展

一、中共太空部隊的組建

在習近平在 2015 年 9 月 3 日國慶大閱兵宣佈裁軍 30 萬時，當時美國國際評估與戰略中心（International Assessment and Strategic Center）資深研究員查德・費舍爾（Richard D.Fisher,Jr.）表示，很多媒體與分析人士談及不多的是建立第五總部，也就是太空部隊的可能性，這在當代軍事行動中對控制訊息起到很大作用，我們知道中國解放軍一直考慮打造一個專門的太空部隊，既能控制太空活動，又能在需要的時候進行太空戰。[20]

習近平強調，「戰略支援部隊是維護國家安全的新型作戰力量，是我軍新質作戰能力的重要增長點。戰略支援部隊全體官兵要堅持體系融合、軍民融合，努力在關鍵領域實現跨越發展，高標準高起點推進新型作戰力量加速發展、一體發展，努力建設一支強大的現代化戰略支援部隊」。[21]

新成立的戰略支援部隊頗為神秘，這究竟是一支什麼樣的軍事力量呢？軍事專家尹卓[22]在接受人民網採訪時表示，戰略支援部隊主要的使命任務是支援戰場作戰，使我軍在航天、太空、網絡和電磁空間戰場能取得局部優勢，保証作戰的順利進行。具體地說，戰略支援部隊的任務包括：對目標的探測、偵察和目標信息的回傳；承擔日常的導航行動，以及北斗衛星和太空偵察手段的管理工作；承擔電磁空間和網絡空間的防御任務。「這些都是決定我軍在未來戰場上能否取得勝利的新領域」。[23]

---

20 楊念軍，〈習總大閱兵，過把癮再說〉，《明鏡月刊》，第 68 期，2015 年 10 月，頁 15

21 新華社，〈陸軍領導機構火箭軍戰略支援部隊成立大會在京舉行習近平向中國人民解放軍陸軍火箭軍戰略支援部隊授予軍旗並致訓詞〉，新華網，2016 年 1 月 1 日，http://www.xinhuanet.com/politics/2016-01/01/c_1117646667.htm。

22 尹卓（1945 年 9 月 -），中華人民共和國政治人物與將領，第十一屆全國政協委員。中國人民解放軍海軍少將。央視《軍情連連看》等節目嘉賓主持。早年加入中國共產黨，擔任海軍信息化專家諮詢委員會主任。2008 年，當選第十一屆全國政協委員，代表科學技術界，分入第二十九組。維基百科，2017 年 8 月 28 日，https://zh.wikipedia.org/wiki/%E5%B0%B9%E5%8D%93。

23 邱越，〈專家：戰略支援部隊將貫穿作戰全過程是致勝關鍵〉，人民網，2015 年 12 月 31 日，http://military.people.com.cn/n1/2016/0105/c1011-28011251.html。

在 2015 年 11 月 2 日，也就是高津[24]在成為戰略支援部隊司令員的前 60 天，在《解放軍報》上發表了題為「深化國防和軍隊改革是強軍興軍的必由之路」的署名文章。文章中特別提到：世界新軍事革命是當今世界大發展大變革大調整的重要構成和關鍵變數，其速度之快、範圍之廣、程度之深、影響之大前所未有。面對這場繼冷兵器、熱兵器、機械化軍事革命之後的又一次劃時代軍事革命，世界主要國家競相調整軍事戰略，加緊推進軍事轉型，以資訊化為核心重塑軍隊組織形態、重構軍事力量體系。戰爭形態處於由機械化向資訊化躍升的質變期，核威懾條件下陸海空天網電一體化聯合作戰日益成為現實，戰場從傳統空間向極高、極深、極遠物理空間和虛擬空間拓展，非對稱、非接觸、非線式作戰樣式更趨成熟，制資訊權成為奪取戰場綜合控制權的核心，戰爭制勝機理深刻改變。高津司令員所說的核威懾條件下陸海空天網電一體化聯合作戰，其中的「天」，就絕不是指的火箭軍部隊的導彈，而是外太空，「網」則是網路空間，「電」則是電磁空間。現代化高技術局部戰爭，陸海空軍部隊仍然是決勝的關鍵，而爭奪的戰場也主要在陸海空空間進行，但資訊化戰爭的一個典型特色，就是非對稱、非接觸、非線式作戰，而實現這一轉變的關鍵環節，就是在外太空、網路空間和電磁空間的「這三個特殊戰場」。電磁空間由於受距離和物理隔絕的影響很大，通常只具有戰術意義。而外太空和網路空間，則是極高、極深、極遠物理空間和無處不在的虛擬空間，從而具有戰略上意義。在這兩個戰場上的拼殺，沒有血肉橫飛，但兇險卻比常規戰場有過之而無不及，軍隊能否擁有對應的力量建設和健全的指揮反應機構，其結果將對常規戰場的勝負產生重要甚至決定性影響。這就是「戰略支援」的含義所在。從邏輯上推理，中國戰略支援部隊，很可能就是針對外太空和網路空間的「新型作戰力量」，賦予「新質作戰能力」新增長點。[25]

---

24 高津，1985 年進入中國人民解放軍第二炮兵指揮學院。擔任中國人民解放軍第二炮兵部隊第五十二基地司令員。2011 年 12 月，任中國人民解放軍第二炮兵部隊參謀長。2014 年 7 月，調任中國人民解放軍總參謀長助理。2014 年 12 月，任中國人民解放軍軍事科學院院長，並取代比其年長 1 歲的副總參謀長乙曉光中將，成為最年輕的正大軍區級將領，也是中國人民解放軍軍事科學院歷史上最年輕的院長（2015 年兼任中國軍事科學學會會長）。2015 年 12 月 31 日，出任新成立的中國人民解放軍戰略支援部隊首任司令員，2017 年 7 月 28 日，晉升上將軍階。參考維基百科，2018 年 4 月 19 日，https://zh.wikipedia.org/wiki/%E9%AB%98%E6%B4%A5。

25 鄭文浩，〈獨家解讀解放軍戰略支援部隊：哪些能耐脫穎而出？〉，鳳凰軍事，2016 年 1 月 3 日，http://www.360doc.com/content/16/0103/11/4295303_525096912.shtml。

戰略支援部隊將包括情報偵察、衛星管理、電子對抗、網路攻防、心理戰等五大部分，是綜合了這個時代最先進的天軍、網軍等看不見硝煙的戰場上的作戰部隊，更確切地說是專門擔負軟殺傷使命的作戰力量。這也是此次軍隊改革在軍種層面的調整，將解放軍分為傳統的陸海空軍、戰略威懾和打擊的火箭軍和最現代化的「天－網」軍，三個層次分工明確，又能強化相互之間的協同配合，最大限度地發揮出整體戰的優勢，無疑是本次軍隊改革的重頭戲。戰略支援部隊的偵察主要就是這方面的技術偵察，可以通過偵察衛星、偵察機、無人機、感應器等現代化裝備來進行。衛星管理就是所謂的「天軍」，這是隨著航太技術的飛速發展，特別是衛星資訊偵察、跟蹤監視、制導導航及航太兵器的廣泛應用，應運而生的新型部隊。美國在 1982 年成立了美國空軍太空司令部，標誌著世界上第一支天軍的誕生。俄羅斯隨後也將軍事航天部隊和太空導彈防禦部隊從戰略火箭軍中分離出來，成立了一支總兵力約 9 萬人的航天部隊。那麼隨著解放軍這次軍事改革，隸屬于戰略支援部隊的「天軍」也正式組建，成為今後專門承擔航天作戰的部隊。[26]

成立「太空部隊」是中共新時期軍事戰略的要求。早在 2004 年，軍委就確立了「空天一體、攻防兼備」的空軍戰略。2009 年，時任空軍司令員的許其亮也要求，必須發展空間防禦和打擊能力，只有實力才能保護太空。2014 年 4 月，軍委主席習近平就強調了加快建設空天一體，攻防兼備的強大空軍。在軍事航太另一個重要領域─軍用衛星方向，則涉及到總參謀部、海軍、火箭軍（原第二炮兵），這些部門都有自己的衛星計畫。正因為資源分散，難以有效管理，這才是這支「太空部隊」成立的初衷。因此，這支部隊是將原屬於解放軍總裝備部各大航太發射基地、測控站以及原隸屬解放軍總參謀部第三部航太偵查局等力量整合到一起。[27]

二、中共太空部隊的指揮

目前就公開資料顯示就戰略支援部隊之下航天系統部，是中國人民解放軍

---

26 名家專欄，〈戰略支援部隊其實就是天網軍：將改變戰爭〉，網易軍事，2016 年 1 月 4 日，http://war.163.com/16/0104/08/BCFMF4HF00014J0G.html。

27 桑梓地，〈秘而不宣 成立解放軍「太空部隊」〉，個人圖書館，2016 年 1 月 5 日，http://www.360doc.com/content/16/0105/08/15447134_525555539.shtml。

戰略支援部隊下屬部，負責航天系統工作。在深化國防和軍隊改革中，2015年底新組建中國人民解放軍戰略支援部隊。此後中國酒泉衛星發射中心等航天系統單位由中國人民解放軍總裝備部轉隸中國人民解放軍戰略支援部隊。2016年成立了中國人民解放軍戰略支援部隊航天系統部，負責領導航天系統工作。[28] 在入選十九屆中央委員的軍人名單之尚宏[29]為中國人民解放軍戰略支援部隊副司令兼航天系統部司令員，尚宏長期在解放軍裝備系統任職，曾任原總裝備部司令部參謀長、酒泉衛星發射中心主任等職務。[30]康春元航天系統部政委[31]，先後在原北京軍區和原蘭州軍區服役，曾任原北京軍區政治部宣傳部部長、陸軍第65集團軍政治部主任、原北京軍區政治部副主任等職，並於2014年12月調任原蘭州軍區副政委。副司令員為郝衛中曾任太原衛星發射中心主任[32]。費加兵任航天系統部參謀長，曾任中國衛星海上測控部主任，中國人民解放軍總裝備部司令部參謀長。[33]

　　由司令員、副司令員及參謀長均是在原總裝備部下航天部隊主管歷練，因航天部隊因屬科技專業部隊，而總裝備部撤銷後，2016年在深化國防和軍隊改革中改組為中央軍委裝備發展部，人員會是在裝備發展部及航天系統部間流動，尤其在2016年全國人大代表、軍委裝備發展部副部長張育林[34]中將接受新

28 〈中國人民解放軍戰略支援部隊航太系統部〉，維基百科，2018年3月27日，https://zh.wikipedia. org/wiki/%E4%B8%AD%E5%9B%BD%E4%BA%BA%E6%B0%91%E8%A7%A3%E6%94%BE%E5% 86%9B%E6%88%98%E7%95%A5%E6%94%AF%E6%8F%B4%E9%83%A8%E9%98%9F%E8%88%A A%E5%A4%A9%E7%B3%BB%E7%BB%9F%E9%83%A8#cite_note-czs-1。
29 〈十九屆中央委員，軍隊出身的都有誰？〉，鳳凰網，2017年10月24日，http://news.ifeng.com/ a/20171024/52777643_0.shtml；〈中國共產黨第十九次全國代表大會代表名單〉，維基百科，2018 年3月28日，https://zh.wikipedia.org/wiki/%E4%B8%AD%E5%9B%BD%E5%85%B1%E4%BA%A7 %E5%85%9A%E7%AC%AC%E5%8D%81%E4%B9%9D%E6%AC%A1%E5%85%A8%E5%9B%BD% E4%BB%A3%E8%A1%A8%E5%A4%A7%E4%BC%9A%E4%BB%A3%E8%A1%A8%E5%90%8D%E 5%8D%95。
30 〈解放軍戰略支援部隊副司令員尚宏已晉升中將軍銜〉，新浪軍事，2017年10月31日，http://mil. news.sina.com.cn/china/2017-10-31/doc-ifynhhay9436954.shtm。
31 〈副大軍區職軍官康春元已晉升中將軍銜，曾任原蘭州軍區副政委〉，澎湃新聞，2016年8月29日，https://www.thepaper.cn/newsDetail_forward_1521187。
32 〈郝衛中任戰略支援部隊航太系統部副司令員〉，超級大本營軍事論壇，2017年4月25日，https:// lt.cjdby.net/thread-2380676-1-1.html；〈中共政治菁英資料庫〉，國立政治大學中共政治菁英資料庫，2017年5月1日，http://cped.nccu.edu.tw/node/1197101。
33 〈中國戰略支援部隊航太系統部陣容曝光〉，萬維讀者網，2017年4月25日，http://news.creaders. net/china/2017/04/25/1816325.html。
34 張育林（1958.1）是陝西千陽人，早年曾在國防科技大學學習，後留校工作，曾任國防科技大學航太技術系副主任、主任，國防科技大學研究生院院長。2002年，張育林出任裝備指揮技術學院院長，

華社記者專訪時建議「近地軌道到月球軌道的地月空間，將是未來人類發展的戰略空間，我國建成空間站、完成載人航太三步走戰略後，應該把地月空間開發利用作為歷史性目標進行戰略規劃」[35]，由談話中可推測裝備發展部負責航天部隊裝備發展規劃計畫、研發試驗鑑定、採購管理、資訊系統建設等職能。

三、中共太空部隊的單位

有關戰略支援部隊航天系統部在已知任務中，相關單位被組織到航天系統部。以總裝備部移轉的太空部門為基礎，還包含前總參謀部的單位。航天系統部涵蓋解放軍太空作戰的每個方面，包括太空發射、支援、遙測、跟蹤和控制（TT 和 C）以及情報、監視和偵察（ISR）。[36]

以下列航天系統部知所單位組織：

（一）發射基地：

1. 中國酒泉衛星發射中心／第 20 試驗訓練基地（原 63600 部隊）

2. 中國太原衛星發射中心／第 25 試驗訓練基地（原 63710 部隊）

3. 中國西昌衛星發射中心／第 27 試驗訓練基地（原 63790 部隊）

4. 文昌航太發射場（海南島）（隸屬西昌衛星發射中心管轄）[37]

（二）太空遙測、跟蹤和控制站：

1. 北京航太飛行控制中心

2. 中國西安衛星測控中心／第 26 試驗訓練基地（原 63750 部隊）

3. 航太測控站

後調任酒泉衛星發射中心司令員。2008 年，張育林出任國防科技大學校長，2011 年調任總裝備部副部長。參考資料〈原總裝備部副部長張育林中將出任軍委裝備發展部副部長〉，澎湃新聞，2016年 2 月 18 日，https://www.thepaper.cn/newsDetail_forward_1433330。

35 李宣良、孫彥新、王經國，〈空間站的下一步是什麼 中國瞄準地月空間〉，中華人民共和國國防部，2016 年 3 月 7 日，http://www.mod.gov.cn/topnews/2016-03/07/content_4646235.htm。

36 John Costello, "The Strategic Support Force:Update and Overview," The Jamestown Foundation, December 21, 2016, https://jamestown.org/program/strategic-support-force-update-overview/.

37 〈中國西昌衛星發射中心〉，維基百科，2018 年 4 月 5 日，https://zh.wikipedia.org/wiki/%E4%B8%AD%E5%9B%BD%E8%A5%BF%E6%98%8C%E5%8D%AB%E6%98%9F%E5%8F%91%E5%B0%84%E4%B8%AD%E5%BF%83。

4. 中國衛星海上測控部[38] ／第 23 試驗訓練基地

（三）航天偵察局[39]（原 61646 部隊）

（四）衛星通信總站[40]（原 61096 部隊）

（五）航太研發中心

（六）工程設計研究所

（七）航天員大隊

（八）航天工程大學[41]

　　中國航天員大隊原先隸屬中國人民解放軍總裝備部。在深化國防和軍隊改革中，2016 年 1 月撤銷解放軍總裝備部，組建中央軍委會裝備發展部，航天員大隊隨即成為裝備發展部直屬單位[42]。2017 年，中國航天員大隊轉隸中國人民解放軍戰略支援部隊航天系統部。[43] 經過本輪軍改，解放軍由原來的陸、海、空三軍，變成了陸、海、空、火箭軍和戰略支援部隊五支獨立軍兵種。航天員大隊這支在中國軍隊和科技界都十分特殊的隊伍，隨之轉隸戰略支援部隊。航天員是中國太空探索的中堅力量，堪稱中國「天兵」的先鋒。戰略支援部隊則是在本輪軍改中誕生的最神秘、最年輕的部隊，從航天員大隊的隸屬關係來看，太空必定是戰略支援部隊與強敵對手過招的一個重要戰場和領域。[44]

---

38 亓創、魏龍，〈中國衛星海上測控部組織陸海視頻親情連線〉，中華人民共和國國防部，2018 年 2 月 14 日，http://www.mod.gov.cn/big5/photos/2018-02/14/content_4804880_4.htm。

39 〈周志鑫任戰略支援部隊航太工程大學校長〉，搜狐軍事網，2017 年 10 月 8 日，http://www.sohu.com/a/196776059_495232;〈中國交通通信資訊中心與戰略支援部隊航太偵察局交流高分遙感「一帶一路」應用需求〉，中國交通通信資訊中心，2017 年 2 月 22 日，https://www.cttic.cn/info/1613；〈遙感衛星十號〉，愛航太網，http://www.aihangtian.com/hangtianqi/yaogan10.html。

40 The Satellite Main Station (衛星通信總站；SMS) from the former GSD Informatization Department (總參信息化部；INFOD) has also been incorporated into the SSF.John Costello, "The Strategic Support Force:Update and Overview," The Jamestown Foundation, December 21, 2016, https://jamestown.org/program/strategic-support-force-update-overview/.

41 〈周志鑫任戰略支援部隊航太工程大學校長〉，搜狐軍事網，2017 年 10 月 8 日，http://www.sohu.com/a/196776059_495232。

42 彭湃新聞，〈航太員劉洋已佩戴軍委裝備發展部直屬單位新臂章〉，騰訊新聞，2016 年 3 月 15 日，https://news.qq.com/a/20160315/019650.htm。

43 賴錦宏，〈景海鵬臂章透玄機航太員改隸戰略支援部隊〉，聯合新聞網，2017 年 10 月 19 日，https://udn.com/news/story/11556/2766096。

44 同上註。

肆、中共太空部隊影響

早在 2000 年，中共軍事戰略分析家王虎城在《瞭望》周刊撰文說：「對於那些永遠無法和美國用坦克對坦克、飛機對飛機的辦法打贏仗的潛在敵人來說，攻擊美國的太空系統可能是一個誘惑力無法抗拒的選擇。其部分原因在於五角大樓在軍事行動中對太空有極大的依賴性」。[45] 美國國防部前副助理部長拉維羅（Doug Loverro）認為，這個看法已經成為中共和俄羅斯發展反介入和區域拒止戰略的穩固基礎。在未來，隨著我們的對手在戰爭領域縮小與美國的差距，隨著他們繼續部署和擴展反太空能力。[46] 五角大廈分析人員曾起草報告指出，解放軍戰略支援部隊的太空戰任務可能分為太空支援、太空進攻任務。未來，解放軍將不斷發展太空支援能力，例如，天基通信、天基定位、導航和定時服務，天基情報偵察，天基導彈預警、發射探測和定性，以及環境監測能力。同時，解放軍將擁有或正在發展天基進攻性武器系統，如動能反衛星導彈、天基共軌武器系統和地面定向能武器。[47] 中共太空部隊發展之運用及威懾略述如下：

一、中共太空部隊的運用

（一）921 工程

神舟十一號載人太空船 2016 年 17 日上午 7 時 49 分在酒泉衛星發射中心，由「長征二號 F」載運火箭推送進入航行軌道，成功把 2 名太空人景海鵬和陳冬送到太空，預計在進入太空 2 天後，「神舟十一號」會與「天宮二號」太空實驗室在 393 公里高的軌道上連接，太空人會在實驗室內工作及生活 30 天，總計 33 天的太空旅程將創下中國載人太空船在軌道飛行時間的最長紀錄。[48]

---

45 葉林，〈川普要建太空部隊是戲言還是必然？〉，美國之音，2018 年 3 月 15 日，https://www.voacantonese.com/a/trump-spacearmy-20180314/4299668.html。

46 同上註。

47 北國防務，〈美報告披露中國這支神秘部隊：將有太空武器能打衛星〉，新浪軍事，2017 年 12 月 13 日，http://mil.news.sina.com.cn/jssd/2017-12-13/doc-ifyppemf6557281.shtml。

48 簡恒宇，〈中國「神舟十一號」發射成功 挑戰中國載人太空船最長飛行時間記錄〉，奇摩新聞，2016 年 10 月 17 日，https://tw.news.yahoo.com/%E4%B8%AD%E5%9C%8B-%E7%A5%9E%E8%88%9F%E5%8D%81-%E8%99%9F-%E7%99%BC%E5%B0%84%E6%88%90%E5%8A%9F-%E6%8C%91%E6%88%B0%E4%B8%AD%E5%9C%8B%E8%BC%89%E4%BA%BA%E5%A4%AA%E7%A9%BA%E8%88%B9%E6%9C%80%E9%95%B7%E9%A3%9B%E8%A1%8C%E6%99

該計畫於 1992 年 9 月 21 日建立，計劃分為三個步驟：第一步驟目標是將太空人送入太空天地往返（神舟一號至神舟六號），第二步驟建立第一階段短期的天宮一號目標飛行器（神舟七號至神舟十號）、第二階段中期三十天為基準的天宮二號太空實驗室（神舟十一號、天舟一號），第三步驟建立長期天宮太空站。[49] 有學者評論中國分別於 2008 年、2011 年發射了天鏈一號和天鏈二號數據中繼衛星，它們不僅可為天宮一號提供資料中繼和測控服務，還可作為其他衛星的衛星，將其他衛星的資料（低軌道衛星）24 小時連續不斷即時傳回地面接收站，增強了解放軍對敵人的監視和預警。動用 6 艘遠望號測量船、10 餘個地面監測站，才能為神舟飛船提供 12% 的全球測控覆蓋率。而一顆天鏈中繼衛星即可覆蓋衛星或飛船 50% 的飛行弧段，兩顆幾乎就可以覆蓋所有的面積。天鏈系列衛星可以集成各類情報、作戰、監偵等資訊集中到指揮平台，實現各級作戰部隊的任務調遣、火力協調、射控、防空、情報、戰子作戰、作戰、後勤，同時更能保證各類聯合作戰的實現。例如當前美軍三軍聯合資料鏈接為 TADIL-J（Tactical Digital Link 戰術資料鏈，或稱 Link-16），所有戰術平台（包括水面、水下、地面、空中、宇宙）都將集成在 Link-16 系統之中。雖然天宮一號仍處於實驗階段，預計在 2015 年之前天宮二號與天宮三號宇宙站開始運行後，將可全面實現這一軍事目的。中國在天宮一號發射成功後，緊接著改變了天宮一號的軌道。衛星變軌表示衛星所有國已經掌握基本的空間作戰能力，因為地面控制人員可以全天候監測衛星以及「危險空間物體」，並監測戰場、並為飛行器和導彈導引。另外，機動變軌可以打擊敵人衛星、攔截敵人導彈、飛行器，甚至對地面攻擊，同時，還可以躲避敵人的攻擊。因此，天宮一號等於告訴中國的潛在敵人：未來中國可以進行星際作戰、還可以在宇宙直接實施核報復。[50] 由上述中共其運用太空科技發展在軍事戰略意圖非常明顯。

%82%E9%96%93%E8%A8%98%E9%8C%84-062700473.html。

49 〈中國載人航天工程〉，維基百科，2018 年 4 月 9 日，https://zh.wikipedia.org/wiki/%E4%B8%AD%E5%9B%BD%E8%BD%BD%E4%BA%BA%E8%88%AA%E5%A4%A9%E5%B7%A5%E7%A8%8B。

50 譚傳毅，〈天宮一號有何軍事意涵？〉，今日新聞，2011 年 10 月 20 日，https://www.nownews.com/news/20111020/446464。

（二）建立太空 C4ISR 系統

太空 C4ISR 系統係由指揮（Command）、管制（Control）、通訊（Communication）、電腦（Computer）、情報（Intelligence）、監視（Surveillance）、偵察（Reconnaissance）的相關硬體與軟體、整合而成的龐大且複雜體系，它是軍事指揮當局對戰爭作出決策、對其部隊與武器實施指揮與管制、進行對目標打擊的重要組織，是軍隊戰鬥的神經中樞，是兵力的倍增器。不論是強國或是弱國，透過部隊和武器對敵方目標實施打擊，都必需具備上述的 C4ISR 系統。其間的差異在於強國的 C4ISR 系統非常完整與先進，基本涵蓋了所有維度、所有空間、所有頻段，主要是依靠「地、天一體」的 C4ISR 系統，具備太空化、資訊化、自動化等特質；弱國的 C41SR 系統則受制於其國力與科技實力，其特質為不先進與不完整。中國火箭軍要進行遠距離作戰，其 C4ISR 系統中的預警、監視、偵察與通訊等必須依賴太空中的人造衛星來執行，才能消除距離上的障礙，即時運用其戰力，因而、須具備太空 C4ISR 系統。[51]

1. 偵察衛星

中國發展太空科技近五十年，已逐漸建構了其太空 C41SR 系統。在偵察衛星方面，中國在近地球軌道運行的有尖兵三號與六號等系列光電偵察衛星，尖兵五號[52]與尖兵七號等系列雷達偵察衛（建構中的「高解析度對地觀測系統」簡稱「高分專項」，其衛星部份共有 8 枚不同功能之對地觀測衛星，目前已有高分一號、二號與四號 3 枚衛星在軌運行。

2. 通訊衛星

通訊衛星方面，中國在軌道運行的有烽火戰術通訊衛星系統與神通戰略通訊衛星系統，以及一組由 3 枚天鏈一號衛星組網的中繼衛星網。[53]

---

51 應天行，〈武器庫與太空 - 中國火箭軍 C4ISR〉，《全球防衛雜誌》，第 64 卷第 2 期，2016 年 4 月，頁 106。
52 〈遙感衛星十號〉，愛航太網，http://www.aihangtian.com/hangtianqi/yaogan10.html。
53 應天行，〈武器庫與太空 - 中國火箭軍 C4ISR〉，頁 106。

3. 定位導航衛星

在 1999 年的車臣戰爭和 2008 年的俄羅斯－喬治亞戰爭中，俄軍使用的 GPS 導航系統就遭到蓄意的干擾和關閉，導致俄導彈和戰機的導航失靈，多枚導彈未能命中目標。1996 年台海危機，當時大陸二炮部隊有兩發導彈未落入 10 海里見方的演習目標區內。解放軍研判認為，可能是美國突然中斷 GPS 訊號所致。美國導航系統干擾威脅使得中俄兩國下定決心，發展衛星導航系統。[54] 所以在衛星定位導航方面，目前中國的北斗衛星導航系統已覆蓋中國全部，和西太平洋以及印度洋大部的亞太大部份地區，2020 年左右北斗系統將建戌由 35 枚衛星組成的覆蓋全球、三維定位、全功能的衛星導航系統。[55]

4. 海洋偵監衛星

針對反艦彈道飛彈打擊移動性目標航艦，中國已建構的反航艦偵察監視體系中，主要由岸基長程超視距雷達與太空軌道上的衛星組成。在軌衛星方面，中國已經發射了 5 組類似美美國「白雲」海洋偵監衛星星座（每一星座由 1 枚主星與 3 枚副星組成），並已組網運作構成一個可在 45 至 60 分鐘重訪頻率的全球海洋監視系統，透過對航艦艦隊輻射出的電磁信號進行連續監測，可以提供艦隊目標的初始航向和速度，能夠在遼闊的海洋上搜索艦艦隊機動目標。2015 年 12 月 29 日中國發射的高分四號衛星，定位於地球同步軌道，其凝視相機的海面解析度達到 50 公尺，足以發現艦體長度 200 到 300 公尺的大型軍艦目標，具備任意時間對特定海域（主要係西太平洋）實施機動偵察與監視的能力。[56]

火箭軍是中國對抗美國、執行「反介入」戰略的威懾力量，中國並不想對美國使用其火箭軍，但希望能將美軍「拒止」於西太平洋之外。由於核彈頭的彈道飛彈不是能輕率與隨意使用的，因而中國火箭軍的彈道飛彈配置了傳統彈頭（先決條件是飛彈的彈著必須精準），以增加其運用的靈活性與威懾力量。不過未來強國間的戰爭可能將始於網路戰與攻擊太空 C4ISR 系統中的衛星，因

---

54 郭匡超，〈甩開美國 GPS 依賴中俄軍力更驚人〉，中時電子報，2018 年 3 月 29 日，http://www.chinatimes.com/realtimenews/20180329000033-260417。

55 應天行，〈武器庫與太空 - 中國火箭軍 C4ISR〉，頁 106。

56 同上註，頁 107。

為這樣不會產生大規模的核污染，也不會激起全球譴責。中國有鑒於此，日前已同時成立了戰略支援軍，負責網路作戰與太空作戰，以建構新型態的戰略威懾力量，與提升中國軍隊的戰略保障能力。[57] 這些在軌的軍用人造衛星能提供作戰必需的偵察、監視、情報、定位、氣象、通訊、指揮與管制，在電腦系統的整合下，已成為太空 C4ISR 體系（目前中國尚缺飛彈預警衛星系統），再與陸基的 C4ISR 體系整合，成為中國完整的 C4ISR 體系。[58]

二、中共太空部隊的威懾

　　中國大陸的東方紅系列通訊衛星、風雲系列氣象衛星、北斗系列導航衛星都是輔戰裝備，不是太空武器。在太空部署了輔戰裝備當然比沒有好，唯愈多未必就愈好。慣用太空輔戰裝備的軍隊，一旦訊源消失，會立即造成軍事行動不順暢；依賴衛星愈深、軍事行動就愈不順暢，甚至翻轉戰局。[59] 未免自身衛星系列為敵人所損毀或致盲，先期致敵人衛星致盲戰術上，必然會研發下列武器，作為反衛星武器。

（一）直升式反衛星導彈

　　2007 年，中國成功試射了反衛星飛彈，一枚開拓者一號火箭攜帶動能彈頭，擊毀了軌道高度 865 公里的一顆已報廢的氣象衛星「風雲一號 C」[60]。其後又多次試射了具備擊落中高軌道衛星能力的飛彈。[61]（表 1）儘管中共對外宣稱彈道飛彈可以攻擊低軌道的美國間諜衛星或飛彈防禦衛星，但還是打不到中軌道的全球定位衛星、高軌道的通訊衛星或小衛星。值得關注的是，這個距離恰好是美、日兩國影像衛星所運行的高度。2010 年除再度以中程飛彈成功擊殺衛星外，其所進行的陸基中段反彈道飛彈攔截試驗的成功，意味著也能夠攔截飛過其上空的衛星。解放軍目前研發地面基地試射一款能力更強的新型反衛

---

57 同上註。

58 同上註，頁 106-107。

59 鍾堅，〈共軍加快發展太空武器系統：極音速滑翔彈〉，《展望與探索》，第 14 卷第 11 期，2016 年 11 月，頁 28。

60 Brendan Nicholson, Foreign Affairs Correspondent, "World fury at satellite destruction," *The Age*, January 20, 2007, https://www.theage.com.au/national/world-fury-at-satellite-destruction-20070120-ge416d.html.

61 BBC 中文網，〈這不是科幻小說！馬斯克發射衛星提供覆蓋全球 WiFi，最擔心被中國擊落〉，風傳媒，2018 年 2 月 27 日，http://www.storm.mg/article/403598。

星武器－DN-2直升式反衛星導彈。美國情報機構指稱，該導彈能夠破壞位於高軌道的戰略衛星，如GPS衛星和間諜衛星。情報機構表示，只要擁有24枚反衛星導彈，北京便能經由破壞全球通信和軍事後勤、限制高科技武器所使用的空中導航系統，嚴重削弱美國的軍事行動。中共亦希望在強化地區性的戰略中占有優勢地位，特別是中共一貫的不對稱理念，認為追求拒外空戰略具有跨越性創意優勢，遠比建航母在海面追趕有效。[62]

表1：中國的反太空行動和測試，包括具有反太空影響的測試

Chinese Counterspace Operations and Tests, Including Tests with Counterspace Implications	
Year	Technology
	Directed Energy
2006	Chinese laser reportedly paints U.S. satellite（雷射對美國衛星致盲）
	Kinetic Energy
2007	China destroys FY-1C meteorological satellite with direct-ascent KKV（動能獵殺載具在超摧毀1枚老舊的中國氣象衛星）
2010	China conducts midcourse ballistic missile defense test（陸基中段反彈道飛彈試驗）
2013	China conducts direct-ascent KKV test to GEO
2013	China conducts midcourse ballistic missile defense test（中段彈道導彈防禦試驗）
2014	China conducts direct-ascent KKV test
2015	KKV test of undeclared purpose（不明KKV試驗）
	Co-orbital
2010	Two Shijian satellites involved in close proximity operation（兩個衛星密切接近操作）
2013	Three satellites involved in close proximity operation to test robotic arm technologies（三顆衛星近距離操作，用於測試機器人手臂技術）

62 張蜀誠，〈中共太空不對稱作戰戰略之研析〉，《清流月刊》，2013年10月號。

Chinese Counterspace Operations and Tests, Including Tests with Counterspace Implications	
Year	Technology
2016	Launch of satellite equipped with robotic arm for space debris removal （發射裝有機器人手臂的衛星來拆除太空碎片）
2016	Launch of satellite to test in-orbit refueling technologies （發射衛星測試在軌加推進劑技術）
Cyber	
2012	Cyber intrusion reported against Jet Propulsion Laboratory （對噴氣推進實驗室的網路入侵）
2014	Cyber intrusion reported against National Oceanic and Atmospheric Administration （對美國國家海洋和大氣管理局的網路入侵）

說明：KKV=kinetic kill vehicles（動能獵殺【攔截】載具）；GEO=geostationary earth orbit.（地球靜止軌道）。

資料來源：Kevin Pollpeter, Eric Anderson, and Fan Yang, China Dream, Space Dream: China's Progress in Space Technologies and Implications for the United States, Institute on Global Conflict and Cooperation, March 2015,p.86.

（二）神龍太空梭

　　香港《亞洲時報》網站 2016 年 1 月 25 日發表報導稱解放軍軍改推動中國的太空計畫。文章稱，在 1 月上旬，曾有中國軍事專家透露稱中國的「神龍」號太空梭可能被部署，作為新成立的戰略支援部隊的一部分，這一新成立的部隊是解放軍新的高科技作戰單位。在香港《東方今報》1 月 8 日的報導中曾援引中國軍事專家宋忠平的話稱戰略支援部隊將由互聯網、航太和電子戰等部分組成，而這支新建部隊未來將裝備神龍太空梭。該型太空梭可以在太空和地球間自由飛翔，相當於中國版的 X-37B。「神龍」將擁有很高的速度和機動性，並且可以隱身，可以進行遠程飛行。根據宋忠平的說法，神龍是正在開發的無人太空飛船，可以進行太空武器發射、監視、偵察情報、和早期預警任務。美國戰略司令部司令塞西爾 - 哈尼（Cecil D.Haney）曾在不久前表示，中國正在開發一系列太空戰爭武器。哈尼稱，對手在太空帶來了多方面的挑戰，有可能威脅國家主權和生存。哈尼還表示中國已經裝備了先進的太空武器，可以致盲

衛星，[63] 而且中國還在 9 月發射搭載 20 枚衛星的火箭，未來可用於太空戰爭。而在不久前還爆出中國成功進行了第 6 次高超音速飛行器試驗，據稱該飛行器速度可達 10 馬赫。美國國會美中經濟與安全評估委員會在最新的年度報告中警告稱，中國追求一系列廣泛和強大的太空反擊能力，包括直接攻擊型反衛星導彈、共軌型反衛星系統、電腦網路、地面衛星干擾器和太空武器。而軍事分析人士也表示解放軍計畫在戰略支援部隊中使用神龍突出了解放軍正在建設太空作戰的能力。[64]

三、中共太空部隊組織特殊性

早在 2014 年習近平在北京接見空軍第十二次代會黨代表時，習近平強調，建設「空天一體、攻防兼備」的強大人民空軍，是時代賦予空軍的重大使命，是新形勢下維護國家主權、安全、發展利益的必然要求。[65] 而中國的軍制介乎蘇美之間，成立初期並未效仿蘇聯分為進攻性的空軍和防禦性的防空軍兩個空中力量軍種，卻與蘇聯一樣，成立了專門的戰略導彈軍種—第二炮兵部隊。同時，中國幾乎所有衛星測繪和航太發射工作，都歸屬於總裝備部管轄。因此，在涉及到空天一體戰方面的問題時，中共軍隊同樣存在 3 個獨立的大機構—空軍、二炮和總裝備部。[66] 從習近平的講話來看，未來中國共軍制改革的方向將會更趨向於美國體制，也就是空軍主導未來空天一體作戰，甚至可能組建空天軍。[67] 但是從現在軍改下戰略支援部隊之太空部隊之的建置，目前所知為航太系統部，屬副戰區級，它並非獨立軍種，也未規劃給空軍，而是直接隸屬於軍委戰略支援部隊，是進行整合完全達到支援海、陸、空軍之聯合作戰體系，但未來有關太空作戰部分，其權責是否提升由空軍負責或者是戰略支援部隊下太空部隊，有待未來世界各國情勢發展，及未來戰爭型態是否改變而有所調整。

---

63 Franz-Stefan Gady, "US Admiral Warns of China's and Russia's Growing Space Weapons Arsenal," *The Diplomat*, January 26, 2016, https://thediplomat.com/2016/01/us-admiral-warns-of-chinas-and-russias-growing-space-weapons-arsenal/.

64 子午星座，〈神龍飛船將成為戰略支援部隊的一部分〉，個人圖書館，2016 年 1 月 26 日，http://www.360doc.com/content/16/0126/16/5169602_530706235.shtml。

65 朱江明，〈媒體：習近平提空天一體戰是鞭策下一步軍事體制改革〉，鳳凰網資訊，2014 年 7 月 1 日，http://news.ifeng.com/a/20140701/40968317_0.shtml。

66 同上註。

67 同上註。

伍、結語

一、中共之太空部隊建立是支援軍事作戰持續建立必要條件。尤其而太空中之衛星所建立 C4ISR 系統，使中共解放軍計劃成為一支能夠獨立執行廣泛、長程戰略打擊任務的現代化戰略空軍及遠海海軍，如海軍向太平洋、印度洋擴張，成為藍水海軍，而其航母、潛艇、飛彈必須有極度良好的藍海資訊掌握，才能握有其優勢。

二、中共戰力目前非美國對手，不對稱作戰是解決戰爭問題的方法論。近年來中共積極提高「反介入 / 區域拒止（Anti-access/Area-denial）」強度訓練，未來東海、台海或南海等主權問題而與美國遭遇時，在軍力仍處劣勢情況下，有效運用中共太空部隊監偵能力削弱美軍的優勢並獲得勝利。

三、我方應深入研究其中共太空部隊改革後現代化監偵及其他創新式戰具、戰法及戰略，以其對我防重層嚇阻防衛影響之因應。

四、中共軍事學者一直強調須爭奪制天權而進行天戰準備，成為太空戰略的最高軍事目標，但此次軍改似乎定調中共太空部隊為支援軍事作戰的角色，而未來戰爭型態若不斷改變，是否在此軍改之下「摸著石頭過河」，而中共太空部隊的角色改變為軍事作戰的獨立兵種或軍種，有待日後的觀察。

全國全民國防教育

從社會化功能探討全民國防教育的創新作為

陳振良 *

壹、前言

為使國人體認全民國防與國家安全的相互關係，透過各種教育宣導管道，增進全民國防知識與愛國意識；進而使全民在心理上認同國防，在行動上支持國防。換言之，國家安全的維護可以從主動與被動防衛角度思考，可以透過直接軍事力量、外交談判、經濟制裁的方式；或是從國內自我防衛的措施，例如增加自我防衛的心裡、建立全民國防理念，來增進國家安全係數。[1]

由於兩岸長期微妙的政治關係，造成國人對國家定位及國防意識多所混淆缺乏共識，此與政府在國家認同、國家定位及國家意識方面未能清楚型塑意念，凝聚國人向心力，應有重大關聯。[2] 面對中共的威脅，國人應建立全民國防的概念，現代戰爭已無前後方之分，應該透過教育的方式，讓全民都具備高度的憂患意識，然後再結合先進的國防建設，達成捍衛國家的目標。

基此，2006 年 2 月政府推動《全民國防教育法》施行，從社會化的概念逐步提升全民國防為政府與民眾的共識。全民國防是一個國家總體戰力的表現，它就是要運用國家整體資源、軍民一體，結合有形和無形的力量，進而突破經濟發展瓶頸，開創嶄新的競爭優勢，有助國防發展建立深厚基礎，並創造軍民雙贏效益。[3]

不可諱言的，全民國防教育歷經十多年來已有一定成效，也累積了相當的學術研究與成果。尤其，以往全民國防教育研究較著重於全民國防教育法規、制度的分析及其檢討與精進。但涉及社會化功能層面較少著墨，本文先探討社

---

* 淡江大學公共行政學系兼任助理教授

1 翁明賢，《解構與建構：台灣的國家安全戰略研究（2000-2008）》（台北：五南，2010 年），頁 161。

2 監察院，《全民國防教育政策執行成效之檢討專案調查研究報告》（台北：監察院，2015 年），頁 48。

3 戴振良，《國防轉型發展趨勢》（台北：幼獅文化，2007 年），頁 7。

會化與政治社會化的概念，並進一步說明全民國防教育推行理念與窒礙，最後提出如何強化全民國防教育的創新作為，據以提升全民國防教育施行的成效。

貳、社會化與政治社會化的概念

本文先將社會化與政治社會化的概念，以理論基礎整理相關文獻後，進一步討論政治社會化的媒介，期望藉由個人經由這些媒介來學習政治文化，並分析從個人及社會觀點達到政治社會化功能。

一、社會化與政治社會化的概念

社會化（Socialization）是個體對社會的認識與適應。它是運用個體與社會環境相互作用而實現的，是一個逐步內化的過程。因此，在社會化過程中，政治社會化（Political Socialization）在於獲取對政治社會知識，形成情感取向與判斷與評估的準則。一言之，社會化與政治社會化的概念，彼此有聯動性，相互影響。然而學者的定義各有不同觀點，兩者的概念必須加以分析，謹就社會化與政治社會化的概念分析，如表1、表2。

從表1及表2概念敘述，可以瞭解社會化過程也不僅止於幼年時期。人生初期或成長期的社會化叫基本社會化，人生後期的社會化叫次級社會化。後者指在人生的各階段須適應不同環境之社會化而言，如入學時的新生訓練、新兵訓練、公司員工訓練等，均是一種社會化過程。年老退休，亦須經一番社會化過程，才能適應退休後的新環境。因此，社會化可以說是終生的歷程。[4]

換言之，社會化在於是個人尋求「自我」的歷程，而政治社會化則是說明個人尋求「政治自我」實現的歷程。政治社會化不僅包括顯示的政治學習在內，同時包括會影響政治行為之隱示的社會化學習在內。[5] 因此，社會化和政治社會化是連續不斷的學習歷程的概念，從「自我」做起達到「政治自我」實現，必賴持續與進行的學習歷程。

---

4 宋明順，〈社會化 Socialization〉，國立編譯館主編，《教育大辭書》（台北：文景書局，2000年），國家教育研究院雙語詞彙、學術名詞暨辭修資訊網，http://terms.naer.edu.tw/detail/1306745/?index=4。

5 蔡政廷，〈社會化歷程探討全民國防教育之推廣〉，(論文發表於2016年全民國防教育學術研討會，台北，2016年10月19日)，頁3。

表1：社會化的概念

作者	概念觀點
Robert Redfield Ralph Linton Melville J. Herskovits	社會化最早提出此一觀點的學者，說明具有不同文化的群體，持續的進行面對面的接觸，使得接觸的任何一方產生文化變遷的現象。
Arnon E. Reichers	社會化是一個有系統的過程，把人員引入機構的文化中，這一過程加以輔導、訓練、獎懲等機制，並接受組織的規範、目標和價值。
林清江	社會化是個人接受文化規範，以形成獨特自我的過程。
宋明順	社會化指將「生物我」轉化成「社會我」的過程。人出生時只是生物人，生物人無法構成社會，只能結成獸群。任何社會均須先將其成員轉化成社會人，社會的共同生活才能維持下去的過程。

資料來源：整理自 Robert Redfield, Ralph Linton and Melville J. Herskovits, "Memorandum for the Study of Acculturation," *American Anthropologist,* Vol.38, No.1, January-March 1936, pp. 149-152; Arnon E. Reichers, "An interactionist perspective on newcomer socialization Rates," *The Academy of Management Review*, Vol.12, No.2, April 1987, pp. 278-287；林清江，《教育社會學新論：我國社會與教育關係之研究》（台北：五南，2002 年），頁 89；宋明順，〈社會化 Socialization〉，國立編譯館主編，《教育大辭書》（台北：文景書局，2000 年），國家教育研究院雙語詞彙、學術名詞暨辭修資訊網，http://terms.naer.edu.tw/detail/1306745/?index=4。

表2：政治社會化的概念

作者	概念觀點
David Easton & Jack Dennis	政治社會化使個人漸漸形成對政治事物的認知、感情與判斷標準：對政治事項的處理之道；並對自我在政治社會中的地位與角色有了認識與看法，並依據此種認識與看法，形成其政治態度與行為。
G. A. Almond & Lucian W. Pye	政治社會化包括「顯示」（manifest）的政治社會化與「隱示」（latent）的政治社會化。

作者	概念觀點
呂亞力	政治社會化是指政治社會成員經歷的調適過程，經此過程，其遂能成為政治社會的一份子，而政治社會也藉此維持其存在。
袁頌西	政治社會化就是指個人成為社會一份子的歷程；這種學習是會終其一生持續學習與繼續存在的，不論是與政治直接相關或是間接相關均包含在內。

資料來源：整理自 David Easton and Jack Dennis, "Children in the Political System：Origins of Political Legitimacy," *The American Political Science Review*, Vol. 64, No1, March 1970, pp.189-191.G. A. Almond, "A Functional Approach to Comparative Politics," in Gabriela. Almond & James S. Coleman, ed., *The Political of the Developing Areas* (Princeton, N.G.：Princeton University Press, 1960), pp. 26-33；Lucian W. Pye, Politics, P*ersonality and Nation Building* (New Haven：Yale University Press, 1962), pp.26-33；呂亞力，《政治學》（台北：三民書局，2002 年），頁 388；袁頌西，《政治社會化理論與實證》（台北：三民書局，2004 年），頁 5-20。

二、政治社會化的媒介與功能

社會化是一連串的人際互動產物，人們學習與自己有關之角色表現和態度之方法。社會化的達成是在不同的情境、經由不同的人員、採用不同的方法、長期而持續作用的結果。因此，個人在接受外在社會期望與規範，以形成獨特自我的過程，期間也受到政治社會化一定程度的影響。謹就政治社會化的媒介與功能，分述如下：

（一）政治社會化的媒介

美國政治學者李文（Herbert M. Levine）認為社會化的管道指的是個人透過哪些手段或工具來學習政治文化，通常這些管道也是一般通稱為媒介、途徑或機構，為利於論述本文通稱為媒介，有關政治社會化的媒介如表 3。

表 3：政治社會化的媒介

媒介方式	內容
家庭	每個人接受政治社會化的洗禮早從幼年時期就開始，在多數社會中，家庭對幼童的成長發揮重要影響力。
同儕團體	包括朋友、同學、同事、教友等經常面對面互動的成員。當同儕之間的家庭性格相近時，對於原來家庭所塑造出的政治態度往往有增強的作用；同樣地，同儕的互動結果也可能將一個人的政治態度社會化成與其原來家庭大異其趣。
學校	現代社會中，學校教育無疑是提供政治社會化的重要媒介，每個國家無不利用教育來灌輸學生對政治制度運作的瞭解與價值。
社會團體	個人因為商務往來，宗教信仰、專業背景等因素所加入的社會團體，對其政治態度的養成也有舉足輕重的影響，例如，公會成員的投票抉擇往往會受到公會的態度所影響。
大眾傳播媒體	主要包括平面與電子媒體。前者以報紙為首，後者則以電視、收音機為主，人們的判斷常受其所接受到的訊息影響。在獨裁社會中，媒體多為政府所控制，並嚴格管制真實訊息在社會大眾間的流通；在民主國家中，資訊自由被視為一個社會是否自由的先決條件。
選舉或其他政治性場合	對許多公民而言，選舉期間參與助選或聆聽候選人的競選言論，是具有教育意義的。尤其候選人的辯論與演說，有助於一般民眾瞭解國家的政治問題。另外，其他政治性場合，如聆聽議會內辯論、參與政黨內討論等，也具有政治社會化的作用。

資料來源：整理自 Herbert M. Levine. 著，王業立、郭應哲、林佳龍譯，《最新政治學爭辯的議題》（台北：韋伯文化，2003 年），頁 192-194；呂亞力，《政治學》（台北：三民書局，2002 年），頁 394-399。

從表 3 所述，每一個人接觸到政治社會化的媒介大致有家庭、同儕團體、學校、社會團體、大眾傳播媒體、選舉或其他政治性場合等。其中家庭是每個人從出生到成年的成長過程是分不開的重要居家場所，家庭對一個人的影響力是重要的。[6] 同時，學校也是傳授政治教育的重要場域，也可以增強人們對國家認同的價值觀。然而，人類是團體群聚的生活互動模式，尤其在現代社會中同儕團體或是社會團體，也會逐漸增加團體認同感影響力，因此，家庭的媒介影響力也漸漸式微。

事實上，學校教育是非常重要的政治社會化的媒介場域，象徵愛國意識的全民國防教育推廣，是任一國家推動學校教育的重要一環。特別是，藉由全民國防教育理念提升公民學習國家安全和國土防衛的基本概念。同樣的，運用同儕團體、社會團體、大眾傳播媒體及選舉或其他政治性場合等媒介，也可以達到推動全民國防教育觀念或灌輸愛國思想。

不過，最重要的政治社會化媒介是家庭、同儕團體與學校。在個人社會化過程中，這三種單位的影響因階段而異。幼年時期家庭的影響最大，其後同儕團體及學校逐漸加強其影響力量。這三種團體的價值與期望，究屬協調一致或相互衝突，對於個人價值的形成及行為的表現，受一定程度的影響。各種社會化媒介都有其文化規範，都對個人形成某種文化期望。個人就在接受各種規範及期望的過程中，形成其人格特徵。[7]

（二）政治社會化的功能

台灣政治學者呂亞力認為社會化的功能，可從個人與社會兩種角度來探討，謹摘述個人與社會的觀點加以說明，有關政治社會化的功能如表 4。

---

6 Hyman, Herbert Hiram, *Political Socialization: A Study in the Psychology of Political Behavior* (Glencoe: Free Press, 1959), p. 69.

7 林清江，〈社會化單位 Agency of Socialization〉，國立編譯館主編，《教育大辭書》（台北：文景書局，2000 年），國家教育研究院雙語詞彙、學術名詞暨辭修資訊網，http://terms.naer.edu.tw/detail/1306818/?index=1。

表 4：政治社會化的功能

項目	項次	內容
個人觀點	政治自覺的形成	由於政治社會化過程中，漸漸瞭解自我在政治社會中的角色與地位。在民主國家，由於此種政治自覺，產生了某種程度的政治效能感，感到自我可以影響政府的政策與行為。
	政治興趣與參與慾望的培養	在民主的社會或其他對人民政治參與加以鼓勵的社會，基本上，政治社會化是有助於人民政治興趣與參政慾望培養。相較的，社會化過程中，也可能壓抑政治參與的興趣。
	政治知識提供	在社會化過程中，個人的政治知識逐漸增加，對政治環境的認識及視野漸漸擴充，個人政治態度亦隨之形成。
	政治能力的栽培	社會化過程中，人的政治能力逐漸被栽培，得以扮演不同的政治角色。
	政治態度的形成	在社會化過程中，人的基本政治態度逐漸形成。如意識形態被灌輸給人民，在傳達收訊較不明確情況下，人民的政治態度即為此一意識形態的支持者。
社會觀點	政治體系「共識」之維持或獲致	政治社會化的成員對於政治競爭規則應有共同的看法，對於基本的規範應一體接受，如此政治社會化方能凝結，而成為一個能追求共同目標的整體。
	新一代認同感之培養	政治社會化在使新一代的成員對其產生認同感，此種認同感之培養是透過象徵性符號（如國旗等）的情感，對基本價值與規範之接受而達成。
	創新精神的維繫（或壓抑）	過分強調「共識」，忽略個體的差異，常導致創新活動的減少，人們創新慾望的壓抑，與社會的僵化、停滯。在社會化過程中，不僅需要重視「共識」的培養與認同感的維持，也應顧及個人創新精神的維繫與批判能力的發展。

資料來源：整理自呂亞力，《政治學》（台北：三民書局，2002 年），頁 394-399。

　　從表 4 所述，政治社會化的功能包括有個人觀點及社會觀點，個人觀點由政治自覺的形成、政治興趣到政治態度的形成都是社會化的過程的一環，相較的，社會觀點政治體系「共識」、認同感及創新精神的維繫，也都必須個人觀點為基礎，從教育著手，凝聚社會大眾的「共識」才能達成。不過，在整體社會化共有認知背離時，國家全體「共識」較難實現，必須從個人及社會觀點同時考量，才有利於政治社會化的功能推行。換言之，個人社會化過程中藉由社會化媒介進行教育與學習，藉以培育公民意識，但是，由於全民國防教育執行，由政府經由法規、教育體制之規劃，因此，在整體社會化過程中，仍有待討論空間。

參、全民國防教育推行理念與窒礙

　　社會化過程中，只有歷經社會化的教育過程，學習中才能獲得嶄新的國家安全觀念和全民國防的理念。換言之，全民國防教育推行理念與作為有賴於政治社會化的功能相配合才能達其效用。

一、全民國防教育推行理念與作法

　　全民國防教育推行與機制源自於 2006 年 2 月《全民國防教育法》第 5 條規定，全民國防教育之範圍包括學校教育、政府機關（構）在職教育、社會教育、國防文物保護、宣導及教育。[8] 另依據《中華民國 106 年國防報告書》說明，為貫徹全民國防及全民防衛之理念，持續深化全民國防教育、軍民互動、營區開放及國軍歷史文物展覽，藉以凝聚全民國防共識。有關全民國防教育執行作法如表 5 所示。

---

8 法務部，〈全民國防教育法〉，全國法規資料庫，2005 年 2 月 2 日，http://law.moj.gov.tw/LawClass/LawAll.aspx?PCode=F0080014。

表 5：全民國防教育執行作法

項次	內容
以多元活動深耕學校教育	1. 走入校園：提供青年學子與軍事院校、部隊間的交流管道，規劃多元活動及融入全民國防教育及募兵制說明，增進對國防事務的認同與支持。 2. 射擊觀摩：配合地方、社會團體與學校，辦理射擊觀摩，讓民眾及青年學子接觸國防事務，體驗軍人射擊實況與執行過程。 3. 寒假、暑期戰鬥營：於軍事訓練場地及設施，採多元教學與寓教於樂的方式，辦理戰鬥營、科學體驗營、新聞研習營、軍樂及儀隊體驗營等。 4. 南沙研習營：配合海軍偵巡任務，由國內各大專院校遴選博碩士生組隊參研，進行生態研究及學習海洋事務，展現捍衛南海主權決心。 5. 國防體驗之旅：開放國軍營區，申請入營參訪，讓學生透過實際體驗，近距離接觸現役裝備與武器，配合專業導覽人員解說，將國防融入生活。 6. 全國高中職校儀隊競賽：展現學子青春活力，激發創意巧思、團隊精神與自我實現，帶動青年學子支持國防、加入國軍之熱情。
培養全民國防教育師資	為落實《政府機關（構）全民國防教育實施辦法》，國防部接受中央各部會及地方政府的申請，選派國防大學專業師資授課。並配合行政院「e等公務園學習平臺」，開設「全民國防教育學堂」深化全民國防理念。
增進國防認知	國防部舉辦陸軍官校等6場次營區開放活動。另辦理全民國防教育網際網路有獎徵答，國軍文藝金像獎擴大到社會人士，達全民國防理念深植人心。
史蹟推廣	為推動國防文物保護，結合政府觀光政策，協調相關部會及各縣（市）政府，提供具國防文物保存價值之景點，納入旅遊規劃，並拍攝「各縣市國防文物及軍事遺蹟介紹」系列專輯，建置網站，結合觀光導覽。
獎勵傑出貢獻	依《全民國防教育法》規定，考核全國直轄市、縣（市）政府全民國防教育執行成效，績優單位給予獎勵。

資料來源：整理自中華民國106年國防報告書編撰委員會，《106年國防報告書》（台北：國防部，2017年），頁140-156。

從表 5 說明全民國防教育在主管機關多年努力初步已見成效，對於全民國防教育推行貢獻良多。不過，整體而言偏重於顯示的政治社會化功能，就如學校教育及相關輔教活動即對全民國防知識、信念的凝聚與建立，發揮相當程度的效果；相較的，隱示的政治社會化的功能施行成效較為不足，就如從某階段開始學習，到某一階段完成，忽略隱示的政治社會化的學習效果，就如社會教育、國防文物保存與宣教等功能，實有精進空間的必要。因此，全民國防教育必須由「個人觀點」潛移默化的學習歷程，再配合「社會觀點」顯示的社會化途徑，才能發揮整體的全民國防教育功能。

二、全民國防教育執行窒礙

基本上，政治社會化的功能首應強化全民國防教育推廣的作為，然而，檢視現行中央主管機關及地方各級政府相關部門因其權責及業務執行不同，加以政策與執行面向執行欠佳，導致全民國防教育推行推廣執行產生窒礙，宜進一步探討產生問題。謹就全民國防教育執行窒礙，分述如下：

（一）制度變革、影響成效

由於募兵制的實施，學生役期折抵高中至大學選課需求降低，直接影響學生選修全民國防教育意願。自 99 學年度起，高級中等學校（包含五專前 3 年）全民國防教育必修課程已減為 1 學年，並且只安排在高一實施，而大專校院部分，相關課程近年來亦多由「必修」調整為「選修」，形成眾多學生高二起即未再接觸全民國防教育課程，減損高級中等以上學校推廣全民國防教育成效。[9] 然而，原全民國防教育係以高級中等學校為核心重點，一方面透過全民國防教育課程實施基礎軍事訓練，以為服兵役預做準備；另一方面，也結合《全民防衛動員準備法》與《民防法》的防衛動員之青年服勤任務，施以相關的認知與教育。[10] 因此，學校教育在面對國防制度變革與募兵制的實施同時，必須考量，全民國防學校教育做好向下延伸及向上擴展的必要。

---

9 監察院，《監察院一零三年度全民國防教育政策執行成效之檢討專案調查研究報告》（台灣：監察院，2015 年），頁 66。

10 王先正，〈理論與轉型：全民國防學校教育之檢視與前瞻〉，（論文發表於 2016 年全民國防教育學術研討會，台北，2016 年 10 月 19 日），頁 17。

（二）在職教育、執行不佳

國防部為協助各機關辦理全民國防在職教育，依所訂之「全民國防政府機關（構）在職教育巡迴宣導實施計畫」，受理各級機關申請協助遴派合格專業師資；監察院調查報告說明：103 年度 9 月 30 日止，參與的中央機關比率甚低；地方政府機關之參與雖為數較多，亦僅有 13 個直轄市或縣（市）政府申請專業師資，公營事業機構、直轄市議會及縣（市）議會部分則未參與，顯然目前機關（構）普遍輕忽全民國防教育之職責。[11] 因此，在職教育必須中央與地方機關應加強要求，將參與率列入單位及人員績效與考核。

（三）社會教育、執行僵化

依《全民國防教育法》第 8 條規定：「各級主管機關及目的事業主管機關應製作全民國防教育電影片、錄影節目帶或文宣資料，透過大眾傳播媒體播放、刊載，積極凝聚社會大眾之全民國防共識，建立全民國防理念。」復依《107年國防部全民國防教育工作計畫》辦理社會教育項目包括：國防知性之旅 - 營區開放、結合觀光導覽、強化網際網路平台、全民國防教育海報甄選、全民國防教育傑出貢獻獎選拔表揚等成果展示。[12] 事實上，社會教育比起政府機關（構）在職教育及學校教育，對象較難特定，參與意願更不確定，須要有足夠的「創意」作為，方能吸引更多民眾主動參與。[13] 換言之，社會教育推行成效在於中央與地方主管機關齊心努力去執行，以「創意」思維才能吸引民眾，帶動社會教育全面及多元化的目標。

（四）文物推廣、事權不一

國防部擇定軍事遺址，供地方政府納入旅遊景點之做法，除少數熱門景點外，絕大多數皆未能達到預期成果；極具教育意義之金門、馬祖戰地文化，則僅侷限於觀光功能，其全民國防教育功能未能被凸顯；另國防部辦理軍事主題特展均集中於「國軍歷史文物館」一處，且宣傳活動多有不足，參觀人數有限；

---

11 監察院，《監察院一零三年度全民國防教育政策執行成效之檢討專案調查研究報告》，頁 60。
12 國防部，〈民國 107 年推展全民國防教育工作計畫〉，2018 年 1 月，國防部政戰資訊服務網，https://gpwd.mnd.gov.tw/Download_File.ashx?id=2895。
13 監察院，《監察院一零三年度全民國防教育政策執行成效之檢討專案調查研究報告》，頁 63-64。

軍事歷史文化發展，未能善加利用。為達成全民國防教育之政策目標，非僅歸責於國防部及教育部等少數部會，當統籌並協調中央與地方各級機關，方能發揮最大綜效。[14] 事實上，國防文物管轄權幾乎都屬於中央政府，執行地方事權又屬於直轄市政府及縣政府，因此，中央、地方宜事權統一及相互協調與支援，建立相關規範並加以整合。[15] 特別是，為彰顯政府推廣全民國防教育之決心，必須釐定權責、事權統一。

肆、如何強化全民國防教育的創新作為

事實上，社會化、政治社會化都是經由個人到社會化的媒介，進行教育學習，培育公民意識的連續不斷社會化歷程；相較的，個人也易受到社會化政治體系的文化、價值、信念之影響。因此，政治社會化的功能如何結合全民國防教育影響甚巨，謹就政治社會化的功能個人觀點與社會觀點提出創新作為分述如下：

一、個人觀點

（一）提升政治自覺的國家整體意識

政治自覺包括民族主義、階級等自，在兒童成長過程中，漸漸感到自我民族受到歧視，遭到不平等待遇，這種經驗使他們較傾向激進的政治立場。[16] 相較的，將政治立場能轉化為加強國人產生了某種程度的政治效能感，藉由全民國防教育可以培養人們對社會與國家的認同感，養成行動力以實踐對國家的認同與支持，並建立認同國家安全與發展之相關概念。[17]

事實上，世界各先進國家所見，瑞士公民深具國防觀念，家家戶戶都有國防宣傳書籍，國防教育內容在小學就編入教科書，並實施終身國防教育；以色

---

14 同上註，頁 51，68-69。
15 戴振良，〈從國防文物管理與推廣探討全民國防教育的作為〉，（論文發表於 2017 年淡江戰略學派年會 - 全國全民國防教育論文發表會，台北，2017 年 5 月 6 日），頁 334-335。
16 呂亞力，《政治學》（台北：三民書局，2002 年），頁 394。
17 教育部，〈十二年國民基本教育課程綱要－全民國防教育〉，臺教授國部字第 1070015028B 號令，2018 年 3 月，國家教育研究院，https://www.naer.edu.tw/files/15-1000-14146,c1179-1.php?Lang=zh-tw。

列採全民皆兵兵役制度，從小學即開始加深國土認識，強化愛國思想和國防觀念；美國國防教育將重點放在青少年教育上及未來國防人才培育上，使青少年能跟上時代的步伐，且中小學課程充滿形形色色的軍事與國防元素，課外活動專案也多與國防事務相關。[18] 因此，為因應學校教育全民國防教育的衝擊與挑戰，主管機關也應配合國情及參考瑞士、以色列、美國等國全民國防教育精隨，結合已通過十二年國教全民國防教育課綱施行，採滾動式修正，將學校教育向下延伸以小學教育作為起點，並銜接及國中及高中教育，向上擴展至大學及研究所等建構完整全民國防教育體系網。

（二）國家認同與參與慾望的培養

所謂「國家認同」，係指一個人為了滿足其自身安全和歸屬感之需要，而自認為本身屬於某一個業經建立或期望建立的國家之成員，同時自願為這個國家奉獻自己的一切，並以這個國家為榮；相對的，此國家亦會透過一切手段，來獲取其所有成員的認同。[19] 事實上，兩岸長期以來，我已逐漸淡化敵我觀念，缺乏憂患意識，政府在國家認同、國家定位及國家意識方面未能清楚型塑意念，凝聚國人向心力。兩岸關係無法避免的將觸及國家定位問題，惟政府機關處理兩岸事務時，卻經常刻意予以忽視或者加以迴避，其結果，不僅造成國軍「為誰而戰」、「為何而戰」的信念迷失；社會各界對於國家意識的觀念亦日趨淡薄。[20] 因此，國家認同建立國人的共識，必須從推廣及宣教全民國防教育才能克竟全功。

政治社會化是有助於人民政治興趣與參政慾望培養。藉由全民國防教育配合參與者認知發展，並整合全民國防、國際情勢、國防政策、防衛動員及國防科技等五大主軸，全民國防教育推廣上力求結合生活經驗、社會時事及國內外議題，以引發國人參與的學習動機與興趣。[21] 就如自 2012 年淡江大學開始舉辦「高階決策模擬營」，以創新的研討方式，採互動式決策，使參與者反應非常

---

18 孫國祥，〈先進國家「全民國防教育」成功案例之比較－以瑞士、以色列、美國為例〉，（論文發表於 2007 年全民國防教育學術研討會，台北，2007 年 8 月 10 日），頁 55-76。

19 林明煌，《憲法與立國精神》（台北：華立圖書，2014 年），頁 103。

20 監察院，《監察院一零三年度全民國防教育政策執行成效之檢討專案調查研究報告》，頁 48。

21 教育部，〈十二年國民基本教育課程綱要－全民國防教育〉，https://www.naer.edu.tw/files/15-1000-14146,c1179-1.php?Lang=zh-tw。

熱烈，並激盪出更多的理路與思維；並再以此良好基礎上，引入「政策決策」模擬系統，採即時（real time）方式模擬政府之決策，目標是讓參與者能同步、水平思考式的總體決策，避免侷限於單一議題、或見樹不見林的思維。[22] 因此。此一「政策決策」模擬系統，可說是首次成功推出全民國防參與決策的案例，是值得國人支持及政府主管機關推廣的決策模式。

（三）全民國防政治知識提供

政治知識是實行民主政治的基底，政治知識程度較高的公民，比較能夠根據政治事實，而非依據個人好惡來討論政治。而政治知識豐富的人，意識形態也比較能夠維持一致性，不會在面對各類政治議題時，產生自相矛盾的說法。[23] 就全民國防教育推廣而言，從安全的角度連結個人與社會及國家，增進學習者對於國家相關作為之瞭解，思考與領會個人與團體之參與對於國家的發展與影響；另透過各類防災演練與實作活動，強化團隊合作的精神，培養同理關懷與溝通互動的能力。[24]

事實上，全民國防學校教育功能，著重在青年編組與訓練，依教育部頒布《105 年度學校青年服勤動員準備計畫》規定：高級中等以上學校在學男女青年，平時運用全民國防教育相關教學及社團活動時間培養專長教育，以交通管制、簡易急救、宣慰、消防、行政支援等為主要服勤內容，戰時或遇重大災害立即於學校基地開設緊急應變及收容中心成立社區服務，結合全民總動員，展現全民防衛意志。[25] 因此，青年服勤動員準備計畫主要在於提升全民國防教育與全民防衛動員 者之間的 帶關係以及相互影響的程 。[26] 透過全民國防教育的推展，使其具備基本全民國防政治知識的體會，俾利於建立保國衛民的理念，以確保國家安全的目標。

---

22 淡江大學國際事務與戰略研究所、整合戰略與科技中心，《第三屆高階決策模擬營政經電腦模擬成果報告》，2013 年 12 月 8 日，excellent.tku.edu.tw/DlFile.aspx?fid=B1B0EC46AA5C98B7。
23 邱師儀，〈公民教育的知識效果：一項定群追蹤的探索性研究〉，《人文及社會科學集刊》，第 29 卷，第 1 期，2017 年 3 月，頁 128。
24 教育部，〈十二年國民基本教育課程綱要－全民國防教育〉。
25 教育部，〈105 年度學校青年服勤動員準備計畫〉，《教育部 105 年度學校青年服勤動員計畫》，2015 年 5 月，頁 2。
26 戴振良，〈從全民災害防救與防衛動員探討全民國防教育的實踐與作為〉，李大中主編，《龍鷹爭輝：大棋盤中的台灣》（新北：淡江大學出版中心，2017 年），頁 199。

（四）防衛動員政治能力的栽培

在社會化過程中，個人學習扮演不同政治角色。最基本的是扮演一般性政治角色的能力，兒童期的社會化即是栽培此種能力。特殊性角色的能力則待成年期社會化的過程中獲得。[27] 社會化過程中，兒童期建立全民國防教育基本理念，成年期灌輸全民國防扮演不同的平戰轉換角色，主要透過防衛動員體制，整合軍民資源與功能，充分運用全民資源，達成平戰結合目標。[28] 其意義在於「納動員於施政、寓戰備於建軍、藏熟練於演訓」，使全體國民建立「責任一體、安危一體、禍福一體」的共識，達到全民支持、全民參與、全民國防的最高理想。[29]

事實上，為強化動員準備具體作為，驗證動員效能，政府每年均辦理「全民防衛動員（萬安）演習」，結合「全國災害防救演習」，置重點於複合式災害防救演練。[30] 持續強化地區戰力綜合協調會報跨域整合功能，及直轄市、縣（市）政府動員準備、戰力綜合協調及災害防救三會報聯合運作，配合汛期及戰況模擬，以天然災害及作戰災損搶救為主軸，納入定期「兵棋推演」研討，並召開動員、災防、戰綜三合一會報等定期會議，以發揮協調整合功能。[31] 並設定狀況模擬演練，以驗證地區內動員、災防、戰綜等會報的作業與執行能力，藉由參與演訓的驗證過程中，達到全民國防教育的體驗成效。

（五）文物推廣政治態度的形成

在社會化學習過程中，人的基本政治態度逐漸形成。政治態度的形成主要是透過不同型態的教育內涵，提高全民憂患意識，整合全民總體資源，強化國防建設、增強國防實力，並結合災害防救，達到維護國家總體安全的目的。[32]

---

27 呂亞力，《政治學》，頁396。
28 沈明室，〈兩岸新情勢下的全民國防教育〉，（論文發表於2010年兩岸關係與全民國防學術研討會，新北市，2010年4月10日），頁83。
29 朱士君，〈全民關注國防建構安全基石〉，青年日報，2015年2月25日，http://news.gpwb.gov.tw/news.aspx?ydn=026dTHGgTRNpmRFEgxcbfZXmgIZG56InU4h6FA541oBi2jPtC1N0AcxXccfdcCvvonyD2mpz9595FliDk%2fQLl0dnRdT8ZHMN18CrNm4q65s%3d。
30 中華民國「國防報告書」編纂委員會，《中華民國104年國防報告書》（台北：國防部，2015年），頁155。
31 同上註，頁153。
32 教育部，〈十二年國民基本教育課程綱要－全民國防教育〉。

尤須藉由全民國防教育推廣，提升國人認知與共識，特別是現代的國防已不僅只是軍事力量，也涉及到整體國力、軍事武力與民力，尤其是精神戰力的總和。因此，愈是先進國家愈是重視全民國防，而國防事務也不再只是國防部的事，更不是只有軍人才有保衛國家的責任，而是全體國民共同的責任。[33]

為了彰顯全民國防教育功效，依《全民國防教育法》律定各級主管機關應妥善管理各類具全民國防教育功能之軍事遺址、博物館、紀念館及其他文化場所，並加強其對具國防教育意義文物之蒐集、研究、解說與保護工作。就如中央及地方政府宜將位於軍事陳展及史蹟之處所，全盤規劃建置以多媒體及導覽系統（Navigation System），藉由參觀時隨身攜帶的個人數位助理器（Personal Digital Assistant, 簡稱 PDA）等行動設備結合網路連結軍事史蹟及遺址的所在地，讓觀光旅遊者或學生能依陳展地點接受參觀導覽及引導，提升其軍事陳展的動機與興趣。因此，推動全民國防教育精神，在於各部會及地方政府及其相關部門宜共同參與國防文物融入新媒體中，結合創新及科技的成果將軍事裝備及戰史簡介呈現國人眼前。[34]

二、社會觀點

（一）政治體系全民國防「共識」之維持或獲致

「共識」共同認可的觀念或想法。政治體系全民國防教育「共識」，必須結合軍隊與全民的力量，才能確保國家安全與永續發展。面對中共文攻武嚇的威脅，政府應統合運用全民國防力量，來推動全民防衛動員理念，運用全國人力、物力、財力及其它各種力量保障國家安全的必要措施與作為，其需以透明合法化的國防事務，爭取全民瞭解與支持；以全民化國防，建立憂患意識，加強國人心防，防制敵人心戰恫嚇之威脅；以有效動員機制，整合全國有利資源，處理立即危機效能。

事實上，為建立全民國防教育「共識」的維持，主管機關也舉辦各項「寓教於樂」的營區開放參觀、學生暑期戰鬥營、實彈射擊訓練等，以及舉辦各種

---

33 蔡政廷，〈社會化歷程探討全民國防教育之推廣〉，頁 5。
34 戴振良，〈國防文物管理與推廣之探討：以金門軍事遺址為例〉，（論文發表於 2016 年全民國防教育學術研討會，台北，2016 年 10 月 19 日），頁 19。

大型而具創新性活動，殊值肯定。為能獲致成果呈現，應藉由舉辦各類社會教育活動，能結合時下青年喜愛的動漫、海報、微電影等熱門主題或競技性活動，納入相關配套機制。[35] 同時，配合漆彈場、攀岩場、垂降場、三索吊橋等設備，讓國小至高中的學生透過實地的體驗，模擬戰爭或危險時可能遇到的情況（如渡河時需使用三索吊橋），從中學習彼此信任與合作的團隊精神，更重要的是，讓他們培養自信及勇氣。[36] 因此，全民國防教育推行必須結合展示、競賽、探索等活動，才能進一步獲得全民參與的「共識」與支持。

（二）新一代學校國防教育認同感之培養

培養新一代對社會與國家的責任感，著眼於培養未來之國防人才，為國防建設與國家安全奠定基礎。在傳統社會，承平時期生活表現出高度的持續性，人們的價值信念一代傳至一代，很難察覺其改變。但在現代社會產生文化價值、社會風尚，乃至科技發展，處於經常的變遷中，對社會化、政治社會化產生不可避免的影響。[37] 換言之，新一代學校教育學習歷程與變遷，也受到所社會環境的影響。由於全民國防隨著環境變遷意味著新的情勢、新的價值觀也為產生變化，將可能面臨改變的壓力。

另外，藉由再社會化（Re-Socialization）[38] 概念去考量，就如成人移民至文化不同的社會、領土割讓以及社會發生革命或改朝換代等，都會使人們既有的文化習慣，國家認同以及價值觀面臨某種程度挑戰，進而使人們主動或被迫進行相對應的改變。因此，對於新住民的再社會化是台灣社會新力量，全民國防教育新力量的相關教育、輔導措施實不容輕視。[39] 因此，全民國防教育中央、各縣市政府主管機關，應將新住民全民國防教育增列為社會教育範疇，設計全

---

35 監察院，《監察院一零三年度全民國防教育政策執行成效之檢討專案調查研究報告》，頁 64。
36 林虹伶，〈挑戰自我　全民國防教育資源中心〉，閱讀臺北，2011 年 1 月 21 日，http://tcgwww. taipei.gov.tw/ct.asp?xItem=1669202&CtNode=41465&mp=100021。
37 蔡政廷，〈社會化歷程探討全民國防教育之推廣〉，頁 4-5。
38 再社會化是指社會化的個人在生涯發展或人際互動歷程之中，因為新的角色、情境或經驗而對自我的價值、態度與行為產生調整，重新學習或改變，以獲得適應，表現新的角色與行為的作用或歷程。參閱高強華，〈再社會化 Re-Socialization〉，國立編譯館主編，《教育大辭書》（台北：文景書局，2000 年），國家教育研究院雙語詞彙、學術名詞暨辭修資訊網，http://terms.naer.edu.tw/ detail/1306818/?index=1。
39 蔡政廷，〈社會化歷程探討全民國防教育之推廣〉，頁 16。

民國防教育相關課程與多元化活動，達成新住民成為全民國防新力量之目標。

（三）國防創新精神的維繫（或壓抑）

創新是指將創意轉化為具體產品或成果，為大眾認可帶來新的價值。[40]換言之，創新宜將國防創意理念轉化為具體成果價值觀，為因應環境之變化，宜活化與深化議題內涵，並依適性地設計具創新、前瞻與統整之全民國防教育的規劃。因此，在社會化的過程，不僅需要重視全民國防共識的培養與認同感支持，也應顧及個人創新精神的維繫與批判能力發展。[41]

事實上，全民國防創新精神的維繫可以由保守（或壓抑）心態轉向積極的創新作為。就如：台灣在歷經解除戒嚴及民主化後，國軍地位明顯下降；反觀美國軍人無不以穿上軍服為榮，即使在機場。火車站、公車站等大眾捷運系統，都能看見軍校生或是軍人穿上制服的英挺的身影，當地民眾會當面給予起立及鼓掌，軍人地位備受尊重，是值得國人學習之處。另外，過去凡是參加從軍報國或是入營服役人員，地方政府都主動在公共場所辦理歡迎入伍儀式，是值得在持續推廣，也可表彰對國軍保國衛民的敬意，提升國軍官兵榮譽感與責任心。

伍、結論

本文說明個人受到社會薰陶與洗禮而接受社會規範，成為社會一份子過程，也是成長中的一種學習接受的過程。在社會化的歷程中，全體國民應建立政治社會化功能的全民國防的意識，這對我國國家安全而言，實具重要的意義。推動全民國防教育是強化整體國防安全的基礎，也唯有支持全民國防的政策，才能確保國家安全與生存發展，也唯有強化「全民關注、全民支持、全民參與」的共識，進而對「綜合國力」的提升，發揮最高效益，真正保障國家安全、社會安定與人民幸福。[42]

---

40 邱茂城、王昭明、吳育昇，《職場倫理與就業力》（新北：普林斯頓，2013年），頁280。
41 呂亞力，《政治學》，頁398。
42 文宣心戰處，〈全民國防教育簡介〉，政戰資訊服務網，2016年11月18日，https://gpwd.mnd.gov.tw/publish.aspx?cnid=519。

　　事實上，如果一個國家政治體系是穩定，其國家安全較易確保，反之，這個國家將陷入動盪不安。[43] 因此，國家面對危機其政治系統若不能採取有效因應作為，則一國將會形成國家及社會動盪不安情況發生。由於兩岸自開放民眾探親以來，民間交流日趨熱絡，但是，在軍事與政治上，兩岸仍是敵對立場，在面對中共不放棄武力犯台之企圖，國人應有同舟共濟的精神，才是捍衛國家生存發展和人民生活福祉的最大屏障。因此，這一事實，恐怕許多國人都已忽視並淡化了，這才是建構全民國防教育最需要努力之處。

---

43 戴振良，《中國國家安全戰略研究》（台北：華立圖書，2014 年），頁 74。

國防科技產業發展與德國二元制技職教育之研議

許衍華 *

壹、前言

從深藍到 AlphaGo 象徵著人工智能（Artificial Intelligence, AI）的深化，也標幟著信息化趨向智能化的演進。[1] 當智能化穿透國防科技產業將深刻影響著軍事技術、作戰概念與組織形式之重大變革。[2] 國防科技發展須前瞻科技發展趨勢及聯合作戰需求並藉由國防科技學術合作計畫，大力度投入「基礎研究」及「應用研究」。[3] 然「基礎研究」及「應用研究」有賴於教育。教育不僅如杜威（John Dewey, 1859-1952）所云，為促進社會變遷進步與奠定民主政治之基石；也正如費希德（Johann Gottlieb Fichte, 1762-1814）所強調，乃是復興民族文化的核心與開創國家富強康樂局面的動力。[4] 德國教育家費新（Jügen Wissing, 1897-1988）強調：「教育必須幫助經濟建設及配合國防需要，他給台灣建議是國教延長八或九年且對國中畢業生實施有效的技職教育」。由於推行技職教育蓬勃發展，促使基礎建設的興立與人力素質育成，提供了十大建設欣欣向榮之養分。[5]

依據中華民國 106 年《四年期國防總檢討》（Quadrennial Defense Review）國防產業發展以目標為導向，航太、船艦及資安三大領域為核心，藉擴大國防需求及結合民間產能，配合科技發展機制，扎根基礎研究，提升國防科技水準，以突破關鍵技術，推動武器自研自製與全壽期支援，進而引領相關產業發展，達成「以國防帶動經濟，以經濟支持國防」之政策目標。[6] 盱衡資訊化趨向智能化之潮流，須在國防科技產業領域以彎道超車之創意方能因應；也

---

* 國立彰化師範大學通識教育中心兼任助理教授

1 王天一，《人工智能革命：歷史、當下與未來》（北京：北京時代華文書局，2017 年），頁 66。
2 Robert O. Walker, Sean Bremley, Paul Scarlett 著，鄒輝等譯，《20YY：機器人時代的戰爭》（*20YY: Warfare in the Robotic Age*）（北京：國防工業，2016 年），頁 5-6。
3 中華民國 106 年國防報告書編撰委員會，《106 年國防報告書》（台北：國防部，2017 年），頁 94。
4 許智偉著，袁志晃編，《白沙薪傳：許智偉博士教育文選》（彰化：彰化師範大學，2013 年），頁 1。
5 同上註，自序。
6 國防部，《中華民國 106 年四年期國防總檢討》（台北：國防部，2017 年），頁 1。

必須結合台灣優良之技職教育基礎上向下扎根、中端銜接、橫向聯通、迎上創新。是以本文構思，先以資訊化邁向智能化時代趨勢作為重要背景；其次，國防產業現今發展與未來目標結合技職教育實力與基礎並提升大學研究所；藉由軍民融合之銜接與注入，達成學、產、官之區塊鏈結，提升國防科技產業自主能力。最後，提出研議與對策。期能達成研究目的。

貳、姍姍到來的智能化時代

未來學研究有五大特定任務：1. 目標和價值的釐清；2. 趨勢的描繪；3. 現況的解釋；4. 如果目前的政策是持續的對其可見的（probable）和可能的（possible）、可欲的（preferable）未來預測；5. 各種替代性政策的創造與評估和選擇。未來研究者要描繪未來趨勢並能同時說明現況，以架構一個他們可能正在引領的未來圖像。未來學者將針對創造一個被認為最為想要的未來，而創造選擇行動的可能結果。[7]現今所面對的趨勢是工業 4.0 的經濟型社會、大數據的預測、雲端科技創新、人工智慧、物聯網革命等，將會進化人類社會生活與顛覆企業運作模式。[8]在《2016-2030 全球趨勢大解密》書中，作者馬修‧巴洛斯（Mathew Burrows）也有類似的看法：奈米、生物科技、3D 列印、人工智慧、機器人等幾種科技結合，產生驚人綜效，使科技變革到達新境界。[9]

時代快速變遷下，全球製造業受到新技術革新和產業變革的挑新戰。新一代智慧芯片快速發展並與製造技術深度融合，引發製造業發展理念、製造模式、製造方法、技術體系和價值鏈的重大變革。以雲端、大數據與物聯網所形成「工業 4.0」正逐漸成形。美國、德國等先進國家提出製造業戰略規劃。[10]例如，德國、歐盟定義為「工業 4.0」、美國表述為「先進製造夥伴計畫」（Advanced

7 Wendell Bell 著，陳國華等譯，《未來學導讀：歷史、目的與知識》(*Futures Studies: History, Purposes, and Knowledge*)（台北：學富文化，2004 年），頁 48-49。

8 Jeremy Rifkin 著，陳儀、陳琇玲譯，《物聯網革命：共享經濟與零邊際成本社會的崛起》(*The Zero Marginal Cost Society: The Internet of Things, the Collaborative Commons, and the Eclipse of Capit alism*)（台北：商周，2015 年），頁 20-21。

9 Mathew Burrows 著，洪慧芳譯，《2016-2030 全球大趨勢解密：與白宮同步，找到失序世界的最佳答案》(*The Future, Declassified: Megatrends that will Undo the World Unless We Take Action*)（台北：先覺，2015 年），頁 158-199。

10 韋康博，《工業 4.0：從製造業到「智」造業，下一波產業革命如何顛覆全世界》（台北：商周，2015 年），頁 8。

Manufacturing Partnership, AMP）、日本稱為「機器人新戰略」、「中國製造2025」、台灣「生產力4.0」等（如圖1）。台灣「生產力4.0」是依據行政院科技會報之重大科技策略會議中，凝聚產官學研的意見與結論共識，進而研擬「行政院生產力4.0發展方案（Taiwan Productivity 4.0 Initiative）」。爰此，行政院所推動的「智慧型自動化產業發展方案」為基礎，整合商業自動化、農業科技化發展進程，提出生產力4.0發展規劃，期能開發人工智慧、物聯網、大數據、雲端運算等技術來引領製造業、商業服務業、農業產品與服務附加價值提升，同時發展人機協同工作的智慧工作環境以因應高齡化社會工作人口遞減的勞動需求。「生產力4.0」願景：1.產業面：促進產業創新轉型；2.技術面：掌握關鍵技術能力；3.人才面：培育產業實務人才。[11]（如圖2）

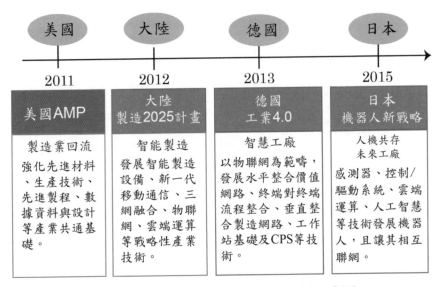

圖1：各國政策及提出製造業戰略規劃示意圖。

資料來源：行政院科技會報辦公室，〈行政院生產力4.0發展方案Taiwan Productivity 4.0 Initiative民國105年至民國113年（核定本）〉，行政院全球資訊網，2015年9月，頁5，https://www.ey.gov.tw/Advanced_Search.aspx?q=/行政院生產力4.0發展方案Taiwan Productivity 4.0 Initiative。

---

11 行政院科技會報辦公室，〈行政院生產力4.0發展方案〉，行政院全球資訊網，2015年9月17日，https://www.ey.gov.tw/Upload/UserFiles/%E8%A1%8C%E6%94%BF%E9%99%A2%E7%94%9F%E7%9
4%A2%E5%8A%9B4_0%E7%99%BC%E5%B1%95%E6%96%B9%E6%A1%88(1).pdf。

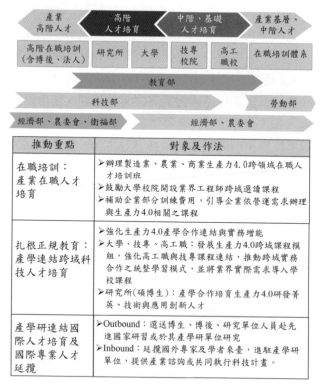

圖 2：生產力 4.0 實務人才培育架構圖

資料來源：行政院科技會報辦公室，〈行政院生產力 4.0 發展方案 Taiwan Productivity 4.0
Initiative 民國 105 年至民國 113 年（核定本）〉，行政院全球資訊網，2015 年 9
月，頁 5，https://www.ey.gov.tw/Advanced_Search.aspx?q=/ 行政院生產力 4.0 發
展方案 Taiwan Productivity 4.0 Initiative。

　　這三項重要戰略規畫正與台灣國防科技產業規劃及深耕技職教育有著契
合之處並能與時俱進。今天人工智能姍然來到，但僅限於狹隘的具體任務，並
沒有展現出人類的一般智慧。盡管如此，人工智能對世界的影響力不斷增長。
我們所看到的進步速度將對以醫療保健到語音識別等領域，產生不可估量的影
響。在醫療保健方面：大量醫療計畫將依靠人工智能來尋找醫療數據模式，提
供醫生診斷治療；在教育方面：人工智能有潛力幫助教師根據學生的需求制訂
教學方案；在交通領域：人工智能是自動駕駛技術的關鍵，這些無人操控的車

輛與飛機可能會在幾十年內改變全球物流與交通模式；在經濟方面：人工智能可以提高勞動生產率使許多發達國家經濟增長率翻倍；在環保議題方面：在零廢棄循環經濟創新下，更能打造永續經營的潔淨地球生活。人工智能技術將使人們能夠更有效地專注更具價值與意義的分析決策、評價並檢視未來之舉。樂觀願景與美好事物之成功關鍵依然是「以人為本」的核心。[12] 隨著人工智能在軍事領域的不斷穿透這些不需要人類干預的自動武器自主決定將那些作戰對象作為襲擊目標，襲擊的強度又會達到何種程度。繼火藥與核武器之後，致命性自主武器系統可能第三次改寫戰爭的面貌。[13]

參、台灣國防科技產業發展與未來目標

戰略大師鈕先鍾在其《孫子三論》原論〈計〉篇：「兵者，國之大事也……故經之以五，校之以計，而所其情……」其中「計」之詮釋為：對國家安全戰略須精密計算、計畫、分析、評估；[14] 在〈作戰篇〉：「篇名曰作戰，而所載乃，完車馬、利器械、運糧草、約費用者，何也？」、「亦已行師必先備乎此，而後可作而用之耳」。照這樣解釋「作戰」的意義即為「作戰準備」。衍生其意義乃為「國防建設，建軍備戰」。同時也內涵著戰爭與經濟之密切互動因素；[15] 在〈謀攻篇〉解釋「謀攻」為如何計畫攻擊。用兵之法：十則圍之—不戰而屈人之兵；五則攻之—優勢攻擊；倍則戰之—重點打擊；敵則能分之—分進合擊；少則能守之—避實擊虛；不若則能避之—間接作戰。從這三篇的闡釋，實屬大戰略的範疇，一切以「計」為開端。[16] 映照現今國防政策與國防戰略乃至國防科技產業發展策畫，雖有時空之遙，卻無意義之別，應所知也。「夫將者，國之輔也，輔周國必強，輔隙則國必弱」。[17]

國防部為發展有效因應未來作戰武器系統與裝備，依《國防法》第22條規範，結合民間產、學、研技術能力，積極建立武器研發生產與全壽期支援能

12　王天一，《人工智能革命：歷史、當下與未來》，頁158。
13　同上註，頁1-2。
14　鈕先鍾，《孫子三論：從古兵法到新戰略》（台北：麥田，1997年），頁41-42。
15　同上註，頁53。
16　同上註，頁66。
17　同上註，頁67。

量,增進國防自主科技能力,並藉向國外採購,進行工業合作及技術移轉,以達成振興國防產業及國防自主目標。國防自主的核心概念是建立自主國防科技能力,也就是必須掌握研發設計製造、測試及後勤支援,才能達成實質的國防自主。國防部前瞻國防需求,以國機、國艦等指標專案,結合部會資源,以需求引導民間產業參與,帶動國防產業發展,營造軍民互利雙贏的良性循環。[18]

國防自主能力,必須前瞻科技發展趨勢及聯合作戰需求,依目標導向之期程,逐步突破關鍵技術,並發展先進科技,以厚植國防自主研發能量;也必須兼顧國防產業發展。國防產業發展是支撐國防自主的重要力量;

一、國防產業發展策略指導

（一）前瞻國防科技趨勢,依聯合作戰需求,完成國防科技規劃確保科研項目結合作戰需求。

（二）擇定航太、船艦及資安三大領域,投入相關資源,振興國防產業,創造經濟成長動能。

（三）建立國防科技發展機制,結合產、學、研多元科技能量發展先進國防科技。

（四）突破武器關鍵技術,建立自主設計、製造、測試至整體後勤支援之自主能量,逐步達成武器系統自研自製目標。

（五）結合產業設計或製造能量,引導廠商投入國防產業,並協助與國際接軌,促進產業升級轉型。另投入相關建案預算並納入資源釋商,創造國防與產業發展的雙贏局面。

（六）制定國防產業發展相關法案,鼓勵與獎助國內產業參與國防科技研發與製造,以提升國防產業技術能量。

（七）強化跨部會協調合作機制,轉化國防科技能量,創造國防產業衍生效益。

（八）建立參與國防事務之廠商安全管控機制,防杜國防科技成果及關鍵技術遭竊或不當移轉。[19]

---

18 中華民國 106 年國防報告書編撰委員會,《106 年國防報告書》,頁 93-94。
19 國防部,《中華民國 106 年四年期國防總檢討》,頁 25-26。

二、國防產業發展策略規劃

（一）航太產業

以國造新式高教機為起始，貫穿後續各型空中載具之研製發展航太關鍵技術，並將技術移轉民間產業，逐步建立完善航太產業供應鏈，帶動航太產業整體發展。

（二）船艦產業

在現有的船艦建造基礎上，執行國艦國造，透過國內研製及認證方式，提升船艦裝備自製率。另藉由設計研發到承造整合的發展過程，厚植國內建造高階船艦關鍵技術，提升造艦能量。

（三）資安產業

推動新世代資安科技研發，協助建立國內資安產業策略聯盟並配合國家級資安測訓場域的建置，強化與產、學、研各界資安技術合作，促進資安產業升級轉型。

如前所述，精密策畫國防科技產業能量，提升國防自主能力，必須加速工業化與資訊化融合。易言之，要從科技密集型向科技質量型邁進。隨著無人系統遠程化、精確化、智能化、隱身化等趨勢日益加快以及先進技術的百花齊放。有三個核心的趨勢：不斷增加的透明度、不斷增加的互聯互通、機器載具更加智能化。[20] 未來將會出現新技術發展與作戰樣式的跨越式變化。[21] 前瞻科技新趨勢我軍聯合戰力規劃及籌劃新裝備：

1. 強化資電攻防及關鍵基礎設施防護等資通電綜合戰力，並結合民間資安潛力，發揮加乘效果。

2. 規劃籌獲具垂直或短場起降、匿蹤功能之新式戰機；強化防空飛彈能量；啟動各型艦艇更新計畫；提升地面部隊之遠程打擊火力及快速應變能力，以精進防衛作戰基本戰力。

---

20 Jason Ellis,Paul Scharre 著，鄒輝譯，《20YY：新武器概念與未來戰爭》，頁 92-93。
21 同上註，頁 6。

3. 推動潛艦國造；整建高速匿蹤艦艇、岸置機動飛彈、快速布雷、反空機降與無人飛行系統等不對稱武器裝備，打造多層次、全方位之防禦陣線與安全防護網。

4. 強化「戰力保存」，降低戰爭損害，發揮持久防衛之聯戰效能增進設施隱匿偽裝、抗炸防護與搶修能量；精進 C4ISR 通信指管備援系統，建構高存活戰力。

5. 以作戰需求驅動科研投資持續投資導彈與雷達系統之性能提升，並運用雲端、微型化等先進科技，推動抗干擾、無人匿蹤載具、精準打擊等武器系統科研案。[22]

「善守者，藏于九地之下；善攻者，動于九天之上」。[23]

工業化加資訊化加智能化融合，便能跨越式發展，達成國防自主能力與國防產業能量提升，實現彎道超車。備戰於未來聯合作戰，發展一套彈性海空拒止戰力，多層式地面防禦戰力，及反指揮、管制、通信、情報、監視與偵查打擊系統，阻止共軍戰力登陸。[24] 我軍聯合作戰能力將會以整合性之聯合作戰的成熟，邁向融合性之聯合作戰能力的證成。[25]

肆、德國二元制技職教育的由來及特點

許智偉博士於《白沙薪傳》的序言中提到：「1967 年 4 月，省教育廳長潘振球先生經美來德考察教育，我陪同去拜望職業教育大師費新（Jürgen Wissing, 1897-1988）。當時費新博士強調：教育必須幫助經濟建設及配合國防需要，亦謂他給台灣的建議是國教延長至八或九年，且對國中畢業生實施有效的技職教育。潘廳長答以行政院長嚴家淦先生已經裁示，將他的建議納入

---

22 國防部，《中華民國 106 年四年期國防總檢討》，頁 29-30。

23 強調防禦必須力求深密，攻擊必須充分發揮機動。善守又能善攻，採取守勢應能自保而轉移攻勢又能獲得勝機。參閱鈕先鍾，《孫子三論：從古兵法到新戰略》，頁 76。

24 美「戰略暨預算評估中心」分析：中華民國應充分執行一套可增加遂行侵略難度的戰略：彈性拒止戰略、重層打擊、精進 C4ISR 反制系統。參閱 Timothy A. Walton 著，章昌文譯，《確保「第三次抵銷戰略」：下任美國防部長的優先要務》，頁 20。

25 許衍華，〈前瞻思考台灣國防戰略：2030 聯合作戰願景〉，（論文發表於 2017 淡江戰略學派年會研討會，台北，2017 年 5 月 6-7 日），頁 22。

『國校畢業生志願就學方案』。同年 6 月，先總統蔣公宣布：『繼耕者有其田政策推行成功之後，加速推行九年義務教育。』立法院迅即通過以上方案為基礎的『九年國民教育實施條例』。使九年國教能於 57 學年度（1968）便可實施。費新博士聞訊非常興奮，要我立即返台協助。不僅安排我到柏林『政策研究所』參加一項以『義務延長教育』為主題的政策演習活動，並且介紹我與陶聲洋秘書長選送來德接受師傅訓練的韓顯壽等 10 位工職優秀教師認識。1969年我應聘政大教育研究所任教，課餘隨潘廳長巡迴全台演講並培訓國中校長及有關師資。1970 年費新門人范海波（Herbert Fleckenstein）及文思漢（Hans Winkelhausen）也應經合會之聘來台。1971 年台灣省立教育學院成立。省主席陳大慶聘我為創校校長，1972 年蔣經國組閣，謝東閔擔任台灣省主席我亦受命繼潘公之後出任教育廳長。因國中第一屆學生已經畢業，故以發展職業教育為使命。不僅根據費新師生建議，設立三重商工實驗『階梯訓練』輔導沙鹿高工加強『建教合作』；同時也試辦農校礫耕機械化及乳牛飼養等實驗。教育部也及時洽得世銀貸款，改善職校校舍及設備，又成立『技術及職業教育司』加強指導，使職業教育蓬勃發展，剛好供應了 1973 年開始的十大建設之人力需求」。[26] 此即，德國二元制技職教育於台灣創辦的由來。

「教育是人與人之間，也是自己與自己之間發生的事，它永不停止就像一棵樹搖動另一棵樹，一朵雲碰觸一朵雲，一個靈魂喚醒另一個靈魂」。[27] 在 19世紀晚期，為了因應工商業的需要，德國各地已經產生了許多職業學校，在這些學校畢業的學生，職業既不專精，公民及品格教育又完全缺乏。凱欣斯泰納（Georg Kerschensteiner, 1854-1932）乃在慕尼黑市教育局長任內創設「職業學校」配合各行業公會的學徒訓練，凡屬同業公會的工商機構，只要擁有訓練師資格之同仁，便可招收學徒施以技術訓練及有關專業課程。這些學徒及國校畢業而未升學的青年男女每周均應至政府所辦職業學校上課 8 至 12 小時，為期 3年。課程中包括專業課程、製圖或筆記、數學或經濟學、工程或園藝、德文、宗教及公民學科。因係由政府及企業雙方各設機構分途共進，故被稱為「二元制」（Dural System）。由於成效良好，逐漸推廣於全國，凱氏也被譽為「職

---

26 許智偉著，袁志晃編，《白沙薪傳：許智偉博士教育文選》，自序。
27 同上註，郭艷光，序文，〈教育的哲學家：許智偉校長〉。

業學校之父」。凱欣斯泰納不僅重視工作中的手、腦、心的陶冶與實用訓練，把職業教育與公民教育加以合一。[28] 德國技職教育之所以備受稱道，乃因其保存了手工業時代學徒制的優良傳統且採用緊迫教育法的辦法，使每一青年，除繼續升學者外，均能接受嚴格的職業訓練而學得一技之長。與之相輔而行的職業補習學校更有效地發揮職業知能，充實公民常識的功能。[29] 易言之，德國二元制技職教育，也就是將傳統學徒制培訓與現代學校教育結合讓學生進入企業實習工作，藉此盡可能完成技術訓練，學科則在學校中學習，是一種「企業與學校合作的職業教育制度」。[30] 這樣的制度有 4 種特色：

一、注重基礎訓練：除專門性的工作技藝外，先使學徒接受技術性的基礎訓練。具體實施的方式，工廠是令學徒在職業學校全日上課並接受有規律的基本操作訓練。然後再到生產工廠去實地學習，以達「先知後行，由簡入繁」的成效。

二、理論與實習並重：學徒訓練注重實際學習與操作，然後研習與實際有關之科目。以實際學習心得理論，幫助理論科目之了解而後又以理論改進實習情境。容易培養專門工作能力。甚者，把工人自身視為工程師，更能培養對生產事業及國家社會有強烈的責任感。[31]

三、能力本位：學徒制訓練出來工作者，有實務經驗及學理能力，畢業後取得證照，即可進入職場或創業自營企業。即使需服役而進入國防義務教育，軍隊亦有技職系統以銜接，不使中斷；一可發揮所長增進軍事科技領域的成就；二能為退役後在社會繼續就職及攻讀深造相關技職領域。可謂是以「二元制」推廣了技職教育的多元化與精緻化。

四、公民文化陶冶：德國不僅重視工作中的人格陶冶價值，把職業教育與公民教育加以合一，強調現代學校應「由養好個人榮譽的場所便為人生全面訓練的場所，由知能獲得的場所化為良好習慣養成的場所。因而將「個人人

---

28 許智偉，《西洋教育史新論：西洋教育的特質及其形成與發展》（台北：三民書局，2012 年），頁 299-300。

29 許智偉，〈西德職業教育發展趨勢〉，在許智偉著，袁志晃編，《白沙薪傳：許智偉博士教育文選》，頁 96。

30 黃清泰，《瑞士學徒制教育在公東：一位老校長引導的學習革命》（台北：圓神，2017 年），頁 172-174。

31 許智偉，〈西德職業教育發展趨勢〉，頁 96-97。

格的陶冶」和「社會應用的知能」合而為一。[32]

教育必須幫助經濟建設及配合國防需要……，中華民國 106 年《四年期國防總檢討》（Quadrennial Defense Review）中強調：「發展國防產業是整合國家資源，為加速國防科技發展及研發成果應用，研議建立發展機制，結合現有資源，以「發展創新技術，推向國防應用」為願景，運用民間科技能量，投入先進科技發展。[33] 國防整體實力在於國防基礎工業，國防基礎工業良窳，在於基層技術人力之培育與養成在整個關鍵鏈條中，必須以技職教育貫穿其中。因此，我國防科技產業發展，可將德國技職教育二元制特色整合，向下扎根堅實國防產業基礎。面向國防自主的提升與創新，可借鏡德國學徒制，培養國防專業人才。本文初議，試以公東高工與空軍航空技術學院及淡江大學航空太空學系，作一聯結，以進參考。[34]

伍、國防科技產業與德國二元制技職教育初議

一、公東高工──低端扎根

公東高工自創立以來，除第一、二屆外，始終保持德國二元制技職教育的精神──學徒制教學。在時代的變遷中，一直秉持著學校與實習工廠技職教育訓練。此二元制教育，隨著本土化教育改革的洗禮，與時俱進的創新。在學徒制的基礎上，也融合群集教育理念，發展出創新的「公東模式」：「群集教育」、「創新的能力本位教學法」、「協同式與分段式教學」。藉此模式自製了 KT-4 車床、氣壓自動控制教學板油壓自動控制教學板；也開發了油壓病床及機器手臂。開啟了自動控制的課程，公東高工機械群的機工科、機械製圖科以及營建群之家具木工科學生都需要接受「自動控制」課程，進行「多能工訓練」。非但可以達到所設定的「現今預期」目標，也加強學生具備發展「未來潛力」的能力，以開拓學生更寬廣的就業能力。[35]

---

32 許智偉，《西洋教育史新論：西洋教育的特質及其形成與發展》，頁 300-301。
33 國防部，《中華民國 106 年四年期國防總檢討》，頁 34。
34 作者畢業於空軍官校，任職經歷飛行部隊，因此以航空為例；台東公東高工創校即實行德國學徒制之工業職業學校。「公東高工」此名其實源於德文 Handwerkersschule 直譯為工藝學校。公東高工的「公」是公教會，亦是天主教的意義。而「公東高工」即是天主教在台東設立的高級工業職業學校，全名為「天主教台東縣私立公東高級工業職業學校」，簡稱「公東高工（公東）」。
35 黃清泰，《瑞士學徒制教育在公東：一位老校長引導的學習革命》，頁 197-200。

二、空軍航空技術學院—中端銜接

本校前身為空軍航空技術學校，民國九十年為因應國防科技之發展及國家技職教育的之變革，經教育部審核同意改制技術學院並附設專科部，於民國九十一年八月一日正式運作並晉名為「空軍航空技術學院」。航技學院負責培訓航空機械、土木工程、技術補給、通信電子、飛航管制及大氣量測等二年制技術學院軍官及專科技職士官，為國軍唯一培訓航空專業人才的技勤幹部學院。該校將於今年6月2日舉辦「2018年亞洲智慧型機器人大賽」。其目的極具意義，為促進各級學校技能教學，提升機電整合技能與人工智慧知識之養成。航空技術學院為國內培訓航空專業人才的軍事學院，畢業後具有國家級評鑑合格之各類技術證照，對未來發動機、航電通訊系統發展等，及維修護管理，具有重大影響與能力。[36]

三、淡江大學航空太空學系—高端聯通

淡江航太航空系所為國內培養航太空專業人員而著名。其課程主要分為基礎數理課程、航太專業課程以及核心課程，教學內容除空氣動力學、飛具結構學、航空發動機、飛具設計、飛行力學、太空力學、火箭工程等，並包括飛機系統、航空電子、航空氣象以及空中交通管制、航空品保、飛行安全等民航技術領域課程。是現今以航太空領域最高學系所之一。並與漢翔公司、長榮航空公司等有著長期產學合作與實習交流。目前也配合政府推動「5+2」產業創新與國機國造及飛機修護中心建立之「國防產業」項目。[37] 是以，在航太空專業領域裡負有聯通關鍵功能。

---

36 空軍技術航空學院，https://www.afats.khc.edu.tw/afats/Home/Default.aspx/ 國防部國軍人才招募中心 / 空軍技術航空學院 /。

37 為加速台灣產業轉型升級，政府打造以「創新、就業、分配」為核心價值，追求永續發展的經濟新模式，並透過「連結未來、連結全球、連結在地」三大策略，激發產業創新風氣與能量。政府提出「智慧機械」、「亞洲·矽谷」、「綠能科技」、「生醫產業」、「國防產業」、「新農業」及「循環經濟」等「5+2」產業創新計畫，作為驅動台灣下世代產業成長的核心，為經濟成長注入新動能。參閱淡江大學航空航太學系，http://www.aero.tku.edu.tw/2-1.html。

四、區塊鏈結合與產官整合

　　技職教育負有配合國家整體發展、促進產業升級、提振產業競爭力、優化勞動條件，並提供社會、產業、國家發展之專業技術人力責任，必須藉由產業、政府各部門與學校協力合作，促使各行各業之專業達人更受到尊重，並能吸引更多人選擇技職教育，提升各行業人才素質。國防產業規劃策略指導說明，建立國防科技發展機制，結合產、學、研多元科技能量發展先進國防科技。在這策略指導下，以中等技職教育體系如公東高工所培育出來的青年人才，可以推薦等多元方式進入高等技職教育體系，如空軍航空技術學院或空軍官校。亦可以進入如淡江航空系所或中正理工學院。形成國防科技產業低端扎根、終端銜接、高端聯通區塊鏈結。（如圖3）

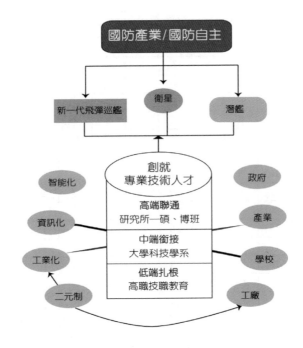

圖 3：區塊鏈結圖

陸、初議與建議

一、根據瑞士世界經濟論壇（WEF）「2017-2018 年全球競爭力報告」（The Global Competitiveness Report 2017-2018），[38] 在 137 個受評比國家，台灣排名第 15，較上年退步 1 名。在亞太地區排名次於新加坡（第 3）、香港（第 6）、日本（第 9）和紐西蘭（第 13）；領先馬來西亞（第 23）、韓國（第 26）、中國大陸（第 27），如表 1。在 WEF 評比競爭力三大類中：

1. 「基本需要」排名全球第 15，較上年退步 1 名。

2. 「效率強度」排名第 16，與去年相同。

3. 「創新及成熟因素」排名全球第 15，較去年進步 2 名。

表 1：世界經濟論壇 (WEF)2017 年「全球競爭力」排名

國家	2017	2016	2015	2014	2013	2012	2011	2010	2009	2008	2007	2006	17-16 變動
瑞士	1	1	1	1	1	1	1	1	1	2	2	1	0
美國	2	3	3	3	5	7	5	4	2	1	1	6	+1
新加坡	3	2	2	2	2	2	2	3	3	5	7	5	-1
荷蘭	4	4	5	8	8	5	7	8	10	8	10	9	0
德國	5	5	4	5	4	6	6	5	7	7	5	8	0
香港	6	9	7	7	7	9	11	11	11	11	12	11	+3
瑞典	7	6	9	10	6	4	3	2	4	4	4	2	-1
英國	8	7	10	9	10	8	10	12	13	12	9	10	-1
日本	9	8	6	6	9	10	9	6	8	9	8	7	-1
芬蘭	10	10	8	4	3	3	4	7	6	6	6	2	0
挪威	11	11	11	11	11	15	16	14	14	15	16	17	0
丹麥	12	12	12	13	15	12	8	9	5	3	3	4	0
紐西蘭	13	13	16	17	18	23	25	23	20	24	24	23	0
加拿大	14	15	13	15	14	14	12	10	9	10	13	12	+1
臺灣	15 (5.33)	14 (5.28)	15 (5.28)	14 (5.25)	12 (5.29)	13 (5.28)	13 (5.26)	13 (5.21)	12 (5.20)	17 (5.22)	14 (5.25)	13 (5.35)	-1
以色列	16	24	25	27	27	26	22	24	27	23	17	15	+8

---

38 Klaus Schwab, "The Global Competitiveness Report 2017-2018," *World Economic Forum*, September 26, 2017, https://www.weforum.org/reports/the-global-competitiveness-report-2017-2018.

國家	2017	2016	2015	2014	2013	2012	2011	2010	2009	2008	2007	2006	17-16 變動
阿拉伯聯合大公國	17	16	17	12	19	24	27	25	23	31	37	34	-1
馬來西亞	23	25	18	20	24	25	21	26	24	21	21	26	+2
韓國	26	26	26	26	25	19	24	22	19	13	11	24	0
中國大陸	27	28	28	28	29	29	26	27	29	30	34	54	+1
泰國	32	34	32	31	37	38	39	38	36	34	28	35	+2
印尼	36	41	37	34	38	50	46	44	54	55	54	50	+5
印度	40	39	55	71	60	59	56	51	49	48	48	43	-1
越南	55	60	56	68	70	75	65	59	75	70	68	77	+5
菲律賓	56	57	47	52	59	65	75	85	87	71	71	71	+1

註：1.( ) 內為評比分數。

　　2. 2017 年受評國家數為 137 個 (2006 至 2016 年分別為 125、131、134、133、139、142、144、148、144、140、138 個。

資料來源：吳明蕙、謝中琮，〈2017 年 WEF 全球競爭力我國排名全球第 15 位〉，國家發展研究會，2017 年 9 月 27 日，https://goo.gl/BoHVT7。

　　這份評比報告，「基礎建設」排名第 15，退步 2 名。主要係港口基礎建設品質第 24 名，退步 4 名、電力供給品質第 41 名，退步 6 名，航空運輸基礎建設品質第 43 名，退步 10 名、行動電話用戶數第 54 名，退步 4 名等項目排名下滑；在高等教育就學率第 13 名，退步 4 名；「創新」排名第 11，與去年持平。其中企業研發支出之投入程度第 10 名，進步 2 名、企業創新能力第 22 名，進步 2 名等方面表現進步、科學家與工程師之充足度第 30 名，退步 2 名、政府購買先進技術第 34 名，退步 9 名等項目排名退步。2017 年台灣在 WEF 全球競爭力排名第 15 (共 137 個國家)，較去年滑落 1 名。WEF 全球競爭力排名可視為對國家經社的總體檢，政府須針對優劣勢項目進行研析，以持續精進及改善施政，維繫全球競爭力優勢。[39] 因此，發展國家科技產業與國防自主及深耕技職教育，台灣實具有雄厚發展與創新的能力。（如表 2）

---

39 吳明蕙、謝中琮，〈2017 年 WEF 全球競爭力我國排名全球第 15 位〉，國家發展研究會，2017 年 9 月 27 日，https://goo.gl/BoHVT7。

表 2：世界經濟論壇（WEF）2017 年東亞六國「全球競爭力」排名

	臺灣		韓國		香港		新加坡		日本		中國大陸	
	2017	17-16 變動	2017	17-16 變動	2017	17-16 變動	2017	17-16 變動	2017	17-16 變動	2017	17-16 變動
全球競爭力指數	15 (5.33)	-1	26 (5.07)	0	6 (5.53)	+3	3 (5.71)	-1	9 (5.49)	-1	27 (5.00)	+1
1. 基本需要	15	-1	16	+3	3	0	2	-1	21	+1	31	-1
(1) 體制	30	0	58	+5	9	0	2	0	17	-1	41	+4
(2) 基礎建設	15	-2	8	+2	1	0	2	0	4	+1	46	-4
(3) 總體經濟環境	5	+9	2	+1	6	+3	18	-7	93	+11	17	-9
(4) 健康與初等教育	15	0	28	+1	26	0	3	-1	7	-2	40	+1
2. 效率強度	16	0	26	0	4	0	2	0	10	0	28	+2
(1) 高等教育與訓練	17	0	25	0	14	0	1	0	23	0	47	+7
(2) 商品市場效率	12	+3	24	0	2	0	1	0	13	+3	46	+10
(3) 勞動市場效率	25	0	73	+4	4	-1	2	0	22	-3	38	+1
(4) 金融市場發展	19	-4	74	+6	5	-1	3	-1	20	-3	48	+8
(5) 技術準備度	25	+5	29	-1	9	-4	14	-5	15	+4	73	+1
(6) 市場規模	20	0	13	0	33	0	35	+2	4	0	1	0
3. 創新及成熟因素	15	+2	23	-1	18	+5	12	0	6	-2	29	0
(1) 企業成熟度	21	+1	26	-3	11	+6	18	+1	3	-1	33	+1
(2) 創新	11	0	18	+2	26	+1	9	0	8	0	28	+2

註：1.( ) 內為評比分數。

　　2. 2016 年及 2017 年受評國家數分別為 138 及 137 個。

　　3.WEF 全球競爭力指標，下分 3 大類、12 中項、114 個細項指標。

資料來源：吳明蕙、謝中琮，〈2017 年 WEF 全球競爭力我國排名全球第 15 位〉，國家發展研究會，https://goo.gl/BoHVT7。

二、政府審視當前國內外環境與兩岸競爭，由行政院主導推動之「生產力4.0」與教育部規劃「技術及職業教育政策綱領」，[40] 均強調技術及職業教育，在提供國家基礎建設人力以及促進經濟發展上，扮演著舉足輕重角色，對締造台灣經濟奇蹟，貢獻厥偉......技職教育體系應成為培育職場就業力之重要養成機構學校及產業應共同深化並落實推動產業實習。配合新型態之技職教育學習方式，並借鏡德國、瑞士及奧地利各國所推動之學徒制模式。[41]

三、面對台灣武器系統的代差及思考未來戰場環境，以國防自主研發是一項重大抉擇。從國防自主研發而言，必須以威脅的來源、技術的成熟、軍隊的信心、充足的預算、人才的智慧與熱誠、學徒制技職所培育的工業基礎。更要的是政策指導必須一致，方能克竟其功。對外採購有其優點，然會受制於人的被動。面對兩相困境，將學徒制精神鑲嵌於技職、產業整個區塊鏈的結合。這可以達到國防科技產業所規劃之戰略目標—國防研發自主與對外採購。這些技職門(類)強調「實作」重視工具性能力，因此，培養「技職力」即形成「戰力」。[42]

四、在國防產業規劃中，為達成落實「國防自主」之戰略目標，須結合政府與民間資源，提升國防產業。能量聚焦航太、船艦及資安三大核心領域，促進國防產業發展，藉由產學合作、轉化研發成果及建立國防廠商安全管控機制，引領產業升級；同樣地，藉由民間的活力與動力，也能促轉國防產業的推陳出新。這是「軍民融合」的概念。因此，成立類似「DARPA」（Defense Advanced Research Project Agency）的「軍民融合」機制，作為引領規劃。[43]

---

40 行政院，〈106年技術及職業教育政策綱領〉，院臺教字第1060165689號，2017年3月2日，教育部全球資訊網，https://ws.moe.edu.tw/001/Upload/3/relfile/6315/52872/d595d36a-1b27-42d3-b50f-95c0f4671296.pdf。

41 同上註，頁9-10，。

42 黃俊傑，〈21世紀大學理念的激盪與通識教育的展望〉，《通識教育學刊》，第20期，2017年12月，頁31。

43 自1958年創立以來，國防部高級研究計畫局（DARPA）已成為美國國防部知名的軍事科研機構，素有「五角大樓之腦」。DARPA的每一項創新都改變了戰爭進程，國家安全乃至最高層面的戰略謀劃。數以千計的科學家和工程師幾十年如一日地投身於技術裝備研發，並將其科研成果投入實戰驗證。他們最終改變了我們作戰方式也改變了我們對世界的認知。參閱Annie Jacobsenv著，〈序言〉，李文婕、郭穎譯，《五角大樓之腦：美國國防部高級研究計畫局不為人知的歷史》(The

五、全民國防教育與公民教育的結合，不是詩歌朗誦，也不是拼湊雜揉理論。
以某些公民教育理念來配合全民國防教育推展，那只是靜態與消極的。[44]
社會化的過程必須了解社會變遷的因素；教育也必須從強制化、型態化中
掙脫，朝實用化及全人教育的發展而邁進。美國杜威曾說，「教育即生
活」；德國凱欣斯泰納也說，將「個人人格的陶冶」和「社會應用的知能」
合而為一，如機械修理、園藝設計等換言之，在學習與生活中是必須以知
識的陶冶與技能的實用與對社會的責任感，所形成的公民教育，才是完整
的。[45] 殊不知，蜂群的啟發造成蜂群無人機的創新是來自大學生的雜耍。[46]
因此，全民國防教育是必須將國防自主與國防科技產業融入，形成文科、
自然與技職的「強—強」聯合。這樣區塊鏈的連結是培養戰略家、國防專
家、科學家與工程師的生命花園。

柒、結語

本文研議，以資訊化邁向智能化是現在、未來進行式的背景描繪，以未來
學的方法是前瞻明日之舉。借重德國學徒制之精神，注重實用之效，因為德國
學徒制曾創造高雄楠梓加工區及十大建設之人力素質育成。以低端高職教育向
下扎根，二專技或軍事院校技職體系為中端銜接，大學或科技大學研究系所，
來實現高端聯通。產業界以社會責任推動協助經濟、科技的助力，使之導向國
防產業，形成產官學的緊密連接而達成國防自主。爰此，台灣整體競爭力排名
15 名，可藉由此實力完成國防科技產業及國防自主能力能量之建成。鑑於國防
議題是艱巨又長遠的運籌與投資，可成立「軍民融合」之機制，作為整合研發
的專門機構。目前雖有困境與短板之處，可藉由工業化資訊化邁向智能化實現
彎道超車。是以，借鏡德國學徒制貫注國防科技產業發展，達成國防自主能力
是可行的。

Pentagon's Brain：An Uncensored History of DARPA, America's Top-Secret Military Research Agency）（北
京：中信，2017 年）。

44 王先正，〈從公民教探討全民國防教育的推展與創新〉，國防大學主編，《103 年全民國防教育學
術研討會論文集》（桃園：國防大學，2017 年），頁 123-144。

45 許智偉，《白沙薪傳：許智偉博士教育文選》，〈慶劉校長白如先生七秩榮壽：論西德公民教育之
理論及實施〉（彰化：彰化師範大學，2014 年），頁 627-647。

46 Annie Jacobsenv 著，李文婕、郭穎譯，《五角大樓之腦：美國國防部高級研究計畫局不為人知的歷
史》，頁 349-360。

國家圖書館出版品預行編目(CIP)資料

川習時期：美中霸權競逐新關係／李大中主編著.--
一版.-- 新北市：淡大出版中心, 2019.08
面； 公分.--
部分內容為英文
ISBN 978-957-8736-30-6(平裝)

1. 國家戰略 2. 文集 3. 臺灣

592.4933   108009999

叢書編號 PS025   ISBN 978-957-8736-30-6

川習時期—美中霸權競逐新關係

著　　者　　李大中 主編

主　　任　　歐陽崇榮
總 編 輯　　吳秋霞
行政編輯　　張瑜倫
行銷企畫　　陳卉綺
文字校對　　徐立中
封面設計　　斐類設計工作室
印 刷 廠　　中茂分色製版印刷事業(股)公司

發 行 人　　葛煥昭
出 版 者　　淡江大學出版中心
　　　　　　地址：25137 新北市淡水區英專路151號
　　　　　　電話：02-86318661／傳真：02-86318660
出版日期　　2019年8月 一版
定　　價　　680元

總 經 銷　　紅螞蟻圖書有限公司
展 售 處　　淡江大學出版中心
　　　　　　地址：新北市 25137 淡水區英專路151號海博館1樓
　　　　　　電話：02-86318661　　傳真：02-86318660